T0262757

INTEGRATED PEST CONTROL IN VITICULTURE

Proceedings of a meeting of the EC Experts' Group / Portoferraio 26-28 September 1985

INTEGRATED PEST CONTROL IN VITICULTURE

Edited by
R.CAVALLORO
Commission of the European Communities, Joint Research Centre, Ispra

Published for the Commission of the European Communities by
A.A.BALKEMA / ROTTERDAM / BROOKFIELD / 1987

The texts of the various papers in this volume were set individually
by typists under the supervision of each of the authors concerned.

Publication arrangements: *P.P.Rotondó*, Commission of the European Communities,
Directorate-General Telecommunications, Information Industries and Innovation, Luxembourg

EUR 10104

Published by
A.A.Balkema, P.O.Box 1675, 3000 BR Rotterdam, Netherlands
A.A.Balkema Publishers, Old Post Road, Brookfield, VT 05036, USA

ISBN 90 6191 749 2

STRUCTURE OF THE MEETING

Scientific Committee

Cavalloro Raffaele, Principal Scientific Officer, CEC, Joint Research
Centre, Ispra

Clerjeau Michel, Maître de Recherches, Station de Pathologie Végétale,
INRA, Pont-de-la-Maye

Domenichini Giorgio, Director, Institute of Entomology, University,
Piacenza

Tzanakakis Minos, Director, Institute of Applied Zoology and Parasitology,University, Thessaloniki

Sessions' Organization   Chairman

Opening session:
Welcome and introductory address   M. Palmieri

Session 1:
Insects and mites   M. Tzanakakis

Session 2:
Diseases and weeds   M. Clerjeau

Session 3:
Strategies of integrated pest management   G. Domenichini

Closing session:
Conclusions and recommendations by   R. Cavalloro
sessions' Chairmen

Local Secretariat

Gentini Umberto
Azienda Autonoma di Cura, Soggiorno e Turismo dell'Isola d'Elba
- Portoferraio

Foreword

The European agricultural economy is more and more oriented towards a production of quality, as requested by the international market.

This necessity is strictly tied to a modern outlook on the problem of productivity. This is considered in the context of interdisciplinarity, and the execution of the various agricultural operations must be with respect of the environment, in particular of the natural equilibrium in the agro-biocenosis.

On this basis the Commission of the European Communities, in the framework of actions carried out on the coordination of agriculture research, took into consideration a programme of research which concerns different cultures, among which is vine growing.

One of the most important actions regards the application of appropriate phytosanitary protection in the light of the most recent knowledge. It is on this line that a meeting of experts was held at the level of the Member-countries of the European Communities.

The principal objective was to better the cultivation of the vines by using means which do not disturb the vine environment and do not cause undesired side effects. In addition, a methodology for the control oriented towards the integrated plant protection was also desired.

The known problem of profound alterations in the ecosystem of the vineyard put into evidence, in a preoccupying manner, the appearance of some harmful phytophages due to the indiscriminate and untimely use of certain pesticides, and the reappearance of serious calamities with the spread of epidemiological elements, which were felt to have been overcome, induced a prudential reexamination of the situation.

Therefore, it was necessary to concentrate on such a field, not only on the principal pests and diseases in the European environment, but also on the prevention of damages and on the treatments in use. This, in the perspective of bettering the interventions on the vine, have stimulated the Commission of the European Communities to organize a specific meeting of experts which came about with a great participation of European specialists.

All the papers presented at this meeting are gathered in this volume and, according to the work sessions, they are subdivided into three large themes: insects and mites, diseases and weeds, and strategies of

Integrated Pest Management. At the end of the text one can find specific recommentations derived from the work discussions and the final debate.

The meeting certainly reached its objectives which turned out to be extremely interesting, vital and stimulating. They indicate valid trends for a better protection of the European viticulture.

R. Cavalloro

Table of contents

Opening session

Session 1. *Insects and mites*

Session 2. *Diseases and weeds*

Session 3. *Strategies of integrated pest management*

Closing session

Opening session

Chairman: M.Palmieri

Opening address

M.Palmieri
Azienda Autonoma di Cura, Soggiorno e Turismo dell'Isola d'Elba

Ladies and Gentlemen,

In my capacity as President of the Azienda Autonoma di Cura, Soggiorno e Turismo dell'Isola d'Elba, it is my honour and pleasure to open your meeting on our island, in the name of which I would like to offer the most hospitable and the least ritual and formal of greetings.

I should like to thank you for having chosen the Island of Elba for your work at the international level and I hope that your work can proceed in the greatest serenity and as fruitfully as possible to achieve the aims that the Commission of the European Communities has proposed in these eventful days. I should thus like to thank Prof. R. Cavalloro who has given Elba the possibility of hosting you and I believe that I can interpret to some extent the importance of your presence here in this land.

In its thousand-year history the Island of Elba has seen agriculture occupying first place in the interests of its population because of vital necessity, because of the vocation of its soil, through economic choice, and also through a choice of lifestyle. For long centuries and for many years in this century, the people of Elba have been first of all farmers. The Elbans preferred the land to the sea, which was in some ways delegated to others, entrusted to others who were not atavically island people or even less Elbans. Elbans saw in agriculture the first source of their own interests, and for long years were proud to produce what they with a little pride and a little over-emphasis said was perhaps the best wine of Italy, certainly the best wine of the Italian islands.

Unfortunately today this is no longer true. This is because of wrong economic and political choices and the intrusiveness, and it may seem strange to you that a President of the Office of Tourism is saying this, the disrupting intrusiveness of tourism, which has meant that Agriculture has been abandoned to itself, losing its capacity to yield profits and its attraction and thus leaving to tourism the duty of sustaining by itself all the weight of a single-product economy and thus of an economy which is certainly not consonant with the ensemble of geography, nature and landscape of an island like ours, which is a productive one, or at least potentially so.

Now there is a revival of interest in agriculture. Today it has been realised that one cannot abandon a tradition of centuries, that one cannot absolutely overturn the country so as to obliterate what was in existence for many centuries. Thus the Public Administrations have understood that it is necessary to integrate the agricultural economy with the tourist economy in a reciprocal vision of interest which we feel is positive.

Today the Public Administrations are represented here by, among others, the Mayor of Portoferraio, whom we will have the pleasure of hearing later. Today the Public Administrations, I would say a Mountain Community which exists also on Elba, is a Mountain Community affected by saltiness, but which however can carry out its institutional function in the agricultural and forestry field, have again taken up these interests and are increasing their political action so that Elban agriculture, in that small amount of territory which now remains to it, can start again in terms of greater rationality, of greater modernity, of greater efficiency and can, above all, come to sustain the social and economic needs of the Elban communities with its contribution which cannot be subordinate.

I thus believe that if you have come to Elba, if you have chosen this our land for your important meeting, it must be due in part to the knowledge that Elba has this vocation, this ancient vocation of being an agricultural land, perhaps before it was a seafaring one, as in its history the Elbans have through necessity, also historical, tended always to go more towards the mountains than towards the sea to defend themselves from the depradations of pirates, because the sea is perhaps more laborious and certainly more dangerous than the land.

Although they are linked to the sea by other bonds, only in agriculture have they found up to the beginning of this century the source of their life, the sources of their wealth.

Look at this island which thus, through its administrators, I believe follows with renewed interest and with a touch of pride your presence here and your work. Certainly you give an essential and fundamental contribution with your studies in the field of vitiviniculture which gives ever-increasing results in the light of new technological research and in the light of new scientific vision.

In our modest capacity as administrators of a small island we are here to thank you for this and we are also here to see if from your acts something of your work can be of use to this island and if we can gain advantages from it.

I do not want to take up any of your time from the brief days at Elba because you must work and study and also, please, must come to know this island. I invite you to travel around our hills and along our beaches and you will see that all in all to the merits of nature one can add, to a certain extent, the merits of the administrators of this island.

We have been able in some way to prevent it being completely overwhelmed by the phenomenon of tourism, a phenomenon which is beneficial in itself but which can induce in the territories where it occurs extremely dangerous negative moments if it is not guided to some extent, and if its capacity for sweeping invasion is not countered. This invasion can as we have seen in agriculture, be overwhelming. It overwhelms traditions and it also often overwhelms the ethnic character of the territory where it exists.

To a certain extent we have succeeded in keeping it within bounds so that it does not lead to the extreme consequences. The island is still livable, is still enjoyable, is an island which this mild September offers itself to you so that you can work serenely and, if you do not know it already, come to know it at least in part and return to visit it. In this spirit I believe that I must thank your organization and must thank all of you and formulate once again my most sincerely felt, cordial, and, if you will allow me, affectionate wishes for your profitable labours.

I should like to read a telegram from the General Secretary of the Italian Academy of Vines and Wine: "Impossible to attend important EC-experts' meeting on integrated control in vine-growing, I offer with my excuses a greeting from Italian vine-wine academy very interested in the topics scheduled. Academy applauds the initiative and is available for any divulgation of meeting results by newsletter sent to all vine-growing countries. Best Wishes for good work. Dr. Arrigo Musiani, General Secretary of the Italian Academy".

Welcome address

G.Pardi

Municipality of Portoferraio, Italy

Ladies and Gentlemen,

The city of Portoferraio, of which I have the honour to be Mayor, in its long and glorious history and through a succession of happy and dramatic hours, has had the fortune to be always a centre of national and international interest and thus to have had the civic pleasure of hosting people from every part of Europe and the world.

Portoferraio, which was a European city before Napoleon illuminated it with his brief island sojourn, is thus particularly happy to be able to host at this time such an important meeting and such illustrious researchers. I should thus like to give them the most cordial and warm greeting of the Portoferraio community and a personal greeting which allows me to thank the C.E.C. most fervently, for having chosen Portoferraio as the location for this meeting.

I would be inopportune and presumptuous of me to talk about the importance of the subject dealt with by such illustrious researchers. It is certain that an administrator of Elba, land of great vine-growing and wine-producing tradition, a land which has been dedicated to agriculture for centuries, cannot be unaware of the subjects which you illustrious people are dealing with.

Now in Elba there is a renewed awareness of the absolute need to implement policies which are suitable for the reconstruction of our agriculture, in the light of modern technologies, in the wake of a glorious tradition, in a modern articulated vision of the various economic compartments, especially that of tourism. With this task and with this hope we administrators feel that we must acquire ever-increasing crop data and that we should always be aware of the work of scientists of whom you are the illustrious representatives.

Thank you, therefore, for your presence among us and sincere wishes for fruitful work and a serene stay in our enchanting island.

The European Communities activity in integrated plant protection and the research programmes in viticulture

R.Cavalloro
CEC, Joint Research Centre, Ispra, Italy

Mr. President, Ladies, dear Participants and Colleagues,

On behalf of the European Communities, in particular of the General Directorate of Agriculture, it is a pleasure for me to present the official welcome and express their pleasure in this meeting where representatives of the ten countries of the European Communities are brought together for the first time to talk about themes of vine-growing and wine-making, from the point of view of the coordination of scientific research. You will not deal with aspects which are commercial in character or of structure characteristics, but with aspects which are scientific, or technico-scientific in character.

This most cordial welcome is added to my personal pleasure at seeing here, in this meeting, many people who are dear to me with a long (alas!) story of interest in our common work. At a meeting of this type it is always heartening to see new young people who have become part of us and who give us the joy of research continuity.

A particular thanks goes to the local oranizers of the Island of Elba because they have enabled us to organize so efficaciously this meeting. I should say that the idea was originated by both the European Communities Commission and the Island of Elba Authorities. In particular I would like to thank the President who wished to outline the character of the island from the agricultural point of view. It is often unknown that the island has a vocation which is more agricultural than seafaring. Our meeting will probably give greater emphasis to this aspect in which I really believe even though agriculture throughout the world is experiencing times which are little difficult.

Our meeting is part of coordination plans for agricultural research at the level of the 10, and from the 1st January 12 members of the European Communities.

The Commission of the European Communities is developing a programme of scientific research in agriculture by coordinating research at the national level.

The themes which are being studied at present are made up of three main threads of activity: - use and conservation of agricultural resources, - structural problems, - improvement of productivity in animal rearing and crop-growing sectors; thus, there are three extremely clear and important threads.

Each of these is being developed, carried out by actions which are mainly oriented under predominant subjects, such as energy in agriculture, the use and management of soils and waters, Mediterranean agriculture - which is a subject with problems which are regional in character -, as well as the rearing of animals and the productivity of crops.

I believe that schematically it is easy to understand the structures and the lines of action through which the European Communities develop the programmes which are entrusted to it by the Council of Ministers.

One of these programmes is particularly concerned with integrated plant protection. It is a theme which is inserted in energy in agricultural actions. One of the qualifying bridges of this larger subject which is the integrated control programme for plant protection.

Integrated control is at present being carried out considering the sectors of protected crops, vines, and cereals, considered in agricultural rotation. And then there is a very particular theme which has come to light recently and that is the research to control the serious, very damaging mite, Varroa jacobsoni Oud., which attacks bees. We are not considering the different aspects of apiculture in this research programme, but the more particular aspect of the defence of bees as pollinators which is of exceptional importance from the agricultural point of view.

This group of activities in plant protection has been chosen for the development of joint activities, which the European Communities concedes on contract to selected institutes so that they can progress with actions which are of interest to all the EC-Member countries.

The programmes are decided by committees of qualified experts, which take into account the activities which exist in the individual countries and which try to complement these by the most incisive joint actions.

In addition, coordination or support activities are also carried out, considering exchange of experts, organisation of meetings, publications and so on. But what are the aims of this action? They are the studies for a more rational use of pesticides or of chemical products, considering them, however, on the basis of a more specific knowledge of the biology and behaviour of phytophagous species. It is the progressive replacement of polluting chemical substances with means which alter the environment less or not at all and which consume less energy.

The criteria for choice are naturally their importance for the EC-Member countries, aspects which are economic in character, as well as the impact on the environment, the influence on consumers and also - and this last is a question mark - the possibility of having results which can be rapidly applied. We do not want our research to go too far into the future so that we do not see at least the first fruits.

In the sector of vines and wine we have found perhaps, with the advice and suggestions of the experts of the Member countries, a more valid point in the sense that it is in this field that the European Communities maiby can operate more rapidly because there are already interesting results. One can more precisely define a final step to produce something which will really be useful instead of starting from the beginning of the research groping a little, as one always does when one begins for the first time.

Today we have important lines which are being followed by many countries and we shall hear of them in the papers which will be presented here but we shall see them above all in the last session on intervention strategies where some countries will refer directly to what they have been doing for several years. This is the passage to an action which is no longer a pilot one but which is a true coordination of integrated control in the open field, which we hope will occur in a short time in the near future.

We hope that with your participation those gaps or important themes will emerge which can be more strongly accented and on which common research can be carried out. We hope that this task of the Commission of the European Communities will be considered as a task which is common to other national and international organisations working in the same sector. But we know that the work carried out by other colleagues and by other experts must be taken into account as well as an action about which we should have a dialogue which will lead us to achieve these common aims more quickly.

Our contacts also allow us to enlarge reciprocally the ideas or the phases which are more structural in character at the level of the European Communities.

A call for offers has led to very interesting results that of having a large number of proposals for research at the level of specialized Institutes of various Member countries. A Committee on which all the countries were represented chose the lines of the programmes which will help us to advance in this research up to 1988. Three countries of the European Communities were chosen for this work, France, Greece and Italy, clearly with the support, collaboration and active interest of the other countries.

The chairmen of our three scientific sessions are in person the movers and scientists responsible for these important programmes which the European Communities support and will try to help as much as possible so that one can achieve a greater diffusion of these main threads to all the vine-growing and wine-producing areas of the European Communities, so that the principles of research which I have described can be achieved.

I thank you cordially for your participation, with many good wishes for your fruitful work, for a better European viticulture, and for full success of our meeting.

Session 1
Insects and mites

Chairman: M.Tzanakakis

Little-suspected conditions possibly affecting the population size of European grape berry moths

M.E.Tzanakakis

Laboratory of Applied Zoology and Parasitology, University of Thessaloniki, Greece

Summary

Data are presented, from the literature, 1) on the occurrence of insect-pathogenic viruses and protozoa on <u>Lobesia botrana</u> (Denis and Schiffermueller) and <u>Eupoecilia ambiguella</u> (Huebner), 2) on the beneficial effect on <u>Lobesia</u> larvae and adults of grapes infected in the laboratory with <u>Botrytis cinerea</u> Pers., 3) on the occúrrence of diapause in <u>Lobesia</u> under summerlike conditions, and 4) on the suitability of host plants other than the grapevine for larval growth. The possible effects of those conditions and/or factors and of certain cultural practices on berry moth population density are discussed within the frame of integrated pest control in vineyards.

1. Introduction

The two European grape berry moths, <u>Lobesia</u> <u>botrana</u> (Denis and Schiffermueller) and <u>Eupoecilia</u> <u>ambiguella</u> (Huebner), have long been important pests of vineyards in most European countries. Research on their biology and control has been extensive, and its results presented in many important papers by French, German, Italian, Soviet, and Swiss scientists. In addition to supervised chemical control in which pheromone traps are used to determine when to spray, microbial control and male confusion have also been tested with encouraging results. The development of integrated pest control often requires a detailed knowledge of the pest's biology, habits and ecological associations. Factors causing relatively low mortality may contribute considerably to the success of integrated control, if a single method is not sufficient. The purpose of this paper is to present some conditions and/or factors which I feel are little known as possibly affecting moth population density, or have been little-studied and, therefore, require further study to determine their possible usefulness for integrated pest control in vineyards.

2.1. Insect pathogens

P. Marchal already in 1911 reported an infection of <u>Lobesia</u> larvae by a sporozoan and Paillot in 1941 described a microsporidian from infected larvae of <u>Eupoecilia</u> (8). Lipa (8) found it in <u>Lobesia</u> larvae, pupae and adults in laboratory cultures and in the field, in France and Czechoslovakia. It is transmitted transovarially and through spermatophores, thus being easily spread. Lipa (8) suggests that the high incidence of infection in a field population he sampled indicates that this pathogen may be a useful agent of biological control of <u>Lobesia</u>.

The fact that no virous diseases of berry moths were known twenty (2)

15

or even ten years ago, may be the reason why research in this field has long been neglected. In 1979, Deseő et al. (3) reported on the occurrence of virous diseases of Lobesia in Italy, and two years later (4) gave further details. Those scientists of the University of Bologna, checking older and recent literature, concluded that discrepancies in the literature suggest that "factors other than pure ecological preference or ecological plasticity of Lobesia adults might play a role in its population dynamics". They subsequently looked into the reasons for abrupt changes in population density of this insect from year to year in central-northern Italy. Eggs they collected gave larvae that died in the first two instars. Their work yielded three pathogens in Lobesia larvae and adults: a cytoplasmic polyhedrosis virus (CPV), a species of Protozoa, Microsporida, and another virus, probably a baculovirus. The first two pathogens are known to cause chronic diseases in insects. In one vineyard, the moth's population collapsed in 3 years. The epidemics were considered to be the result of the joint action of the CPV and the protozoan. Without questioning the important role of climatic factors, Deseő et al. (4) showed that pathogens had a decisive influence on the population dynamics of Lobesia. This could explain the different population flunctuations between different vineyards and grapevine cultivars which are within the same region and have similar weather and topographic conditions. The disease is certainly worth taking advantage of, especially where ambient temperatures are favorable for the expression of the symptoms of the disease. Even where relatively high temperatures do not allow the expression of the disease symptoms, the infected insects may be more susceptible to other mortality factors.

Two years ago, specimens of a Lobesia colony of our laboratory, which had been maintained for years on an artificial larval diet and originated from northern Greece, was found by Deseő and collaborators to be infected by at least two pathogens: a protozoan of the Microsporida, and a virus. Only a small percentage of the eggs laid by moths of that diseased stock gave viable larvae (K.V. Deseő, in litteris). Therefore, there should be little doubt about viruses and protozoa occurring on Lobesia also in other European countries. The fact that some of them cause debilitating diseases and some do not cause symptoms under relatively high temperatures, thus escaping notice, calls for detailed surveys and in-depth studies to find or select strains of high virulence and exploit their potential for biological control of the berry moths.

2.2. Effect of Botrytis on Lobesia

Botrytis cinerea Pers., is a polyphagous fungus, growing on a number of ornamental plants, vegetables, strawberries, and fruits of trees, especially juicy ones. In grapevines it may attack leaves, flowers, shoots, graftings, cuttings in nurseries, and especially grapes including harvested ones. It can enter the plant through the epidermis, stomata and wounds. Plant sap or fruit juice oozing from wounds favors spore germination and allows the entrance of fungal hyphae. Thus, wounds made by the feeding of grape berry moth larvae favor the entry and spread of the fungus (22, 12). In France, erosions caused by 2nd and 3rd-generation larvae and cracks caused by the powdery mildew fungus Uncinula necator (Schw.) Burrill, are responsible for the spread of Botrytis. The percentage of grapes infected, from 5% in the absence of insect attack, rose to approximately 50% in grapes infested by Lobesia larvae (21).

In nature, Lobesia develops well on grapes infected by Botrytis (14), but it was not known whether the presence of the fungus made the grapes better for the larvae or not. Savopoulou-Soultani (17), under laboratory conditions, found that larvae developed faster and yielded more adults when

reared on ripe grapes of the cultivar Razaki or ripe apples which had been
infected with Botrytis. This resulted in a 2.2-2.5-fold population increase
from one generation to the next on infected grapes, and 1.4-2.7-fold increase on infected apples, as compared to uninfected fruits.

2.3. Diapause

Both european grape berry moths are long-day species, with a faculta-
tive automnohibernal diapause occurring in the pupal stage. In Lobesia this
diapause is induced when the embryonic and/or early larval stages are exposed to short photophases (7, 15, 6, 4). Consequently, a generation whose
embrya and/or early larvae occur beyond a certain critical date, should be
destined to give diapausing pupae to overwinter. It may be worth determining the critical date in more regions, and especially where late-harvested
cultivars are grown, to avoid unnecessary trap servicing and control late
in the season.

Recently, Tzanakakis et al. (20) reported that part of a laboratory
population of Lobesia originating in northern Greece and reared on an artificial larval diet, entered diapause also under conditions different from
those one would expect. Diapause occurred under high temperature and a long
photophase during the whole larval stage, in conjunction with high temperature during the embryonic stage. In other words, diapause occurred also
under conditions typical of summer rather than of autumn. The percentage of
the laboratory population which exhibited this diapause from low in some
years, reached 37-50% in others. We do not yet know whether this high temperature long photophase-induced diapause is a typical summer one as defined
by Masaki (11), neither whether it occurs in the vineyard. Work under way I
hope will soon cast some light on this matter.

2.4. Wild host plants

Both European grape berry moths are polyphagous. Lobesia is known to
feed on 25 plant species and Eupoecilia on 32, besides Vitis. Among the
wild hosts are trees, shrubs and herbaceous plants belonging to many families (1, 19, 2) and a number of them occur frequently in or near vineyards.
Bovey (2) mentions that, when a spring frost damages the new growth of vines,
larvae of Eupoecilia become abundant on Ribes, Lonicera, Viburnum, Rhamnus
and some other plants near the vineyards. Resulting adults may subsequently
oviposit on the grapes, but this aspect has not been studied sufficiently
to allow an estimate of the hazard to the vineyard from such neighboring
wild plants. With Lobesia the situation must not be much different, except
that, in addition, there is a preferred wild host, Daphne gnidium. This
plant is known to habour dense populations of Lobesia, whether near vineyards or not (2). Of interest is the observation of Roehrich and Carles(16)
that a high percentage of adult Lobesia was captured in pheromone traps in
such trees as chestnut and peach near vineyards. This phenomenon was noticed
also in other regions. In Sicily, in two consecutive years, Genduso (5)
captured males of Lobesia and Eupoecilia in pheromone traps in vineyards at
the beginning of March. This made him suspect that, in favorable years, the
moths may complete the first generation on host plants other than Vitis.

3. Discussion

Integrated control, same as biological control, of plant pests often
requires a much more detailed knowledge of the life history, habits and e-
cological associations of the pests with the other components of the agro-
ecosystem. The available data on the conditions or factors given above,
leave many questions to be answered by future research. For example, how

widespread and effective can insect-pathogenic viruses and protozoa be a-
gainst the berry moths? Would certain cultural practices favor their spread?
Does the favorable effect of Botrytis on Lobesia, also occur in the vineyard? Does it also occur when the larvae feed on fungus-infected foliage
and flowers in addition to grapes? Does it involve both species of grape
berry moths? Does it occur in most grape growing areas and most cultivars,
or only in specific locations and in certain years? Where such a favorable
effect is manifested does it result in an additional generation of the
moths, or only in greater population densities late in the growing season
and early in the following spring? Do alternative wild host plants, if infected by B. cinerea, produce larger moth populations? Will those, in turn,
produce more damage to neighbouring vineyards? Do other fungi infecting the
grapevine or wild host plants affect moth populations? How important is the
necrotrophic and/or fungivorous ability of Lobesia larvae for the survival
of this insect in nature?

In the laboratory, larvae of Lobesia were reared well on vine leaves,
but with difficulty on grape berries (10). However, leaves of vines are reported not to be oviposited by adults of this species (2). If this is true,
does it mean that in late-flowering cultivars part of the moths' eggs are
laid on other plants nearby? If so, do larvae disperse from the wild host
plants to the vines and vice versa? What happens in the next generation of
the insect? where weeds are abundant, adults of Eupoecilia were reported to
concentrate (2). Could this be due to the moth's preference for high air
humidity created by the dense weed vegetation, or could it be due also to
the moth's preference to oviposit on certain weeds when the vines are not
in a preferred stage? To what extent does berry moth dispersal from grapevines to wild host plants and vice versa affect population density in vineyards? Which factors make the berry moths disperse to and stay in non-host
trees? Whether due to a specific attractant or to accidental landing on
trees which offer abundant shelters, such information might lead to the development of more effective traps. Picard (13) suggested the use of Daphne
gnidium as a trap plant for Lobesia. His suggestion was then considered
dangerous for the vines. Today that the available insecticides are effective enough to prevent the dispersal of moths from this plant to the vines,
is it worth testing Picard's suggestion?

In some regions, the time of pruning vines affects the time of grape
maturity. Could we affect berry moth abundance by changing the time of pruning? Could we affect moth abundance by other cultural practices? In some
regions, systems of vine growing which favor high air humidity in the vineyard are known to result in denser populations of Eupoecilia and sparser of
Lobesia (9). In regions where only Lobesia exists, how is it affected by
high humidity? Is the rare occurrence of Eupoecilia in southern Europe the
result of only low air humidity?

The berry moths are key pests in most European vineyards. Therefore,
spending time and funds to study in more detail some aspects of their life
history which might bear on population density is justified. Some such aspects were presented above. Work on those aspects should be carried out in
addition to further work on the use of Bacillus thuringiensis, entomophagous insects, including egg-sucking Hemiptera, mass trapping, male confusion
and other promising methods of control. The fact that the efficiency of Bacillus thuringiensis, against the grape berry moths was considerably increased, to reach satisfactory levels, by the addition of sucrose (18), shows
how valuable a detailed study can be in improving conventional or new methods of pest control.

Acknowledgement. I thank Prof. R. Cavalloro and Prof. S. Zangheri for providing literature.

18

REFERENCES

1. BALACHOWSKY, A. and MESNIL, L. (1935). Les insectes nuisibles aux plantes cultivées. Paris Vol. 1, 1137 pp.
2. BOVEY, P. (1966). Super-famille des Tortricoidea. In Entomologie Appliquée a l' Agriculture, A.S. Balachowsky (ed.). Tome II, Lepidoptères.
   Masson et Cie, Paris, : 456-893.
3. DESEÖ, K.V., MARANI, F., and BRUNELLI, A. (1979). Virus diseases of Lobesia botrana Den. & Schiff. (Lepidopt.; Tortr.) in Italy. IX Intern.
   Congr. Plant Prot., Washington D.C., Abstr. no. 910.
4. DESEÖ, K.V., MARANI, F., BRUNELLI, A. and BERTACCINI, A. (1981). Observations on the biology and diseases of Lobesia botrana Den. and Schiff.
   (Lepidoptera, Tortricidae) in Central-North Italy. Acta Phytopath. Acad.
   Scient. Hungar. 16: 405-431.
5. GENDUSO, P. (1985). Osservazioni su Lobesia botrana (Den.& Schiff.) ed
   Eupoecilia ambiguella (Hb.) in Sicilia. Atti XIV Congr. Naz. Ital. Entomol., Palermo, Erice, Bagheria: 409-410 (summary).
6. GEOFFRION, R.(1970). Observations sur le troisième vol de l' Eudemis
   dans les vignobles du Val-de-Loire. Phytoma (jan.): 27-36.
7. KOMAROVA, O.S. (1949). Factors inducing diapause in the grape berry
   moth. Rep. Acad. Sci. USSR. 68: 789-792 (in Russian).
8. LIPA, J.J. (1982). Plistophora legeri (Paillot) comb. nov. (Microspori-
   dia) as parasite of Lobesia botrana Den. et Schiff. (Lepidoptera, Tortricidae). Bull. Polish Acad. Sci., Biol. Sci. Ser. 29: 305-310.
9. LOZZIA, G.C. and RANCATI, M.A. (1984). La distribuzione delle tignole
   della vite in Lombardia. Vignevini No. 6: 15-19.
10. MAISON, P. and PARGADE, P. (1967). Le piégeage sexuel de l' Eudemis au
    service de l' avertissement agricole. Phytoma 19: 9-13.
11. MASAKI, S. (1980). Summer diapause. Ann Rev. Entomol. 25: 1-25.
12. NELSON, K.E. (1951). Factors influencing the infection of table grapes
    by Botrytis cinerea (Pers.). Phytopathology 41: 319-326.
13. PICARD, F. (1924). Les origines de la vigne. Feuille Nat. 45: 25-27.
    (from Bovey 1966).
14. ROEHRICH, R. (1967). Elevage des chenilles de l' Eudemis (Lobesia bot-
    rana Schiff.) sur des aliments naturels de remplacement. Rev. Zool.
    Agric. Appl. 66: 111-115.
15. ROEHRICH, R. (1969). La diapause de l' Eudemis de la vigne Lobesia
    botrana Schiff. (Lep. Tortricidae): Induction et elimination. Annls.
    Zool. Ecol. Anim. 1: 419-443.
16. ROEHRICH, R. and CARLES, J.P. (1981). Observations sur les déplacements
    de l' Eudemis, Lobesia botrana. Boll. Zool. Agr. e Bachic. 16: 10-11.
17. Savopoulou-Soultani, M. (1985). The effect of the fungus Botrytis cine-
    rea on the biology of the insect Lobesia botrana (Lepidoptera, Tortri-
    cidae). Doctoral dissertation, Univ. of Thessaloniki, 78 pp. (in Greek).
18. SCHMID, A., ANTONIN, PH., GUIGNARD, E., CACCIA, R. and RAYMOND, J.-C.
    (1977). Bacillus thuringiensis dans la lutte contre les vers de la
    grappe, eudémis (Lobesia botrana) et cochylis (Clysia ambiguella) en
    Suisse romande. Rev. Suisse Vitic. Arboric. Hortic. 9: 119-126.
19. SILVESTRI, F. (1943). Compendio di entomologia applicata. Parte Specia-
    le. Portici, Vol. II, 512 pp.
20. TZANAKAKIS, M.E., OUSTAPASSIDIS, C.S., VERRAS, S.C. and SAVOPOULOU-SOUL-
    TANI, M.C. (1984). Diapause induction in Lobesia botrana under high tem-
    peratures. 17th Intern. Congr. Entomol., Hamburg, Abstr. vol. : 382.
21. VIDAL, J.-P. and MARCELIN, H. (1964). L' Eudemis en Roussillon. Travaux
    et essais 1963. Bull. Techn. des P.-O., no. 29: 7-30.
22. VIENNOT-BOURGIN, G. (1949). Les champignons parasites des plantes culti-
    vées. Masson et Cie, Paris, 755 pp.

Study of the biology of *Lobesia botrana* (Denis et Schiff.) (Tortricidae) in Macedonia (Greece) during 1984-1985

H.Stavraki, T.Broumas & K.Souliotis
Benaki Phytopathological Institute, Kifissia, Athens, Greece

Summary

Lobesia botrana (Dennis and Schiff.) in the Kavala area, and particularly in the biotopes of Nea Peramos, Elevtheres and Eleochori, produces three full generations per year on vines between 15 April and the end of August. There is probably the start of a fourth generation during the period September-November, the grape-picking period. Observations are continuing in these areas. In 1984 the flight of the second generation lasted 37 days and that of the third 42. In 1985 the flight of the first generation lasted 42 days, that of the second 28 and that of the third 34. The second generation laid its eggs on the leaves of the vine rather than on the clusters. This generation was observed between 4-9 June and 3 July 1985 in the above-mentioned areas. Eupoecilia (Cochylis) ambiguella Hb. was observed in Greece for the first time in 1985 in the area of Kavala in the Nea Peramos biotope during June, July and August. The dominant species caught by the sex pheromone traps was L. botrana. 23.3% of the adults captured belonged to the species E. ambiguella.

1. INTRODUCTION

In 1872 (5) L. botrana was observed in the vineyards of the Peloponnese, where it caused serious damage. It was also observed in 1941 (3) as a vine pest throughout Greece. In 1962 (6) it was again observed as a major vine pest. In Kavala, the area under table grapes of the "Razaki" variety has increased in recent years to 3500 hectares. The vines are cultivated in rows. Some 20-26 km to the south-west of the city of Kavala (Macedonia) are the Elevtheres, Nea Peramos and Eleochori areas. These are on the coast and extend for 1500 hectares. Serious damage to the grape clusters from L. botrana has been observed in recent years in these areas. Our country exports a total of 60 000 tonnes of "Razaki" grapes each year (source : Ministry of Agriculture). 40 000 tonnes are exported by Kavala, which makes it the main area for exports of table grapes of the "Razaki" variety. Owing to the major problem in the areas mentioned above posed by L. botrana, it was thought that it would be useful to study the biology of this insect in these biotopes.

2. MATERIALS AND METHODS

This work was carried out during 1984 and 1985 in the Kavala area in vineyards which grow the "Razaki" variety (**Fig. 1**). In 1984 only

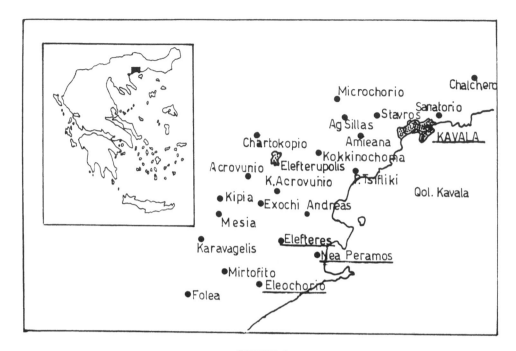

FIGURE 1
Map of the Department of Kavala

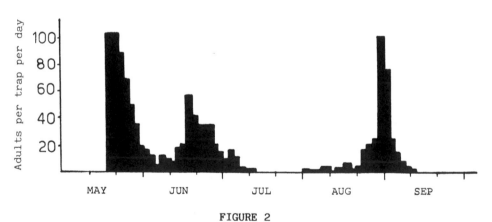

FIGURE 2
Adult <u>Lobesia botrana</u> caught by sex pheromone traps in the vineyards of
Kavala (Macedonia) in 1984.

flights of adult <u>L. botrana</u> were monitored by setting up, in May, four
sex pheromone traps of the zoecon (USA) type. These were set up by the
Kavala Agricultural Directorate of the Ministry of Agriculture in the
Elevtheres area at distances of approximately 1000 m from each other.
This was done on 15 May. Captured adults were counted at one or two-day
intervals up to 8 September, when the harvesting of the grapes began.
Each year the grape-picking begins between 1 and 8 September and
finishes between 10 and 15 November.

22

Studies of the biology of L. botrana in 1985 concerning the flight of the population of the adult insects were carried using sex pheromone traps. The population at the various stages of the insect's life-cycle was estimated by taking samples from various parts of the vine (once a week).

The flight of adult L. botrana was monitored using a network of traps of the same type as was used in 1984, i.e. the zoecon type, in the Elevtheres, Eleochori and Nea Peramos areas. In each area four traps were set up on 15 April 85, i.e. a total of 12 traps for the three areas at a distance of 1000 m from each other. On 10 July, a further two Biotrap type traps (Hoechst) were set up in each area. All the traps were checked at one or two-day intervals.

In the area of Nea Peramos two sex pheromone traps for Eupoecilia (Cochylis) ambiguella Hb. of the Farmoplante (Italy) type were set up. The observations were carried out as for the L. botrana traps. Weather conditions (humidity and temperature) for the three areas were the same.

Simultaneously, on 25 April 1985, the work of taking samples from untreated vineyards began, covering an area of half a hectare at the phenological stages of vine growth, in accordance with the Baggiolini table (Federal Agricultural Testing Station, Lausanne). One hundred (100) growing vine buds were examined per week. Subsequently, one hundred (100) clusters from the untreated vineyard were also examined. On 5 August the work of taking samples from the leaves of the vineyard began. 100 well-developed leaves were examined weekly. It should be noted that in all the areas of the vineyards treatments were carried out against this pest.

This work was carried out by the "Benaki" Phytopathological Institute in collaboration with the Kavala Agricultural Directorate of the Ministry of Agriculture.

3.   RESULTS

The number of adult L. botrana captured in the Elevtheres area during 1984 is given in **Fig. 2**. The number of adults captured by the traps fluctuated between two and 100 adults per trap per day between 15 May and 3 June. These adults are the first generation. A new increase in captures was observed about 5 June, when the number of adults captured fluctuated between 4 and 60 per trap per day. The number of captures reached a maximum between 20 and 22 June and then decreased to zero at the end of June. This is the period of the second generation. At the start of August there was another increase in the number of adults, varying between one and 98 per trap per day. The maximum number was recorded between 30 August to 3 September.

The numbers of adult L. botrana caught in 1985 in the Nea Peramos, Eleochori and Elevtheres areas are given in **Fig. 3**. Recording of data for the traps began on 18 April 1985 in the three areas. The number of adult L. botrana caught per trap per day varied between 4 and 168. The maximum number was recorded on 28 April in the Nea Peramos area. In the two other areas a large number of captures were observed from the outset using the network of sex pheromone traps and these captures did not reach the maximum for the first generation (**Fig. 3**). Captures of adults subsequently dropped to zero in the Nea Peramos and Eleochorio areas, but in the Elevtheres area captures continued during May and June, although at a low level. The second generation appeared on 6-9 June 1985 in the Nea Peramos area with a maximum per trap per day of 42 adults, with 14 for Eleochorio. The second generation lasted from 4-9 June to 28

23

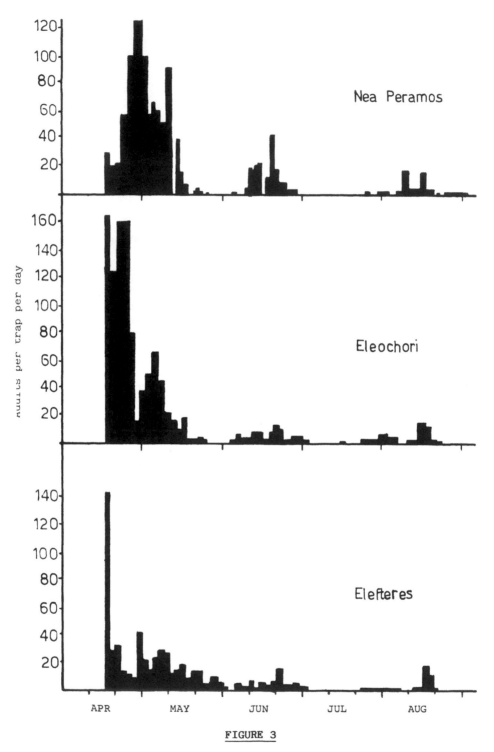

FIGURE 3

Adult <u>Lobesia botrana</u> caught by sex pheromone traps in the vineyards of
Kavala (Macedonia) in 1985.

24

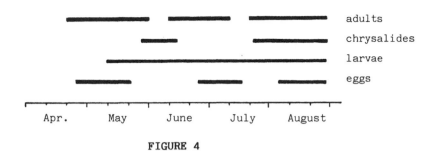

FIGURE 4

Life-cycle of <u>Lobesia botrana</u> in the Nea Peramos district (Kavala) in 1985.

June-3 July. During most of July captures fell to zero in the three areas. The third generation appeared towards the end of July, more specifically on 23-31 July. The maximum numbers recorded in the three areas varied between 14 and 19 adults per trap per day and occurred on 15-20 August 1985. The results of the samplings carried out during 1985 are given in **Fig. 4**. The first laying was observed on 25 April 1985. The eggs were deposited on smooth surfaces (young flower buds, their stalks, bracts and young leaves). On 5 August 1985 fresh eggs were observed on 13% of vine leaves. <u>Eupoecilia (Cochylis) ambiguella</u> was observed during June, July and August at a percentage of 23.3% in the Nea Peramos area. <u>L. botrana</u> accounted for the greatest percentage of the captured population.

4.   DISCUSSION AND CONCLUSIONS

On the basis of the adults captured in the network of sex pheromone traps there were (**Fig. 2**) three generations of <u>L. botrana</u> per year (1, 2) during 1984 in the Elevtheres area. Use of the network of traps was discontinued when grape-picking began on 8 September 1984. It should be noted that the grape-picking period lasts for almost two months. During 1985 three generations per year were observed up to the end of August. We assume that throughout the grape-picking period, which continues until 15 November, there is perhaps a further generation (1/2-1). We are continuing our observations so as to determine the number of generations of <u>L. botrana</u> in the areas mentioned. In Crete (7), 3 1/2 generations of this species were observed in the area of Heraklion which produces grapes of the "Razaki" variety. In Italy <u>L. botrana</u> also produces three generations a year (8). In France three generations of <u>L. botrana</u> were observed in 1985 and also a fourth which is of no economic importance (4). The second generation in 1984 lasted 37 days and the third 42 days. In 1985, the first generation lasted 42 days, the second 28 days and the third 34 days. The variation in the life-span of each generation is influenced by the weather. The <u>L. botrana</u> began laying on 20-25 April 1985 in the Nea Peramos biotope. By this time leaves had appeared on the vine. Some had already unfolded and the clusters were visible.

Eggs were observed on the young leaves, on the bracts and on the stalks of the flower buds. As stated above, at the second generation stage a small number of eggs were observed on the grapes. <u>L. botrana</u> eggs were observed on mature leaves in France (4) by Marcelin. In 1985

E. ambiguella was observed for the first time in Greece (3, 5, 7). The predominant species among the population of insects trapped in the Nea Peramos area was _L. botrana_.

ACKNOWLEDGMENTS

Thanks are due to the Kavala Agricultural Directorate of the Greek Ministry of Agriculture and more particularly to the agricultural engineers, Mr. J. STYBIRIS, Head of the office of phytopathology, and Mr. N. SARAFIS, who helped us with the experimental work in the field.

REFERENCES

1. BALACHOWSKY A.S., 1966, Entomologie appliquée à l'agriculture, Tome II Lepidoptères. 1er vol. Masson et Cie, Paris, 1057 p.
2. BOUVEY R., 1967, La défense des plantes cultivées. Edit. Payot, Lausanne, 1-845 p.
3. ISAAKIDES C.A., 1941, Insects interesting the greek agriculture with some observations on them (in Greek). Proc. Acad. Athens (Meeting Nov. 27, 1941), 1-238-263.
4. MARCELIN H., 1985, La lutte contre les tordeuses de la grappe. PHYTOMA (Défense des cultures) juillet-août, p. 29-32
5. ORPHANIDIS G.T., 1872, About an epidemic larva on grape and currant vineyards, and its control. Geoponica 1 (7), 221-232 (in Greek).
6. PELEKASSIS C.E.D., 1962, A catalogue of the more important insects and other animals harmful to the agricultural crops of Greece during the last thirty-year period, Annls. Inst. Phytopath., Benaki, Nouvelle série, volume 1, p. 5-104
7. RODITAKIS N.E., 1983, Ecological studies on grape vine moth _Lobesia botrana_. Denn. & Schiff. in Heraklion, Crete, Greece, 1st Hellenic Congress on Plant Diseases and Pests, October 5-7, Athens, Greece
8. SILVESTRI F., 1942. _Polychrosis botrana_ Schiff., pp. 407-416. In Silvestri, F., Compendio d'Entomologia applicata, vol. 2, Bellavista Portici.

Observations on biology and control of grape moths in Venetia

S.Zangheri, L.Dalla Montà & C.Duso
Istituto di Entomologia Agraria dell' Università di Padova, Italy

Summary

The Authors refer to some results obtained during 10 years of observations on grape moth (Lobesia botrana Den. and Schiff. and Eupoecilia ambiquella Hb.) in Venetia. The research was carried out on vineyards in the plains and on the hills. Both species normally have 3 generations. They are present throughout, but with varying degrees of frequency. With regard to the damage, experiments confirm that the infestation of the first generation does not reach the economic threshold, except on rare occasions. The second generation almost always requires treatment. Only on the late maturing cultivars is a second treatment necessary, against the third generation. Experiments have been carried out with various insecticides of different toxic classes. Observations have also been made on the role of the natural enemies in the population control of grape moths. Samples taken from numerous vineyards made it possible to note the presence, above all, of Ichneumonids endoparasites of the chrysalids. The percentage of parasitisation is highest (up to 60%) in overwintering chrysalids, while it often appears very low (from 3% to 15%) in the spring and summer generations. Various Arachnida (Trombididae and Araneidae) show predatory activity, mostly against overwintering chrysalids.

1. Introduction

Research on grape moths begun by the Institute of Entomology, University of Padova, in the 1950's (11) in local vineyards, was extended in the last 10 years to many places in Venetia, to gain more knowledge on this problem.

Research work was done on population numbers of these two species (Lobesia botrana Den. and Schiff. and Eupoecilia ambiquella Hb.) and on the course of the generations. Further studies was carried out on sampling methods and damage evaluation, economic thresholds, biological control factors and chemical and microbiological treatments.

Some results have already been presented at meetings of the O.I.L.B. Group "Lutte intégrée en viticulture" at Gargnano (1981) Toulouse (1983) and Bernkastel (1985) by V. Girolami and L. Dalla Monta'.

2. Distribution, Population Density, Biological Cycle

Research was carried out by placing feromone traps for many years in vineyards situated in different parts of Venetia (Padova, Treviso, Pordenone and Gorizia country). Results can be seen from diagrams of catches, chosen as examples of various environmental

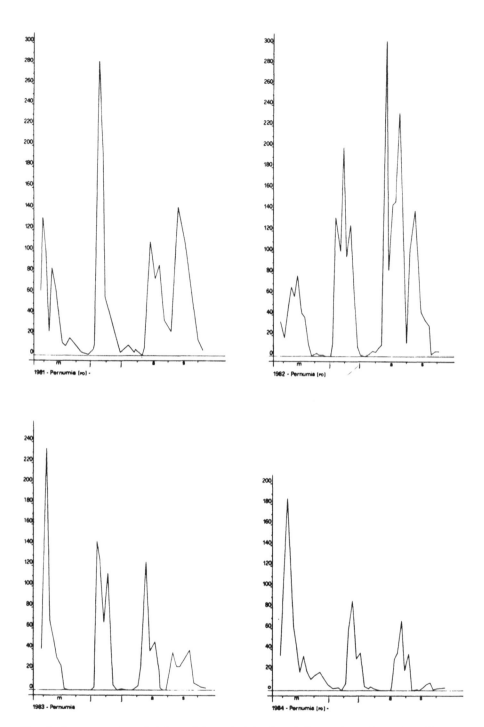

Fig. 1: Average number of daily catches of __Lobesia__ __botrana__ in a vineyard on the Padova plains on Raboso veronese cultivar from 1981 to 1984.

28

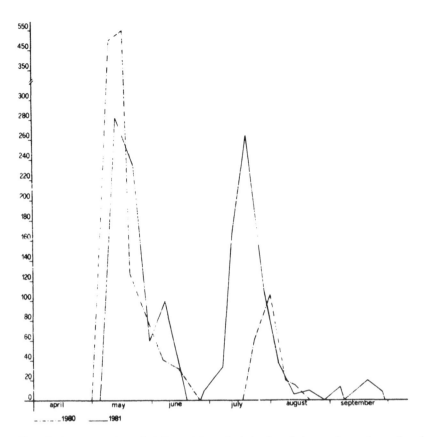

Fig. 2: Average number of daily catches of <u>Lobesia</u> <u>botrana</u> in a vineyard on the Pordenone plains on Pinot bianco cultivar in 1980 and 1981.

situations. Observations have shown that <u>Lobesia</u> is more abundant than <u>Eupoecilia</u> (male catches up to about 200:1) in the Padova plain (Pernumia, Salboro) where research has continued for many years. Peaks of the three generations of <u>Lobesia</u> are clearly visible in some years with continuos catches (Fig.1).

<u>Eupoecilia</u> has nearly always three generations, sometimes with very low catches. This fact can be explained by the very low population density in climatically unfavourable years (very high temperatures); however this requires further research. Third generation larvae can be found mostly on late maturing cultivars (e.g. Raboso, Friularo).

<u>Lobesia</u> is more abundant than <u>Eupoecilia</u> also in the Pordenonese plains (Fig.2). The first two generations of <u>Lobesia</u> are clearly visibile, while the third is either very low or non-existant, on the basis of catches by traps. The very low number of the third generation (without insecticide treatments) as results from three years of experimentation, can lead us to believe it is not a complete generation. Instead, in some parts of the plains around Treviso (Vazzola, San Michele di Piave) <u>Eupoecilia</u> often is more numerous than <u>Lobesia</u> (Fig.3). In Colli Euganei vineyards (e.g. Baone, Rovolon) the population density of <u>Eupoecilia</u> is higher

29

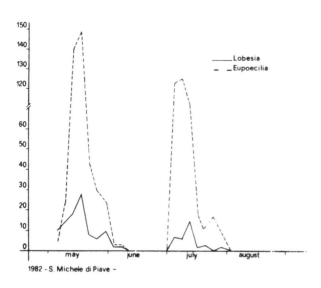

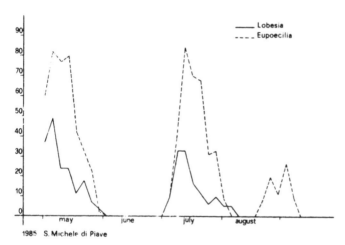

Fig. 3: Average number of daily catches of <u>Lobesia</u> <u>botrana</u> and <u>Eupoecilia</u> <u>ambiquella</u> in a vineyard on the Treviso plains on Merlot cultivar in 1982 and 1985.

compared with nearby plains vineyards, however, this density is still lower than the density of <u>Lobesia</u> . For both species there are three generations, with <u>Eupoecilia</u> having a particularly low density in the third generation (Fig.4). A change in population density between the two species was observed in Baone following the use of insecticide treatments. Up to the time when insecticide treatments were first used, <u>Eupoecilia</u> was clearly dominant. After beginning treatments with organophosphorous compounds there was an clear increase of <u>Lobesia</u> , especially in the third generation, over a two year period. In vineyards of Collio Goriziano <u>Lobesia</u> dominates <u>Eupoecilia</u> in some places, in others the contrary happens, and yet others have only <u>Eupoecilia</u> present (Fig.5). The density between the two species often varies from one year to the next in the same

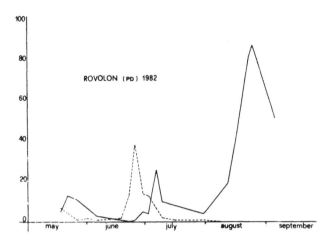

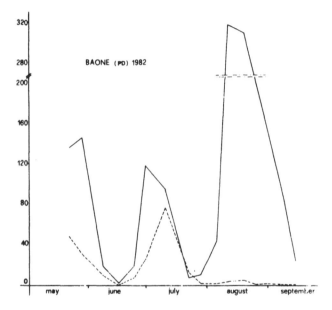

Fig. 4: Average number of daily catches of <u>Lobesia</u> <u>botrana</u> and <u>Eupoecilia</u> <u>ambiquella</u> in a vineyard of Colli Euganei on Merlot cultivar, in 1982.

places. Both species have three generations and can vary in the number of catches of the third generation. While <u>Eupoecilia</u> has a high population density in the first two generations and a decrease in the third generation, <u>Lobesia</u> , on the contrary, has a low density in the first two generations with an increase in the third one. It can be stated that <u>Lobesia</u> , prefers, and is more prevalent in warmer temperatures.

In conclusion, the following general aspects can be observed:

a) <u>Lobesia</u> <u>botrana</u> is the species more abundant in the plains vineyards of Padova, Treviso and Pordenone. There are

31

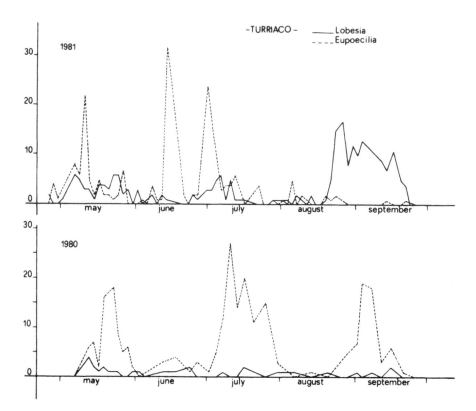

Fig. 5: Average number of daily catches of _Lobesia botrana_ and _Eupoecilia ambiquella_ in a vineyard of Collio goriziano on Pinot bianco cultivar, in 1980 and 1981.

sometimes considerable fluctuations in the population density of the two species in different years.
  b) _Eupoecilia ambiquella_ is more frequent in some hilly regions such as the Colli Euganei and Collio Goriziano.
  c) Both species usually have three generations per year.

3. Damage Evaluation

  The amount of damage of the first generation does not usually require any insecticide treatment, as confirmed by various research. For two years, the weight of bunches of healthy and attacked grapes were compared at harvest, without any important differences resulting. It is generally accepted (1), (6) that a loss of weight in bunches due to a first generation attack is compensated for by the remaining grapes growing larger.
  Research carried out on medium maturing cultivars (Merlot, Cabernet) for two years in different vineyards on the plains around Treviso and Pordenone, shows that the percentage of first generation attacks reached, for example, in 1981 maximum levels of 16-17% (in 4 vineyards) and in 1982 levels of about 30% (in only two vineyards). In other vineyards on the Padova plains with more vulnerable cultivars the first generation shows an attack level of 80-90%.

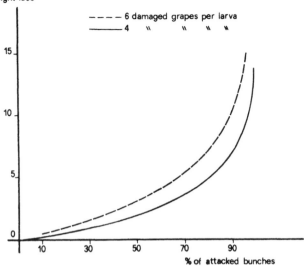

%of weight loss

- - - - 6 damaged grapes per larva
———— 4   "   "   "   "

% of attacked bunches

Fig. 6: Relation between percentage of attacked bunches and weight loss, at harvest, if
4 or 6 grapes are damaged by a single larva.

Experiments carried out in different vineyards with
differential treatments have shown that, even with high levels of
infestation, there is no difference at harvest between vines which
were treated or untreated at the time of first generation attacks.
These results are very interesting as they show the uselessness of
springtime treatments which are overused both before and after
blooming.

Population estimates of second and third generations were always
carried out by periodical sampling of larval stages on bunches.
These samples were taken on the basis of a scheme realised by
Girolami (unpublished data) which as well as simplifying the work,
eliminates any subjective choices in the bunches chosen.In vines
grown by the Sylvoz method the first and last vine branches were
chosen in every couple of vines. On each of these shoots two bunches
were observed. If the first and last shoot had less than two bunches
the next one was considered; if more than two bunches were present
the two first (base) bunches were considered. Sampling was carried
out on the same bunches all season.

Damage evaluation of 2nd and 3rd generations (only on late
maturing cultivars) is the basis for the economic threshold. As it is
not possible to establish amount of damage by weighing the bunches,
the larger ones are also the most vulnerable, it is better to
estimate damage by measuring loss in weight of a certain number of
attacked grapes. Research already published (5) has shown that there
is a relation between the percentage of attacked bunches and the
number of larval nests and bored grapes and, consequently, a relation
between attacked bunches and a loss in weight (Fig.6). For example,
in Pernumia a 27% attack level corresponds to a loss in weight of
1.2%, while at an infestation of approximately 50% the loss visibly
increases.

Table I: Results of chemical control trials using organophosphorous compounds on an early maturing cultivar (Pinot bianco) in 1981 (Prata di Pordenone, PN).

Compounds	Treatment date	% attacked bunches (larval nests/bunch)	Duncan Test
Chlorpyri-phos ethyl (200 g/hl)	20.7	27 % (0.37)	A
	20.7, 18.8	10.4 % (0.10)	A
	18.8	63.5 % (0.91)	B
Not treated		70.8 % (1.61)	B

Table II: Results of chemical control trials using pyrethroids on a late-maturing cultivar (Raboso veronese) in 1980 (Pernumia, PD).

Compounds	Treatment date	% Attacked bunches (larval nests/bunch)	Duncan Test
Deltamethrin (60 g/hl)	25.7	51 % (1.25)	B, b
	25.7, 22.9	27 %	A, a
	22.9	68 % (1.63)	C, b
Not treated		89.1 % (2.91)	D, c

4. Chemical control and Treatment periods

The methods used in cases where chemical control is necessary are different according to the cultivar and compounds.

Regarding early-medium cultivars (Pinot bianco, Tocai, Merlot) it has been observed that the 2nd generations are the most damaging.

Therefore, useful treatment against the larvae of this generation must be done in July while treatment done at a later date (end of August) is superfluous (Tab.I).

On the contrary, for late maturing cultivar (Raboso veronese, Friularo, Corvina) two treatments against 2nd and 3rd generations are always necessary in cases of high attacks (only one treatment is not sufficient) (Tab.II).

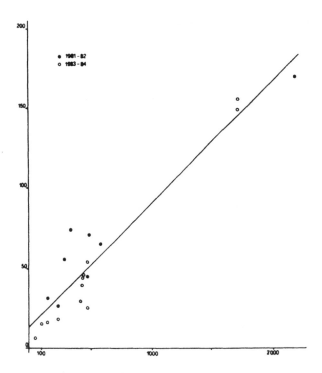

Fig. 7: Relation between the total number of catches and the number of 2nd generation larvae on 100 bunches.

By using ferormone traps, it has been observed that the best treatment period varies according to compounds used. Contact products (e.g. Pyrethroids, _Bacillus thuringiensis_ Berl.) must be used immediately at catches peak, while penetrating insecticides (e.g. methyl-parathion, chlorpyriphos-methyl, fenitrothion) can be used, with good results, even 8-15 days after peak.

The compounds used must be chosen both for their efficiency and also for their side effects (toxic effects in humans, danger for parasites and predators).

Consequently, first class compounds which are very toxic for humans and sometimes for entomophagous species must be excluded; also pyrethroids (deltamethrin, flucithrinate, fenvalerate) are dangerous, especially to Phytoseids, predators of phytophagous mites (3), (4).

Many experiments carried out between 1979 and 1985 with diflubenzuron and some organophosphorous compounds (trichlorfon, phosalone, acephate) gave bad results while good results were obtained with chlorpyriphos-methyl, fenitrothion and pyridafenthion.

Another aspect often discussed by many authors (9), (10) regards the possible relation between total number of catches per generation and percentage of attack on 100 bunches. Observations done at Collio Goriziano (Fig.7) suggest that generally a definite relation does not exist. According to this information it does not seem necessary to treat for very low catches (totalling about 100 males per trap up to peak) but it is certainly necessary if the catch exceeds 500 males. In between these two figures, treatment can be decided on by sampling bunches. Further studies on these

35

above-mentioned relations are in the process of being published by
Dalla Monta' and Cecchini.

5. Biological Control Factors

During many years research the role of entomophagous in limiting
the Lobesia population was studied. To collect as much data as
possible regarding the presence and distribution of these auxiliary
organisms, periodical and extensive samplings were done mostly in the
vineyards around Padova. Vineyards with different conditions were
chosen, both regarding population density of Lobesia , and methods
of treatment used. Sampling was carried out on larval stages and on
overwintering chrysalids 2 or 3 times per generation. Larvae and
chrysalids were isolated in separate containers and observed in a
laboratory until the emergence of any possible parasites; then they
were counted and identified to establish the percentage of parasites
and frequency of each species.

Observations were started in 1982 and are still continuing now,
in many hillside and plains vineyards, but only in a few highly
infested by Lobesia botrana were entomophagous organisms present.

From the data collected, it was possible to establish that
entomophagous fauna consists mostly of Hymenoptera Ichneumonidae ,
and to a lesser degree, Hymenoptera Chalcidoidea . Six species
were identified among the Hymenoptera Ichneumonidae : Pimpla
turionellae L. Itoplectis tunetanus Schm., Dicaelotus
resplendens Holm., Theroscopus hemipterus F. Bathythrix
decipiens Gr. (= meridionator Aub.), Ischnus alternator Grav.,
Gelis cinctus L. Two species were identified among the
Hymenoptera Chalcidoidea : Pteromalid Dibrachys affinis
Masi and Eulophid Colpoclypeus florus Wlk.

All these species, apart from Colpoclypeus florus , are
chrysalid endoparasites and generally polyphagous. Some, such as
Dibrachys affinis , are hyperparasites, while C. florus is a
larval ectoparasite. This species was found occasionally in only one
of the vineyards, situated close to an apple orchard which was highly
infested by Tortricid Argyrotaenia pulchellana Hb. and also
parasitized by C. florus . Samples taken in different periods of the
year have shown, as has been already found by many authors (6), (8),
(7), (2), that the highest percentage of parasites present can be
seen on overwintering chrysalids, where the percentage can reach up
to 60%. In spring, summer and autumn samples, on the other hand, the
percentage is considerably lower, varying betwen 2% and 15%. With
reference to the distribution and the frequence of the
above-mentioned species, it is observed that the most common species
are the following three:

Pimpla turionellae , Itoplectis tunetanus and Dicoelotus
resplendens . The first two species make up 25% of total parasites
present, and the third 20%. The remaining species appear only
sporadically and in very small quantities.

In general, it has been noted that there is a higher number and
more species of parasites present in hillside vineyards as compared
to vineyards on the plains.

As well as the parasites mentioned above, predatory organisms
which can limit Lobesia populations have been found. It is more
difficult to identify the activity of these predators as compared to
parasites, however, their activity is still useful at least against
overwintering chrysalids. In autumn and at the end of winter, it is

possible to see, under the bark of the vine, a high number of Arachnidae , particularly of Trombidiphormes which by eating can destroy a large number of overwintering chrysalids. The presence of these predators, together with parasites, helps to increase the mortality of the juvenile stages during the diapause period.

In conclusion, it seems that entomophagous fauna do not play a determining role in limiting populations of Lobesia . This problem, therefore, requires further research.

Acknowledgements

The Authors wish to thank the specialists Prof. J. Aubert, Prof. G. Viggiani and Prof. G. Delrio for identification of parasites.

References

1. BASSINO, J.P., BLANC, M., BONNET, L., EHRWEIN, B., FABRE, F., RAMEL, J.P., SELVA, M.F. (1978). Les Tordeuses de la grappe. La protection de la vigne en Provence dans l'optique de la "lutte intégrée". La Défense des végétaux, 189.

2. COSCOLLA, R. (1981). Parasitisme de Lobesia botrana Schiff. dans le région de Valencia. Lutte Intégrée en viticulture. IV Réunion plénièere. Rapport. Gargnano (Italie), 10-12 Mars 1981: 11-12.

3. DALLA MONTA', L. and DUSO, C. (1984). Prospettive di lotta integrata in viticoltura. Atti II Convegno S.IT.E. Padova, Giugno 1984.

4. DUSO, C. (1985). Effetti collaterali di alcuni insetticidi sugli Acari Fitoseidi del pesco. Incontro internazionale sull'influenza degli antiparassitari sulla fauna utile in agricoltura: Fitoseidi e Antocoridi, Verona, 27 - 29 Maggio, 1985.

5. GIROLAMI, V. (1981) Evaluation des dégats dus aux vers de la grappe. Lutte intégrée en viticulture. Rapport. IV Réunion plénière, Gargnano. Boll. Zool. Agr. Bachic., (Ser. II), Vol.16: 16-18.

6. ROEHRICH, R. and SCHMID, A. (1979) Lutte intégrée en viticulture.Tordeuses de la grappe. Evaluation du risque, détermination des périodes d'intervention et recherche de méthodes de lutte biologique. Proc.Int.Symp. IOBC/WPRS on Integrated Control in Agriculture and Forestry. Wien 8 - 12 Oct.1979: 245 - 254.

7. SCHMID, A. (1978) Vers de la Grappe 1977 en Suisse Romande. Rapport pour la Réunion OILB "Lutte intégrée en viticulture" Zaragoza, Février 1978.

8. SCHMID, A. (1979) Vers de la Grappe 1978 en Suisse Romande. Rapport pour la Réunion OILB "Lutte intégrée en viticulture" Beaune, Février 1979.

9. TRANFAGLIA, A. and VIGGIANI, G. (1976) Osservazioni sui voli di Lobesia botrana Schiff. (Lep. Tortricidae) con trappole a ferormone sessuale sintetico e prove di lotta. Boll. Lab Ent. Agr. F. Silvestri, Portici, XXXIII: 259 - 264.

10. TRANFAGLIA, A. and MALATESTA, M. (1977) Utilizzazioni di trappole a ferormone sintetico e valutazione del grado di infestazione di Lobesia botrana Schiff. (Lep. Tortricidae) nell'isola d'Ischia nell'anno 1976. Boll. Lab. Ent. Agr. F. Silvestri. Portici, XXXIV, 19 - 24.

11. ZANGHERI, S. (1959) Le "Tignole dell'uva" (Clysia ambiquella Hb. e Polychrosis botrana Schiff.) nel Veneto e nel Trentino. Riv. Vit. Enol. Conegliano, N. 1 - 2.

Damage evolution, larval sampling and treatment period for grape moths

F.Pavan, G.Sacilotto & V.Girolami
Istituto di Entomologia Agraria dell' Università di Padova, Italy

SUMMARY

The economic threshold for the second generation of grape moths based on attack and damage levels at harvest is not useful, since treatment must be given in July.

The relationship between catches on pheromone traps and attacks do not always allow a decision on whether to treat or not.

The "anticipated" economic threshold based on egg-laying on grapes (1 - 10% bunches with visible eggs) presents some sampling difficulties.

The first larval holes on the grapes are easier to observe than eggs. A sampling method based on the first larval holes (10 - 15 days after the emergence peak of male moths) is reported.

The new economic threshold is based on the analysis of the evolution of attacks and damages. In particular, it has been observed that:

1. The groups of rotten grapes, visible on mature bunches, are the consequence of perforations on unripe grapes, made by second generation larvae.

2. A good relationship has been found between early larval attacks (10 - 15 days after the emergence peak of male moths) and final attacks.

3. Penetrating insecticides allow a successful treatment even 10 - 15 days after the emergence peak of male moths.

1. INTRODUCTION

Grape moths are the main entomological problem in vine. On early wine varieties, damage is linked to the first carpophagous generations.

Supervised control of grape moths requires:

-the estimation, for each variety and environment, of the attack level (economic threshold) which causes enough crop damage to justify treatment;

-the availability of simple sampling methods and the establishment of the right moment to treat.

Roehrich (9), in a final report of the O.I.L.B/S.R.O.P. sub group on grape moths, proposes a final damage threshold of 15-20%; however, he states that treatment must be given when 5% of bunches show eggs or larval holes, because there are no indications on the final infestation level when it is still possible to control larval population.

Bassino (1) advises a preventive treatment with contact insecticide 8-10 days after first emergence, and a further treatment 10 days after if 10% of bunches are attacked (20 days after beginning of emergence).

In France the ACTA-ITV (14) recommend the following control strategies:

a) No treatment must be done in absence of catches on pheromone traps (negative threshold).

b) On emergence, it is necessary to sample the first egg-laying on bunches and use economic thresholds which vary from 1% to 10% according to risk of rot.

If the risk of rot is high, sampling and, if necessary, treatment is carried out 8-10 days after the beginning of the actual increase of male emergence (before larval penetration on grape).

If the risk of rot is low, sampling is carried out immediately after mass emergence peak.

If there is further emergence, another sampling is taken observing the first larval holes one week after the first sampling, and a further treatment is required if the threshold is reached.

In central and southern Italy, "early" damage thresholds varying from 2% to 5% of attacked bunches are reported (11, 7).

A better understanding of the following aspects and problems is necessary for a supervised control of grape moths:

1) Treatment must be given in July to prevent harvest damage two months later when practically all larvae will have reached maturity (one or more weeks earlier).

2) It is possible to evaluate attack and damage level at harvest (5) but this is not useful in establishing the economic threshold, since treatments must be done early. The final attack must be predicted at time of treatment (9).

3) The relation between emergence and attack level gives a vague indication of the necessity of treatment (12, 4, 10). However, in North-east Italy (Dalla Monta' and Cecchini, unpublished data) and in Switzerland (3)), the relation between emergence and attack levels advises no treatment for very low emergence; treatment is necessary for very high catches; no indication on control strategies is given for intermediate catches.

4) In any uncertain situation, it is necessary to sample eggs laid or first holes at a time in which treatment can still be useful. Sampling of the very small eggs must be done by expert technicians and requires a long time. The larval holes are larger than eggs and easier to observe, therefore, an economic threshold possibly based on first holes seems more practical.

5) Pheromone traps give a good indication of the beginning of infestation and it is possible to use the emergence peak as a base for sampling.

In this paper it is reported:

a) Study of cronological evolution of attack and damage level, paying particular attention to the relation between early attacks (first holes) and final ones.

b) An estimation of the possibility of delayed treatment (when first holes are already visible) in order to base the economic threshold on the first holes and not the eggs which are difficult to observe.

2. MATERIAL AND METHODS

2.1. Evolution of larval activity and damage

Research was carried out in North east Italy (Pasiano di

Pordenone, at the Pase farm) on the plains with a humid climate and a clayey soil. This farm has a vine area of about 5 ha. with 4 varieties (Chardonnay, Tocai, Merlot, Verduzzo Trevigiano) using "sylvoz" training system.

Data concerned all four varieties in 1983 and only Chardonnay in 1984. For each variety the same 100 bunches were sampled weekly, starting on the week following the emergence peak (when treatment is usually advised). Observations concerned:
- number of larval nests
- bored grapes not yet rotten
- bored rotten grapes shrivelled or fresh
- grapes fallen down
- rotten grapes not bored surrounding bored ones
- rotten grapes not bored and not surrounding bored ones

The attacked grapes which do not become rotten at harvest time, due to the suberization of lacerations, were considered to be not damaged.

In 1984 damage evolution was studied on plots treated with chlorpyriphos-methil (0.2%) 10-15 days after emergence peak (4 August).

In 1985 the effectiveness of delayed treatments was verified on about 50 ha. in 13 different farms.

Four bunches per vine on five vines per row, in five different rows, were observed.

To avoid subjective choices the four bunches observed on each vine were chosen following an a priori scheme. The first vine branch (1 year old wood on which shoots with bunches will grow) near the vinestock was considered; between the shoots of this branch the first one (base) and the last one (top) were considered; on each of these shoots two bunches were observed. If the first or last shoot had less than two bunches the next one was considered; if more than two bunches were present the two first (base) bunches were considered (6).

2.2. Early and final attacks

The early attack level (10-15 days after peak emergence) and final attack level has been observed starting from 1980 (11, 13, 8)). Data was collected by different samplers on various varieties in different farms in North east Italy (tab 1). The number of bunches on which larval nests were counted was 64, 96 or 100 according to experimental designs carried out for other purposes.

Sampling of larval nests must be carrried out at least 10 days after the emergence peak because before this time the number of visible nests is very low and larger sample size is necessary to determine low percentages. The infestation increases rapidly so sampling must be done as late as possible before the last possibility of killing larvae with a chemical spray.

3. RESULTS AND DISCUSSION

3.1. Population dynamics

The emergence peak of the second generation (observed from feromone traps) took place between 9-11 July, 1983 and between 19-21 July, 1984.

The first larval holes were already visible a week after the

41

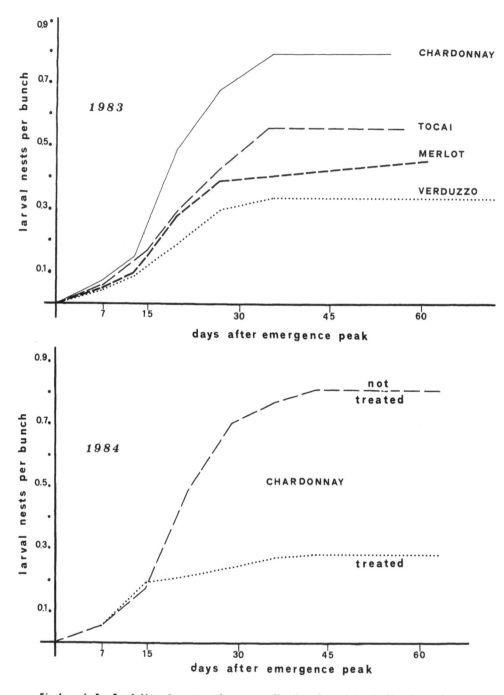

Fig.1 and 2 Population increase of grape moth (larval nests) starting from male emergence peak: Four varieties were observed in 1983 on untreated plots (above); in 1984 only Chardonnay varieties were observed on untreated and treated plots (15 days after emergence peak) (below).

42

Table 1. Years, number of samplers employed, localities, varieties, number of bunches observed for each variety to evaluate the ratio between early attack levels and final attack at harvest.

The early attack corresponds to the number of larval nests observed 10-15 days after the emergence peak, established on pheromone trap catches.

Year	Number of samplers	Localities		Varieties	No.samples for each variety	No. of bunches /sample
1980	1	Prata	(Pn)	Tocai	1	96
1981	2	Prata	(Pn)	Tocai	2	96
1982	1	Pasiano	(Pn)	Merlot	4	96
					7	64
1983	2	Pasiano	(Pn)	Chardonnay	1	100
				Tocai	1	100
				Merlot	1	100
				Verduzzo Tr.	1	100
1984	2	Pasiano	(Pn)	Chardonnay	1	100
1985	3	Pasiano	(Pn)	Chardonnay	1	100
				Tocai	1	100
				Verduzzo Tr.	1	100
		Villorba	(TV)	Merlot	2	100
				Chardonnay	2	100
		Negrisia	(Tv)	Cabernet	1	100
		Rosa'	(Vi)	Pinot B.	1	100

peak, but only between the 15th and 30th day (25 July - 10 August in 1983 and 5 - 20 August in 1984) a rapid increase in larval nests was observed per bunch (Fig.1 and 2).

The population increased until 35 days after peak in 1983 (15 August) and until 40 days in 1984 (30 August). The last second generation larvae were found 45 - 50 days after peak in 1983 (end of August) and 50 days after peak in 1984 (first decade in September).

No third generation larvae have been found in the early (normal) harvesting varieties reported in this work. Data on the infestation on late harvesting varieties will be reported in another paper.

3.2. Early and Final Attacks

Between the early attack level, at 10 - 15 days after peak, and the final attack level a close relation can be seen. The regression (Fig.3) shows an intercept of 0.01 and a slope of 3.53, (R =0.90), this means that the ratio between final number of larval nests and early ones is around 3.5.

This ratio resulted valid in different years, different farms and different samplers and can be extended to all the varieties which have an early harvest.

This data can be considered correct for the vineyards on the

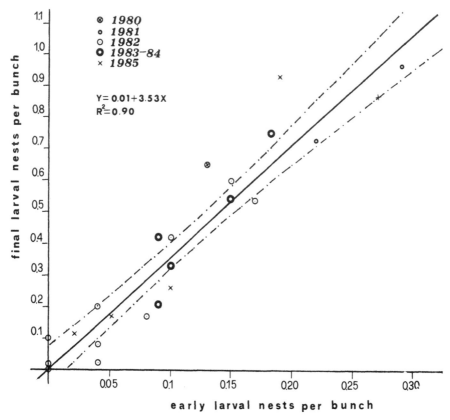

Fig.3 Relation between early attacks (10-15 days after emergence peak) and attacks at harvest. Data refers to sampling on four varieties (Merlot, Tocai, Chardonnay, Verduzzo Trevigiano), carried out by different samplers, in various areas of North east Italy, between 1980 and 1985.

plains in North east Italy; obviously, this ratio must be verified in different climatic conditions, but a definite ratio can be expected in every environment.

3.3. Damage Evolution

The second generation larvae penetrate the grape (which is still green) either near the pedicel or at the point where the two grapes contact, often held together by silk.

Early attacks which consist of a hole in the centre of a small, dark area on grapes can only be seen by moving the grapes which contact. The dark area corresponds to the lacerated pulp and faeces visible under the transperent skin. At the entrance to the holes excrements are often seen.

The damage, starting from the peak, shows increases in quantities and modifications in qualities (Fig.4).

Up to 15 days after peak the number of bored grapes per bunch increase slightly because few larvae are present on bunches and the small larvae have attacked only one or at the maximum two grapes.

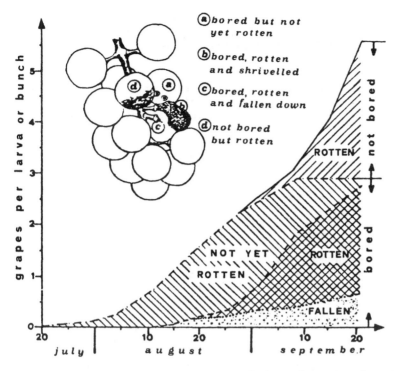

Fig.4. Evolution of damage provoked by the first carpophagous (second annual) generation larvae.

The number of grapes directly attacked by larvae increases until the first week of September when all the larvae have reached maturity; in the second half of August some bored grapes begin to rot. The surrounding grapes which are not bored but in contact with rotten ones, begin to rot. The rot spreads, in the absence of larvae, until harvest, doubling the damaged grapes (3 bored per larva (or per attacked bunch) and about 6 completely rotten).

The data refers to weekly samples on Chardonnay varieties with a particularly rainy season (1984) in a humid region (Pordenone) and therefore the number of grapes indirectly damaged by one larva is relatively high.

The number of bored grapes increase rapidly in the following 3 weeks due to the increase of larvae per bunch and the number of grapes bored by a single larva. The number of bored grapes tends to become constant 40 days after peak because most of the larvae reach maturity.

The findings reported in Fig.4 corresponds to a final mean of one larvae per bunch (fortunately reached in the bunches on the top shoots, according to the sampling methods reported, and correspond to 50 bunches on 100 observed). The data can be therefore expressed at harvest both for larva or for bunch. The data for larva can be extended to different infestation levels; the data for bunch are only valid for a final mean of one.

Every larva during its activity can directly damage from 1 to 6 grapes with an average of 2-4 grapes bored per larva. The number of grapes bored can change depending on the larval survival and on the varieties and the climatic conditions (6).

45

Table 2: larval nests per bunch, bored grapes per larva, bored but not yet rotten grapes per larva, rotten grapes not bored per larva, on plots treated (15 days from peak) and non treated, against the first carpophagous of Grape moths. Data refers to cultivar Chardonnay at harvest in 1985

	Larval nests per bunch	bored grapes per larva	bored but not yet rotten	rotten grapes not bored around larval nests
Treated	0.28 a	1.44 a	0.90 b	0.15 a
Not Treated	0.81 b	2.88 b	0.25 a	3.34 b

Generally as maturity takes place, the grapes bored by Grape moths become rotten. This rot develops from the larval excretions contained inside; the rotten grapes shrivel and sometimes fall.

The grapes which are not bored and in contact with the bored and rotten ones, become infected and themselves rot especially near harvest time when the grapes become ripe. These unbored grapes can spread rot. Even these grapes may split and shrivel so that at harvest it is difficult to distinguish them from the bored ones.

Some bored, unripe grapes can heal and remain whole until harvest time.

Even grapes which are far from bored ones can rot but this is independant from larval attacks. No statistical correlation has resulted between these odd, rotten grapes and larval nests (6).

3.4. Damage evolution and efficacy of delayed treatment

Treatment carried out 15 days after the "peak" has reduced the number of larval nests by about 1/3. The increase was almost stopped by the treatment (Fig 2 and tab.2).

Treatment has also reduced the number of grapes bored by a single larvae due to the penetrating insecticides which killed a large number of newly-penetrated larvae (Tab.2).

In treated plots many grapes heal and develop normally without either a loss in weight or infecting surrounding grapes. Therefore there are fewer rotten grapes not bored per larval nest on these plots (Tab.2). The capacity of the "early" bored grapes to heal and not become rotten, even in rainy weather, is due to the fact that Botrytis is normally unable to invade unripe grapes (2). Probably the suberization around small bored areas, on green tissues before they become dark coloured is sufficient to prevent rot even when the grapes ripen.

The last useful moment to spray seems to be 15 days after the peak. After this period there is a rapid increase in larval nests and in the number of bored grapes per larva; therefore an increase in bored grapes per bunch (Fig.1,2,3). However, a further delayed treatment can result useful because it stops a large number of larvae and the increase in rotten grapes. In the last weeks when larvae have reached maturity treatments are completely useless (11).

For delayed treatments penetrating insecticides must be used.

A further advantage of delayed treatments is that it kills the
larvae derived from the last emerged adults that normally escape
treatments. These larvae are more dangerous than the first ones
because they attack grapes which become rotten near harvest and more
easily infect the nearly-ripe, surrounding grapes. The grapes
attacked by first larvae are already dry and transmit infection less
easily when surrounding ones become nearly-ripe.

The first larvae are already developed when they are killed, but
the grape is far from being ripe and so the number of rotten grapes
is limited.

The efficacy of delayed treatment was confirmed in 1985 on
different farms and varieties, in 3 of these farms on 5 varieties an
unusual early attack level of 30%-60% was found. At least 90% of
larval nests did not develop and the bored grapes were completely
healed.

On a farm, where an untreated control plot was maintained, an
early attack level of 60% on bunches was found; after the delayed
treatment three larvae on 100 bunches survived on treated plots and
197 on 100 bunches on the unsprayed one.

Delayed treatment is not applicable on table varieties because
grapes with holes that are healed are not estetically acceptable.

4. CONCLUSION

1. The % of final attacks which provokes an economic damage
higher than the cost of treatment can be decided for every variety
and environment (5). Even with the same % of attacks, damage is
higher on compact variety grapes which are susceptible to rot, in
humid climates and rainy years (6).

2. Delayed treatment (10 - 15 days after peak) is useful in
limiting damage of larvae already penetrated in the grapes and also
in limiting the increase of further larvae after treatment.

3. The close relation between early attacks (10 - 15 days after
emergence peak) and final ones is, in practice, very important,
making it possible to predict at time of treatment the % of attack
and damage to harvest.

4. Once the final damage threshold has been established and the
relation between early and final attack level has been observed, it
is possible to fix an early economic threshold beyond which treatment
becomes economically convenient.

5. Early larval sampling is necessary only if catches with
pheromone traps are contained in a range superior to a minimum value
under which any treatment is useless and inferior to a maximum value
over which it is always necessary to treat, as already mentioned (3,
Dalla Monta' and Cecchini, unpublished data). The range must be
calculated for every environment, or different climatic conditions.

Attention should be given to the third generation in climates
warmer than those mentioned since this generation seems to cause more
damage than the second one on all varieties.

It is to be hoped that someone will confirm the existence of
definite ratios between early attacks and final damage for the most
dangerous generations (second or third) in different environments.

Acknowledgements
        Thanks to Dr E. Turbian for his help in collecting field data.
        We would like to thank Farmer's s.r.l. Cooperative Service
(Carrara S. Giorgio, Pd) which allowed the publication of data
(1985) and the Consorzio Vini D.O.C. "Lison - Pramaggiore" Venezia.

REFERENCES

        1. BASSINO, J.P., BLANC, M., BONNET, L., EHRWEIN, B., FABRE, F.,
RAMEL, J.P., SELVA, MARIE F. (1978). Les tordeuses de la grappe. La
protection de la vigne en Provence dans l'optique de la "lutte
intégreée". La Défense des Végétaux, 189: 12-26.
        2. BISIACH, M. and VERCESI, A. (19). Aspetti biologici ed
epidemiologici di Botrytis cinerea Pers. su vite
        3. BOLLER, E. and REMUND, U. (1981). Relations entre les
captures de Eupoecilia ambiquella et le pourcentage de grappes
attaquées. In ROERICH R.: Travaux du sous-groupe "Tordeuses de la
grappe". Lutte intégrée en viticulture; IV Réunion plénière, Gargnano
(Italie), 10-12 mars 1981. Boll. Zool. agr. Bachic. (ser. 2) 16:
21-22.
        4. CACCIA, R. (1981). Lotta guidata nella Svizzera Romanda. Atti
3 incontro su "La difesa integrata della vite". Latina, 3-4 dicembre
1981: 145-155.
        5. GIROLAMI, V. (1981). Evaluation des degats dus aux vers de la
grappe. In ROERICH R.: Travaux du sous-groupe "Tordeuses de la
grappe". Lutte intégrée en viticulture; IV Réunion plénière, Gargnano
(Italie), 10-12 mars 1981. Boll. Zool. agr. Bachic.,(ser.2) 16:
16-18.
        6. GIROLAMI, V., PAVAN, F., TURBIAN, E., CECCHINI, A. (1985).
Evoluzione cronologica dei danni di Lobesia botrana Den. e
Schiff. e Eupoecilia ambiquella Hb. - Redia (in print).
        7. MOLEAS, T. (1981). Biologia ed etologia della Lobesia
botrana in Puglia. Possibilita' di lotta integrata. Atti 3 incontro
su "La difesa integrata della vite". Latina, 3-4 Dicembre 1981 :
91-97.
        8. PAVAN, F. (1983). Controllo delle tignole in vigneti del
Pordenonese, pullulazione di fitofagi indotti e verifica di nuovi
criteri per una soglia 'anticipata' di intervento. Doctoral
diseratation, University of Padua.
        9. ROERICH, R. (1977). Recherches sur la nuisibilité des
tordeuses de la grappe E.ambiquella Hb. et L.botrana Den. et
Schiff. Compte-rendu de la Réunion OILB/SROP "Lutte intégrée en
vignoble" (sous groupe "Vers de la grappe"). Nimes 1977.
        10. ROERICH, R. (1981). Travaux du sous-groupe "Tordeuses de la
grappe". In Lutte intégrée en viticulture; IV Réunion plénière ,
Gargnano (Italie), 10-12 mars 1981. Boll.Zool.agr.Bachic. (ser. 2)
16: 7-34.
        11. SACILOTTO, G. (1981). Valutazione dei danni di Lobesia
botrana Schiff. Oppurtunita' d'intervento e conseguenze dei
trattamenti fitosanitari. Doctoral disertation, University of Padua.
        12. TRANFAGLIA, A., WEBER, F., BIANCHI, A. (1981). Prima
esperienza sull'applicazione a livello territoriale della lotta
integrata agli insetti della vite. Atti 3 incontro su "La difesa
integrata della vite". Latina, 3-4 dicembre 1981: 71-89.

13. TRANFAGLIA, A. and MALATESTA, M. (1977). Utilizzazione di trappole a feromone sintetico a valutazione del grado di infestazione di _Lobesia_ _botrana_ (Schiff.) (Lep.Tortricidae) nell'Ischia nell'anno 1976. Boll.Lab.Ent.agr.Portici 34: 19-24.

14. TURBIAN, E. (1983). Stima precoce dell'entita' degli attacchi della prima generazione carpofaga di _Lobesia_ _botrana_ Schiff., evoluzione dei danni e possibilita' di controllo chimico posticipato. Doctoral disertation, University of Padua.

15. AA divers (1980). Protection intégrée controles periodiques au vignoble. ACTA-ITV ISSN 0337-83 49. Imp.Laboureur et Cie, 36100 Issoudum, vol.II: 35-37.

Observations on the biology and risk of grape moths in Piemonte

L.Corino

Istituto Sperimentale per la Viticoltura, Sezione di Asti, Italy

Summary

During the period 1979–1985 a survey on the presence of grape vine moths was carried on by pheromone traps in a great viticultural aera of Piemonte. The research has proved that : Lobesia botrana is the vine moth mostly widespread while Eupoecilia ambiguella was found rare and in cooler environments mainly. Grape vine moths' presence is greatly different according to viticultural zones' features. Moth growth stages are changeable in the years according to climatic conditions, temperatures mainly, and vines development stages. The vine moth risk is rather different from one place to another and may be changeable as well in adjoining vineyards. Training systems as pergola are more suitable for moths development in comparison with trellis which, in the shade side, are normally more affected by the moths. High temperatures and low humidity conditions may affect larvae survival more than eggs hatching. Overwintering chrysalids parasitism is up to 12% and done by Hymenoptera Ichnemonidae and Braconidae. Efforts in order to improve the knowledge of grape moth risk in the different viticultural areas should continue.

1. INTRODUCTION

The grape moths are widespread in Piemonte viticulture area but their risk is changeable from one environment to another. Wine growers are often in difficulties with regard to the real risk, the treatment period and insecticides choice in order to avoid secondary effects. Biological studies on grape moths and an attempt to get some answers for the above problems have been done in the present experience. In the past, different authors have been dealing with similar subjects (1, 2, 3, 8, 9, 10).

2. MATERIAL AND METHODS

Observations on the biology of grape moths have been done with pheromone traps during the period 1979–1985 in a large Piemonte viticultural area (**Fig. 1**). For each year were chosen 40 observation stations differing greatly for climatic and viticultural specific situations.

3. RESULTS

Lobesia botrana Den and Schiff was found the most widespread grape moth and almost single may times. Eupoecilia ambiguella Hb was met less important and mainly in cooler environment. Occasionally was also found Sparganothis pilleriana Den and Schiff; Lobesia and Eupoecilia presence has been very changeable in the viticultural zones and in the years; the ratio between the two species is as well very different from one place to another. Emergence peaks of male moths, useful for possible treatment decision, are definitely determined by climatic conditions which are liable of great changes in the years (**Fig. 2**). It has been proved for both moths that the mass emergence peak of anthophagous generation is almost at the same date while for the first carpophagous generation the Eupoecilia peak comes first as regards to Lobesia emergence peak; such differences are much greater as far as environment temperatures are higher.

Pheromone traps have proved a good evidence of the spring and of the first carpophagous male moths flight but almost useless for the second carpophagous generation which, for Lobesia botrana, has been observed in all the seven years of the experience. The moth adults number catched by pheromone traps in the same places showed a lowering trend in the years. There seems to exist, for extreme values, a connection within traps adults catched and expected risk : over 60-80 and below 10 male moths adults a day at the maximum peak level, we would expect high or low risk respectively. Great amount of moths adults catches were mainly observed in the intensive viticultural area and where grape moths control by insecticides has been widely done over the years. In such environments Lobesia botrana was found almost sole.

Where vineyards are more occasionals and few or any are the insecticides treatments, grape moth risk is almost inexistent. An interesting example is that of Monferrato Casalese region, where at the beginning of the century, the vineyards were rather widespread and the grape moth risk high (5, 6, 7). At the present time in that area, many vineyards have been replaced by cereal fields, the woodland has improved and almost no grape moth risk exists.

The anthophagous generation is almost without risk with the exception of some specific environments like Carema where the damage is important. The first carpophagous generation is often damaging berries whilst the second generation is hardly harmless.

Until 1983 a large grape moths presence was found not over 350 m height; starting with 1984 a great presence of Lobesia botrana was also recorded up to 400 m height in an intensive grape cultivation (Moscato area). In the pergola training systems in comparison with espalier training is recorded a higher damage by moths reason of more suitable microclimatic conditions for larvae survival and fungus infections spreading from perforated berries.

The great amount of observations done with pheromone traps allow to conclude that few moth recording traps are enough for great viticultural areas informations, provided that climatic conditions are similar. In the same hill at the similar height between the south and the north side is recorded a difference in the mass moths emergence peak of 7/8 days or more. Viceversa long distance viticultural zones with the same climatic conditions have a moth peak at the same period.It has been observed in viticultural uniform areas an infestation graduality which,

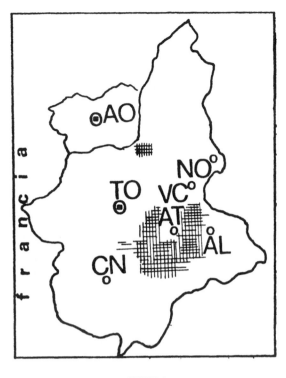

FIGURE 1

Piemonte observation zones on grape moths by

pheromone traps

starting with earliest varieties is extending gradually to latest ripening varieties following plants phenology. In warmer and drier summer conditions like 1983 and 1985 has been shown a low eggs death rate (3%) but a more substantial larvae mortality (30%). Overwintering chrysalids' natural enemies were found in Hymenoptera Ichneumonidae Braconidae. The percentage of parasitisation was about 12%, of collected chrysalids without any relationship with the vineyard insecticides history (never, occasionally or systematically treated).

4.   CONCLUSIONS

The validity of pheromone traps has been proved successful in order to determine the possible date for insecticide treatment but are lacking in forecasting the expected risk.

-   Few pheromone traps, positioned in well representative climatic areas, are good enough for getting extensible informations for wide viticultural zones.

-   Piemonte viticultural zones are very different for grape moth occurrence and risk. Where grape moth control by insecticides has

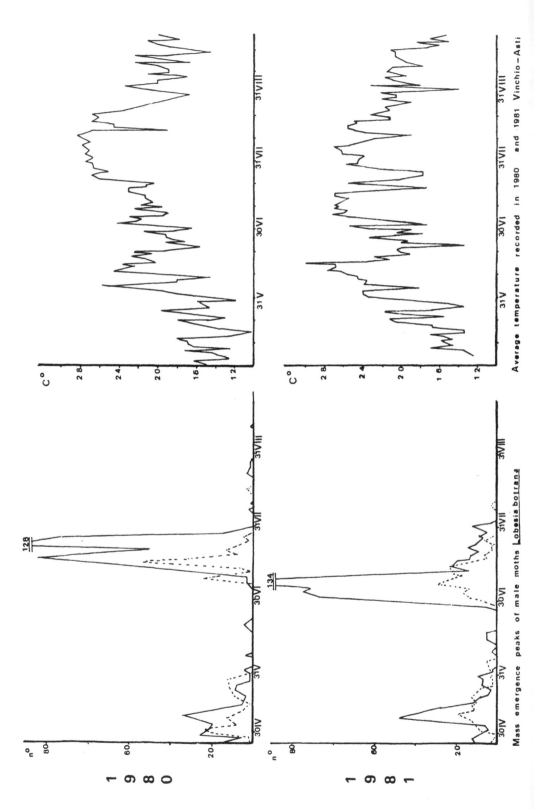

Mass emergence peaks of male moths *Lobesia botrana*

Average temperature recorded in 1980 and 1981 Vinchio-Asti

been more intensive over the years, moths are much more serious than elsewhere and Lobesia botrana is almost the only moth species present.

- Natural enemies of grape moths are lacking in efficiency toward infestations in the high risk zones.

- Very often the insecticide control of grape moths have been liable of noxious growing of spider mites Panonychus ulmi and Eotetranychus urticae mainly, and of the death of Phytoseids with the exception of one case of organophosphorous compounds resistence found out for Amblyseius aberrans in Carema (4).

- It is important to make efforts in order to improve the knowledge of the moth risk following viticultural areas.

REFERENCES

1. CELLI G., CASARINI C., BARBIERI R., BECCHI R. (1980) - Risultati di trattamenti con preparati a base di Bacillus thuringiensis contro la Lobesia botrana Schiff. (Lepidoptera Tortricidae) nel Modenese in rapporto al rilievo feromonico e al danno. Atti Giornate Fit., 431-439
2. CORINO L., MAGNAGHI G. (1982) - Esperienze di controllo delle tignole dell'uva in Piemonte. Atti Giornate Fit., 197-205
3. CORINO L. (1984) - Pullulazioni di Panonychus ulmi (Koch) quale conseguenza dell'impiego di alcuni insetticidi nella lotta alle tignole dell'uva. Atti Giornate Fit., 149-158
4. CORINO L. (1985) - Le specie di fitoseidi (Acarina : Phytoseiidae) presenti in vigneti del Piemonte. Vignevini N.6, 53-58
5. DALMASSO G. (1910) - La lotta contro le Tignole dell'uva - Studi ed esperienze. Le Staz. Sper. Agr. Ital. XLIII fasc. VII-IX, 593-645
6. DALMASSO G. (1921) - La lotta contro le tignole dell'uva - Stato attuale del problema. Novità viticole - Ed. F.lli Marescalchi di Casale Monferrato
7. GABOTTO L. (1920) - I risultati di un esperimento di lotta collettiva contro le tignole della vite. Ed. Cons. Prov. per la difesa della viticoltura di Alessandria. Arti graf. già F.lli Torelli
8. GEOFFRION R. (1977) - Les "vers de la grappe" toujours présents et toujours menaçants. Phytoma - Défense des cultures 9-18
9. SCHMID A., ANTONIN Ph. (1977) - Bacillus thuringiensis dans la lutte contre les vers de la grappe, eudemis (Lobesia botrana) e cochylis (Clysia ambiguella) en Suisse romande. Rev. Suisse Vit. Arb. Hort., 9, 119-126
10. SCHMID A., ANTONIN Ph., RABOUD G. (1977) - Effets des conditions météorologiques particulières de l'année 1976 sur l'évolution des vers de la vigne. Revue Suisse Vitic. Arboric. Hortic. 9 : 131-135

Researches on grape-vine moths in Sardinia*

G.Delrio, P.Luciano & R.Prota
Istituto di Entomologia Agraria dell' Università di Sassari, Italy

Summary

The main wine-grape insect pests in Sardinia are Lobesia botrana (Den.
et Schiff.) and Eupoecilia ambiguella (Hb.). L. botrana was found in
all types of vineyard, producing three generations a year in central
and northern areas and a fourth partial generation in the south.
E. ambiguella was only present in some irrigated arbour-type vineyards;
three generations a year were observed, but with high population densities only in years with moderately hot summers and high humidity.
The most severe infestations normally occurred in vineyards with extended training systems (e.g., arbour) and on compact-cluster varieties.
L. botrana populations varied considerably over the years; limiting
factors were high summer temperatures and mortality of overwintering
pupae from numerous parasites and predators. Infestations were controllable by intervening on the second generation with a single insecticidal treatment, on the basis of pheromone-trap captures and when the
intervention thresholds (10-15% of clusters attacked for loose-cluster
varieties and 5-10% for compact) were exceeded.

1. INTRODUCTION

Sardinian viticulture covers in all about 70,000 ha, the grapes being
grown almost completely for wine production (only a small part of which is
classified as DOC —Denominazione d'Origine Controllata). Low head-training
is still the most common form of culture, but other forms such as the espalier and the irrigated arbour are gradually gaining ground.

Among a number of insect pests, the common grape-vine moths, Lobesia
botrana (Den. et Schiff.) and Eupoecilia ambiguella (Hb.), are predominant.
Other insects causing only sporadic damage, but which can be sometimes
quite substantial, are the termite Kalotermes flavicollis (Fabr.), the
leafhoppers Jacobiasca lybica (Berg.-Zan.) and Zygina rhamni Ferr., the
mealybug Planococcus ficus (Sign.), and the coleopters Triodonta raymondi
Perr. and Apate monachus Fabr. Attacks by Viteus vitifoliae (Fitch) and
spider mites, which have been reported elsewhere in Italy, are extremely
rare in Sardinia.

---

*Work supported by M.P.I., Italy

Control methods employed against the moths vary considerably, but are generally based on interventions repeated at fixed times of the year. With the aim of devising a comprehensive pest management scheme suitable to the island, studies were begun in 1979 on the distribution and population dynamics of the moths and on the damage caused by them. In this paper, only that part of the elaborated data is reported which is considered sufficient for a significant explanation of the situation in Sardinia.

2. MATERIALS AND METHODS

The researches on life cycle, flight periods, natural enemies, damage and control measures were carried out on some representative cultivars in the following localities:

Tempio, 1 ha, headed vines, cv: Moscato and Vermentino;
Bonnanaro, 6 ha, espalier, cv: Pascale di Cagliari and Sangiovese;
Alghero, 350 ha, arbour, cv: Vermentino, Cannonau and Carignano;
Uta, 200 ha, arbour, cv: Nuragus and Monica;
Villamassargia, 30 ha, arbour, cv: Cardinal and Italia.

Lobesia life-cycle data were particularly obtained in the Bonnanaro vineyard; weekly sampling was effected of all stages of development, larval instars being distinguished by measurement. Pheromone traps (Farmoplant) were used in all the vineyards for the adult flight studies; catches were noted weekly and in some areas twice a week or daily, and the traps renewed at the end of each generation. For the infestation data, 100 to 200 clusters were examined of each variety and for each generation. The overwintering pupal stage densities were obtained by barking 200 plants chosen at random in each vineyard under observation, and the pupae found were placed in separate plastic containers for examination as to parasites. Samples of the spring and summer generation larvae and pupae were also examined for parasites. In the control trials, various insecticides were tested as well as two different preparations of Bacillus thuringiensis Berl. (Bactospeine and Thuricide HP).

A further large scale investigation was conducted in 1982 in collaboration with the Technical Assistance Centres of the Sardinian Region. Almost a thousand sites were involved, in each of which 3 traps were utilized for Lobesia and one for Eupoecilia and 100 clusters per vineyard per generation were examined.

3. RESULTS AND DISCUSSION

3.1 Distribution and density of the moths

Lobesia was found in every vineyard studied in Sardinia. Eupoecilia only occurred in some irrigated arbour-trained vineyards, whether in the north or south of the island, and in some northwestern espalier vineyards (Alghero) where the immediate vicinity of the sea creates particular microclimatic conditions. The ratio between the two species as shown by the numbers caught varied considerably from year to year (Fig. 1), at one time the proportions were nearly equal and at another Eupoecilia was almost inexistent. Climatic exigencies, particularly hygrometric, differ widely for

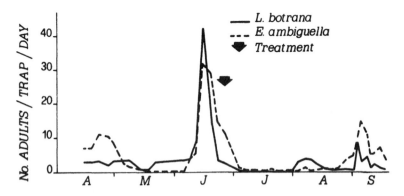

Fig. 1. Grape-vine moth catches by pheromone traps in an arbour-trained vineyard (Alghero, 1981).

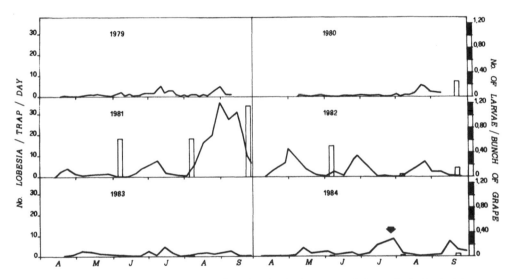

Fig. 2. Pheromone trap catches of <u>Lobesia</u> <u>botrana</u> and larval infestation of grape-clusters (histograms) in an espalier-trained vineyard at Bonnanaro; in 1984, an application of insecticide was effected (indicated by the arrow).

the development of the two species (15, 16). With a greater need for high relative humidity, Eupoecilia populations were larger in the years with humid spring and summer seasons (1980 and 1981) and in the vineyards where irrigation and the arbour system (Alghero and Villamassargia) create much higher local relative humidity than exists in head-trained systems.

3.2 Flight periods
    Pheromone trap catches in the north-central area revealed three flights (Fig. 1 and 2), which corresponded to the same number of gener-

59

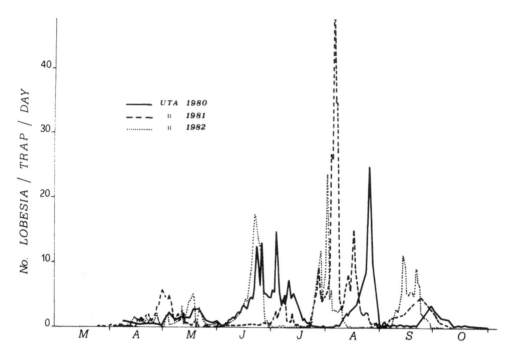

Fig. 3. Pheromone trap catches of Lobesia botrana in an arbour-trained vineyard (Uta).

ations. The flight periods obviously depend upon seasonal climatic trends, but generally the first flight was observed from halfway through April to the end of May, the second from the beginning of June to about the 20th of July, and the third from early in August to the end of September. These findings in the northern and central areas refer equally to both species, but in the south L. botrana produced a fourth flight and generation, the latter being only partial in that the larvae were prevented from completing development by the vintage (Fig. 3). Here, the first catches were recorded at the beginning of April, and subsequent flights all began about ten days earlier than in the north. The third flight always came to an end by the last days of August, the fourth running from early September to the middle of October. Our results for the north of Sardinia are similar to those for Venetia (8) where the two species equally produced three generations. The catches of Lobesia made in the south confirm the existence of a fourth generation, if only partial, as reported in Apulia (12).

3.3 Development of Lobesia botrana generations and mortality factors

The Bonnanaro studies showed three periods for the 1st and 2nd larval stages: (i) the month of May, (ii) from 20th June to 20th July and (iii) the months of August and September. Pupae of the anthophagous generation

were observed all the month of June, those of the first carpophagous generation during the whole of August and those of the second (under the rhytidome) from the end of September onwards. The young larvae of the first generation reached maximum numbers about 20 days after adult catches reached a
peak; in the second generation larval and adult peaks almost coincided; in
the third some 10 days separated the two (Fig. 4).

The incidence of biotic mortality factors on the overwintering pupae
was observed from 1980 to 1982 at Bonnanaro and from 1982 to 1984 at Alghero. Mortality due to parasites (Ichneumonidae, Braconidae and Pteromalidae)
varied from 38% to 48% at Bonnanaro and from 74% to 84% at Alghero. Predators, on the contrary, (Forficula auricularia L., Malachius sardous Er.
and M. spinipennis Germ.), proved to be more active in the former locality
causing mortality which varied from 16% to 25%; at Alghero it was only from
8% to 10%. The principal parasites observed were the pteromalid Dibrachys
affinis Masi and two ichneumonids, Pimpla apricaria Costa and P. turionellae
L. Numerous Cryptinae were taken from the pupae and are probably to be
counted among the hyperparasites of the lepidopter. As in Spain (6), the
extent of parasitism varied considerably; D. affinis, however, proved to be
the most diffuse in both countries (7). Mortality due to other causes (including fungus diseases) varied from 3% to as much as 25%. In the four
years' study of Lobesia overwintering pupae, from 5% to 20% adults succeeded in emerging (Tab. I).

With regard to the larvae, much less parasitism was observed than in
the case of overwintering pupae —10% to 12% of anthophagous generation lar-

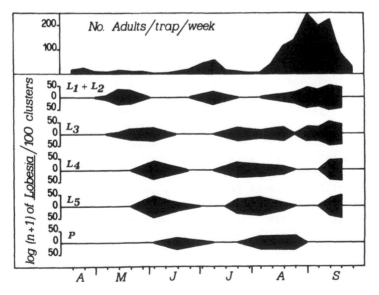

Fig. 4. Lobesia botrana pheromone trap catches of males and larval and
pupal densities in an espalier-trained vineyard (Bonnanaro, 1981).

Tab. I. Lobesia botrana pupal mortality in winter; parasites noted (asterisks indicate relative importance).

		BONNANARO		ALGHERO	
		1980–81	1981–82	1982–83	1983–84
Pupae monitored	(No.)	120	100	110	226
Killed by parasites	(%)	38.4	48.2	74.5	84.0
" " predators	(%)	25.5	16.4	10.0	8.0
" " other causes	(%)	25.6	15.8	7.3	2.7
Adults emerged	(%)	10.5	19.6	8.2	5.3

ICHNEUMONIDAE		Cremastinae	
Pimplinae		Pristomerus vulnerator Panz.	*
Itoplectis alternans Grav.	**	Metopiinae	
Pimpla apricaria Costa	***	Triclistus lativentris Ths.	*
Pimpla turionellae L.	***	Ichneumoninae	
Scambus elegans Woldst.	*	Dicaelotus resplendens Holm.	*
Cryptinae		BRACONIDAE	
Agrothereutes pumilus Krb.	*	Chelonus sp.	*
Bathytrix argentatus Grav.	*	PTEROMALIDAE	
Hemiteles sp.	*	Dibrachys affinis Masi	****
Theroscopus hemipterus F.	**	Dibrachys cavus Walk.	*

vae and about 5% of each of the two carpophagous generations. The parasites were identified as the eulophid Elachertus affinis Masi and several braconid species belonging to the genera Habrobracon, Chelonus and Agathis. In some vineyards, Chrysopa predators probably caused some mortality, since numerous eggs were found on the grape clusters.

The most incisive mortality factor, however, containing the two carpophagous generations was that of high summer temperature combined with low relative humidity due to the poor resistance of Lobesia eggs to these conditions (9, 11). In the head-trained vineyards, where eggs and larvae are more exposed to the effects of the sun, the attacks were normally over 50% lighter than in espalier and arbour types, higher off the ground. The large population and infestation variations occurring between one year and another (also in vineyards left untreated with insecticides) were exactly due to this susceptibility to summer temperature. In six years' observations at Bonnanaro (in espalier-trained vineyards) extremely high infestations were only recorded in 1981 (Fig. 2), a year with summer conditions particularly favourable. In 1982, despite profuse numbers of larvae on the flowers and a substantial second flight of adults, the second and third generation attacks were negligible (Fig. 2), because the daily maximum temperatures during the last week of June always rose above 38°C. A similar situation occurred in 1983, when the daily maximum temperatures superseded 40°C during the last ten days of July.

3.4 Intervention thresholds and control measures

In our experiments, it has usually been possible to disregard infestations by the anthophagous generations of the moths in view of the negligible resultant damage. In corroboration, trials using Cannonau and Nuragus cultivars showed that the experimental removal of up to 30% of the flowers did not diminish the eventual weight of the grape-clusters. An intervention threshold was, therefore, adopted of 200 larvae per 100 flower-clusters (an average of 90% of clusters attacked). During the six years' studies in the Alghero vineyards, this threshold was only exceeded twice (first-flight adult catches had also been over 300 per trap), and intervention with insecticide was considered prudent to contain the expected high levels of second generation populations. Various researchers, using different methods (comparison by weighing attacked and unattacked clusters, inducing artificial infestations, or simulating damage by removing the flowers), have reported that only rarely do first generations cause significant harm. According to the variety, the vines are able to compensate for losses varying from 10 to 30 flowers per cluster; in some cultivars, even up to half their florescence (1, 14, 18).

Damage provoked by the carpophagous generations varied with the cultivar, compact-cluster varieties being the most attacked (e.g., Cannonau, Giro', Carignano, Nuragus, Moscato and Vernaccia). Varieties with loose grape-clusters (Pascale di Cagliari, Vermentino, Monica and Malvasia) were less susceptible. At vintage time in an untreated vineyard, for example, counts of Lobesia larvae per grape-cluster gave an average in 1980 of 0.3 on Vermentino and 0.6 on Giro', and in 1981 1.7 and 5.6, respectively (Tab. II).

The experiments at Bonnanaro in 1981 showed that the number of grape-berries directly damaged by a 2nd generation larva averaged 3.76 and per 3rd generation larva 3.79. Greater damage, however, occurred indirectly, since the number of berries attacked by grey mould and sour rot proved to be in proportion to the number of larvae (Tab. III). Although fairly high levels of damage can be tolerated at the vintage — in Venetia a tolerance threshold of about 50% of grape-clusters infested (10)— , thresholds for intervention against the crucial 2nd generation have to be set lower.

The intervention thresholds established from the control experiments and vintage damage evaluations took into account the dependence of treatment effectiveness upon cluster conformation as well as cultivar susceptibility to attack. For the less susceptible varieties, sampling was done a week after pheromone-trap catches reached their peak and the threshold was set at a larval infestation of 10% to 15% of the clusters. A single insecticidal treatment a few days after reaching this threshold (i.e., about ten days after the peak catches) was sufficient to ensure only limited damage to the vintage, especially when adequate antibotrytic measures were adopted. For varieties with compact grape-clusters, not only did the intervention threshold have to be lower (5% to 10%), but also the decision to intervene required much more urgency. Since waiting to assess the true larval situation might have led to overstepping the moment for effective treatment, the need for intervention was also based on past experience and on the num-

63

Tab. II. Lobesia botrana infestations on different cultivars at vintage-time (Bonnanaro).

Cultivar	No. of larvae per cluster	
	1980	1981
Girò	1.7	5.6
Cannonau	1.4	3.1
Pascale di Cagliari	1.3	1.9
Sangiovese	0.9	1.0
Torbato	0.8	1.0
Cagnolari	0.7	1.4
Vermentino	0.3	0.6

Tab. III. Relationship between Lobesia botrana infestations and grey mould disease (Bonnanaro, 1981).

No. of larvae per cluster	No. of clusters	No. of diseased berries per cluster
0	60	8.4
1	25	17.2
2	11	19.6
3	5	37.8
4	5	59.8
5	5	114.0

bers of adults caught by the traps. Monitoring was necessary at least twice a week, so that treatment (if indicated by catches and/or larval counts) could be applied soon after peak catches were reached, in time to control the larvae. With the use of these thresholds, intervention against the third generation was only necessary in a few exceptional cases, even for late-maturing compact-cluster varieties.

The intervention thresholds established by us differ to some extent from those in other regions with three Lobesia generations. In Emilia-Romagna, for instance, they have been fixed at 4% to 5% of clusters infested, and in Lazio at 3% to 5% (3, 17). In Switzerland and Lombardy, where the moths produce only one carpophagous generation, tolerance thresholds have been fixed at 5 to 10 larvae per 100 clusters and 15% to 25% of clusters infested (2, 18). Most probably, the variations between the thresholds applied in different regions depend upon a combination of many factors, particularly environmental, cultural and economic.

Although no significant correlation between 2nd generation catches and larval infestations resulted from our experiments, it was observed that peak catches (monitoring every three or four days) of no more than 10 to 15

moths per trap always indicated that no infestation of importance would occur. The usefulness of the traps at least for negative prediction, i.e. as
indicators for non-intervention, was thus confirmed (14). Conversely, if
the catches among all the viniferous varieties in many vineyards exceeded
50 males per trap, the corresponding larval attacks generally proved to be
substantial. However, the latter situation did not always signify eventual
serious damage, since at times summer temperatures were high enough to reduce the infestations to insignificance (Fig. 2) —a factor to be given
careful consideration. In vineyards where both species of moth constituted
the infestations, the intervention thresholds referred to the sum of their
numbers.

In the control trials (against 2nd generations), good results were obtained with Trichlorphon and Tetrachlorvinphos if used at the time of peak
catches. Carbaryl gave poor results. Deltamethrin, Dimethoate and Metida-
thion (especially the latter two) proved effective even after larval penetration into the berries. Two separate trials were carried out of the two
Bacillus thuringiensis preparations, Bactospeine and Thuricide HP, used
with 1% of sugar. In each case, the results were comparable with those of
the most effective insecticides. However, there are conflicting reports in
the literature on the effectiveness of this biotic agent, which seems to be
affected by temperature and the date of application (4, 5, 13, 14). Its use
would, therefore, be most tenable against 1st generation attacks heavy
enough to raise doubts or against moderate 2nd generation infestations
(peak catches of 15 to 50 moths per trap, monitoring every 3 or 4 days).

4. CONCLUSIONS

Despite the ubiquity of both species of moth, these researches showed
an almost complete absence of Eupoecilia ambiguella and only occasional
damage by Lobesia botrana in the majority of Sardinian vineyards; that is,
in those employing head-training, which exposed the plants to the action of
high summer temperatures. Heavy attacks, particularly damaging to the compact grape-cluster varieties, were observed in the vineyards using irrigation and forms of training high off the ground. The thresholds proposed on
the basis of our experiments for intervention against the 1st carpophagous
generations have proved to be sactisfactory for rationalizing control of
the moths, but need further verification as well as adaptation to the different environmental and cultural conditions in the many island vineyards.

Pheromone traps helped considerably to minimize pest management costs.
Not only did they indicate more accurately the moment to sample the larval
infestations, but also when not to intervene if the attacks were light and
unlikely to cause appreciable damage. Thus, they rendered obsolete the
expensive and ecologically dubious routine use of insecticides. Further,
their use made it possible to intervene at exactly the right moment against
heavy attacks, particularly important in the case of compact-cluster varieties.

More research work is required to determine conclusively the correlation between trap catches and infestations, and to obtain more comprehen-

sive data on the efficiency and range of the traps. Also, investigations have still to be made into the effects of other nearby host vegetation on vineyard pest populations and their natural enemies.

REFERENCES

1. BAILLOD, M., BOLAY, A., ROEHRICH, R., RUSS, K. and TOUZEAU, J. (1980). La lutte intégrée en viticulture. La défense des vegetaux 33: 91-101
2. BOLAY, A., BAILLOD, M., VALLOTTON, R. and GUIGNARD, E. (1981). La protection phytosanitaire en viticulture. Revue suisse Vitic. Arboric. Hortic. 13: 13-18
3. CASARINI, C. (1980). Lotta guidata in viticoltura. Guida pratica per il viticoltore. Regione Emilia Romagna. Cooptip, Modena: 39 pp.
4. CELLI, G., CASARINI, C., BARBIERI, R. and BECCHI, R. (1980). Risultati di trattamenti con preparati a base di Bacillus thuringiensis contro la Lobesia botrana Schiff. (Lepidoptera, Tortricidae) nel Modenese in rapporto al rilievo feromonico e al danno. Atti Giornate Fitopatologiche 1980: 431-439
5. CORINO, L. and MAGNAGHI, G. (1982). Esperienze di controllo delle tignole dell'uva in Piemonte. Atti Giornate Fitopatologiche 1982: 197-205
6. COSCOLLA, R. (1980). Aproximacion al estudio del parasitismo natural sobre Lobesia botrana Den. y Schiff. en las comarcas viticolas Valencianas. Bol. Serv. Plagas 6: 5-15
7. COSCOLLA, R. (1981). Algunas observaciones sobre el pteromalido Dibrachys affinis Masi, parasito de Lobesia botrana Den. y Schiff. (polilla del racimo de la vid). Bol. Serv. Plagas 7: 57-63
8. DALLA MONTA, L. (1981). Résultats de piégeages sexsuels en Venetie et en Friuli. Lutte intégrée en viticulture, IV Réunion plénière, Gargnano. Boll. Zool. agr. Bachic., ser. II, 16: 24
9. GABEL, B. (1981). Uber den Einfluss der Temperatur auf die Entwicklung und Vermehrung des Bekreuzten Traubenwicklers, Lobesia botrana Den. et Schiff. (Lepid., Tortricidae). Anz. Schadlingskde., Pflanzenschutz, Umweltschutz 54: 83-87
10. GIROLAMI, V. (1981). Evaluation des dégâts dus aux vers de la grappe. Lutte intégrée en viticulture, IV Réunion plénière, Gargnano. Boll. Zool. agr. Bachic., ser. II, 16: 16-18
11. GOTZ, B. (1941). Laboratoriumsuntersuchungen uber den Einfluss von konstanten und variierend Temperaturen, relativer Luftfeuchtigkeit und Licht auf die Embryonalentwicklung von Polychrosis botrana. Anz. Schadlingskde. 17: 73-83; 85-96; 125-129
12. MOLEAS, T. (1984). Biologia ed etologia della Lobesia botrana in Puglia. Possibilità di lotta integrata. 3° Incontro Difesa Integrata Vite, Regione Lazio: 91-97
13. ROEHRICH, R. (1984). Travaux du sous-groupe "Tordeuses de la grappe et insectes broyeurs". Bull. SROP 1984/VII/2: 36-52
14. ROEHRICH, R. and SCHMID, A. (1979). Lutte intégrée en viticulture. Tordeuses de la grappe: evaluation du risque, determination des periodes

d'intervention et recherche des méthodes de lutte biologique. Proc. Int. Symp. IOBC/WPRS on Integrated Control in Agriculture and Forestry, Wien 8-12 oct. 1979: 245-254

15. SPRENGEL, L. (1931). Epidemiologiche Forschungen uber den Trauben-wickler Clysia ambiguella Hubn und ihre Auswertung fur die praktische Grossbekampfung. Z. ang. Ent. 18: 505-530

16. STELLWAAG, F. (1939). Der Massenwechsel des bekreuzten Traubenwicklers Polychrosis botrana in Weinbau. Z. ang. Ent. 25: 57-80

17. TRANFAGLIA, A., WEBER, F. and BIANCHI, A. (1981). Prima esperienza sull'applicazione a livello territoriale della lotta integrata agli insetti della vite. 3° Incontro Difesa Integrata Vite, Regione Lazio: 71-90

18. VALLI, G. (1975). Lotta integrata nei vigneti. Ricerche e valutazioni preliminari sulle Tignole. Notiziario sulle Malattie delle Piante 92-93: 407-419

The grape-vine moths in the framework of IPM in Sicily

P.Genduso

Istituto di Entomologia Agraria dell' Università, Osservatorio per le Malattie delle Piante, Palermo, Italy

Summary

The problem of grapevine phytophagous is being discussed in relation to an integrated control programme in Sicily. Many pests are occasional and, although particularly harmful, they are limited to certain zones and years.

Research carried out in 29 biotopes in Western Sicily have shown that Lobesia botrana (Den. et Schiff.), which is the key phytophagous, causes variable damages.

The population was recorded numerous only in 6 biotopes. Eupoecilia ambiguella (Hb.) is rare but almost always present.

Some biological data on moths are reported in Sicily. They have different behaviour in comparison with other areas of Southern Italy.

The fr equent spraying of insecticides in many areas, are not justified.

For this it is necessary to organize a prevention Service in order to carry out an integrated or at least a guided control that will be able to reduce the spraying.

1.1 Introduction

The grapevine pests, reported by Stellwaag (1928), are a few hundred.

Recently, Pastena (1985) listed 487 species of metazoa, of which 150 are usually harmful. It is not possible to list here the harmful ones in Sicily, because we have very little information of different groups of animals, for example, nematoda, which deserve a more precise study. We can say that in the past, mites did not cause worries, whereas today they are causing much damage.

I will limit this list to insects which have caused and are still causing severe damage.

Among the Isoptera we have Kalotermes flavicollis (F.), very common in the old implantations.

Among the Orthoptera it is frequent to find imfestation of different species of Hensiphera and Celiphera which sometimes invest the vines from nearby uncultivated areas.

Among Hemiptera Heteroptera, Metopoplax ditomoides Costa, have developed in great quantity in some years, causing death of some of the branches or the intire young plant.

Fig. 1 - Map of Sicily indicating the districts where the pheromone traps of the grapevine moth have been installed.

Fig. 2
Pheromone trap.

Among the Hemiptera Homoptera: Planococcus ficus (Sign.) and Viteus vitifoliae (Fitch); this last one have caused problems also on the European grapevines.

Among Coleoptera we can mention Phillognatus excavatus Forst and Schistocerus bimaculatus Ol. and Epicometis hirta and E. squalida Scop. (Mineo 1964).

Among Lepidoptera, besides the two Tortricoidea we can cite Cryptoblabes gnidiella (Mill.) which have caused primary damage to ripe grapes; and Myelois ceratoniae (Z.). Antispila rivillei (Stt.) is to be considered less important because in my opinion the damages caused to the leaves are severe only at the end of the vegetative cycle of the plant.

Among the Hymenoptera we have a lot of problems because of wasps.

The moths Lobesia botrana (Den. et Schiff.) and Eupoecilia ambiguella (Hb.), are the key pests of the grapevine in Sicily like in other paleartic regions. (1)

In Sicily L. botrana have been reported by De Stefani Perez (1889), Costantino (1939) and by Vivona (1935); the latter gives information on the time of flying and the number of generations. Silvestri (1943) noted that in ten years of observations on grapes coming from different zones of South Italy, no E. ambiguella have been found.

This study had the principal aim to know both the diffusion of L. botrana and E. ambiguella, as well as the proceeding of flight of the two species. In some biotopes the entity of damages caused to the grapes have been tried to be evaluated.

1.2 Methods and Materials

As it was necessary, since 1978 in some vineyards in the Province of Palermo, Trapani, Agrigento and Caltanissetta some pheromone traps for both species (Fig. 1) have been installed.

The biotopes with data related to the cultivar and the raising system are reported in tables I, II and III.

In the Province of Palermo the biotopes no. 4, 5, 6, 7, 8, 9, 10 and 13 are in typical grapevine producing zones. The others are in small vineyards interposed with other cultivations. It is to be noted that the biotopes 4 to 9, which are found in the district of Camporeale, are in a single firm of about 80 ha., but exposed to different microclimates; the biotope no. 8 in particular is found near a little hilly lake.

All the biotopes in the Province of Trapani are typical viticultural zone. The biotopes in the Province of Agrigento and Caltanissetta are characterized by the production of table grapes from the cultivar "Italia", whose center is in Canicattì.

(1) As it is known the geographical diffusion extends also to the Eastern areas and as far as E. ambiguella is concerned it extends also to South America.

71

TAB. I - Biotopes of the province of Palermo, where traps with pheromones of L. botrana and E. ambiguella were installed.

Biot. no.	Year	Obser. period	District	Locality	m/ s.l.	Cultivar	Rai- sing syst.	n. traps Lob. bot.	Eup. amb.
1	1979	August	Bisacquino	Crocilla	600	Insolia Catarratto	A 2	3	-
2	"	"	"	Cascia	400	"	A 2	1	-
3	"	"	Contessa Entellina	Realbate	400	"	A 2	2	-
4	1980	"	Camporeale	Marchese	200	"	B 1	1	1
5	"	"	"	"	"	"	B 1	1	1
6	"	"	"	"	"	Nero d'Avola	B 1	1	1
7	"	"	"	"	"	Nero mascalese	C 1	1	1
8	"	"	"	"	"	Nero d'Avola	B 1	1	1
9	"	"	"	"	300	Nero mascalese	A 1	1	1
10	1983	"	Cerda	Canna	200	Grecanico Insolia	B 1	1	1
11	"	"	Ciminna	Pecorone	300	Catarratto	A 2	1	1
12	"	"	Trabia	Burgio	360	Trebbiano	B 2	1	1
13	"	"	San Cipirello	Chiusa	300	Insolia	A 2	1	1
14	"	"	Monreale	S.Domenico	130	Catarratto Insolia	A 2	1	1
15	"		"	Portella Paglia	150	"	A 2	1	1
16	"	"	Corleone	Malvello	520	Catarratto Trebbiano	C 1	1	1
17	"	"	Polizzi Generosa	Cuca	720	Varie	A 2	1	1

A = stake   B = trellis   C = arbor
1 = irrigated   2 = not irrigated

B. II - Biotopes of the province of Trapani, where traps with pheromones of <u>L.</u> <u>botrana</u> and <u>E. ambiguella</u> were installed.

ot. o.	Year	Observ. period	District	Locality	m/ s.l.	Cultivar	Rai- sing syst.	n. traps Lob. bot.	Eup. amb.
8	80-81	May - Sept.	Marsala	Messinello	150	San Giovese	A 2	2	2
9	"	"	"	Mamuna	110	Catarratto	C 1	4	2
0	"	"	"	Samperi Torrelunga	20	Grillo	B 1	3	2
1	82-83	March-Sept.	"	Puleo	125	Catarratto	A 2	3	-
2	1982	"	Petrosino	Scaletta	20	Grecanico Trebbiano	B 2	1	1
3	"	"	"	"	"	Catarratto	B 2	1	1
4	"	"	Mazara del Vallo	Ramisella	"	Grillo	C 1	2	-
5	"	"	"	"	20	"	A 1	2	-

B. III - Biotopes of the province of Agrigento and Caltanissetta, where traps with pheromone of <u>L.</u> <u>botrana</u> and <u>E. ambiguella</u> were installed.

ot. o.	Year	Observ. period	District	Locality	m/ s.l.	Cultivar	Rai- sing syst.	n. traps Lob. bot.	Eup. amb.
6	80-81	End of Feb. - November	Naro	Rocca di Mendola	380	Italia	C 2	3	3
7	"	"	Castrofi- lippo	Babilonia	500	"	C 1	3	3
8	"	"	Delia	Fruscula Scuola	400	"	C 1	3	3
9	"	"	Canicattì	Agraria	380	"	C 2	3	3

= stake   B = trellis   C = arbor
= irrigated   2 = not irrigated

73

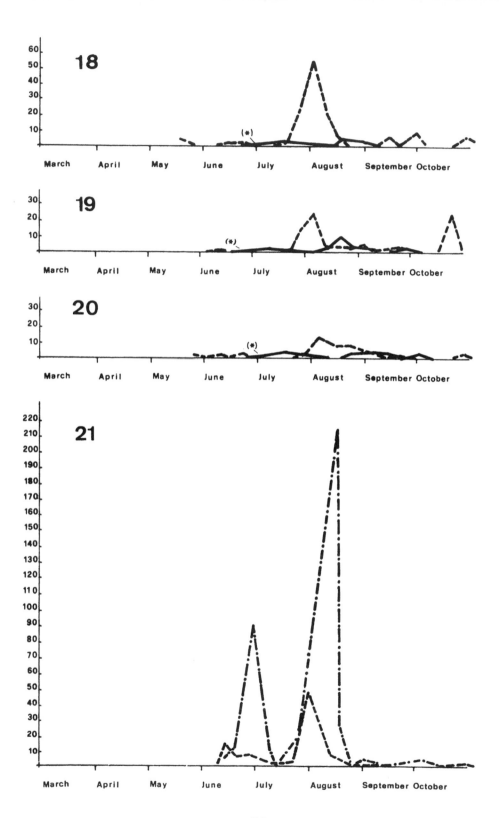

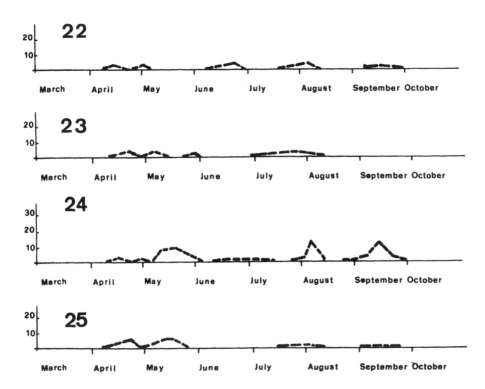

Fig. 3 —Diagram of capturing of males of L. botrana (average/trap) of bio
topes 18 to 25 (Trapani). —— : 1981; - - - : 1982; -.-.- : 1983;
(*) : unreported data.

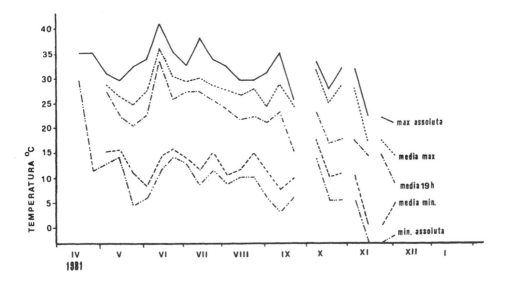

Fig. 4 – Temperatures recorded near biotope 21 in 1981.

During the first years traps commercialized by several firms were used (2). Later, metal traps reproducing a model built in the Institute were used. These last one (Fig. 2) in the form of a hut, varnished in yellow, have the roof with two slopes and four spacers sustaining two binaries in which runs a metallic plate cm 20 x 15, smeared with glue on which the capsule of pheromone rests (3). The plate is substituted in all the biotopes about every 40 days.

The evaluation of the damages is related to the year 1981 and limited to one biotope of the Province of Trapani and to two surveys carried out in the Province of Agrigento. In the first phenological stages, that is from flowering to full size unripe grapes, during which the anthophagous and the first carpophagous generation take place; 200 vines in about 1 ha. have been randomly chosen. The infested bunches were taken from the field, the glomerules with larva being counted, while 200 bunches were harvested when ripe and later observed in the laboratory.

During the sampling period, beside recording the infected bunches, the number of infected berries were also observed. Moreover, an empirical evaluation of the damage was made through a visual observation in the other biotopes.

1.3 Results

Province of Palermo

In the biotopes 1-17, in which it was not possible to follow the trapping regularly, trappings related to the month of August are globally reported.

	Lobesia botrana	Eupoecilia ambiguella			Lobesia botrana	Eupoecilia ambiguella
1979				1983		
biotope n.1	2	(°)		biotope n.10	35	–
" " 2	5	(°)		" " 11	15	6
" " 3	4	(°)		" " 12	26	3
1980				" " 13	28	8
biotope n.4	40	1		" " 14	28	–
" " 5	95	–		" " 15	24	3
" " 6	21	–		" " 16	13	2
" " 7	19	–		" " 17	–	6
" " 8	184	–				
" " 9	19	–				

(°) Uninstalled traps

(2) As it is known, in Italy they are commercialized by Sipcam, by Agrovit-Hoechst-Russel and by Farmoplant. The glue was furnished by the Kollant Firm.

(3) The capsules used have been furnished by Farmoplant, by the University of Southampton (Great Britain) and by the Group of Laboratory and Service La Minière of INRA (France).

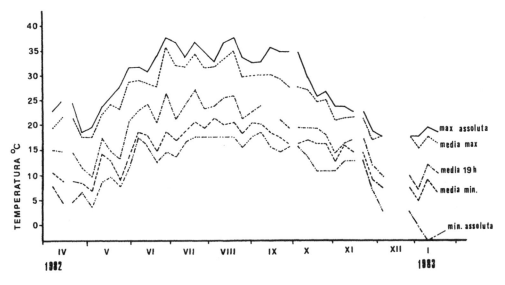

Fig. 5 - Temperatures recorded near biotope 21 in 1982.

Province of Trapani

The capturing of males of L. botrana in the eight biotopes in the Province of Trapani has been reported in the graphics of Fig. 3 and the temperatures in Fig. 4 and 5.

The graphics related to the captures of E. ambiguella have not been reported because limited to only 30 individuals in five of the biotopes in the whole period of observation.

Provinces of Agrigento and Caltanissetta

The capturing of males of L. botrana related to the years 1980 and 1981 are reported in Fig. 6, whereas those of E. ambiguella are reported in Fig. 7 united in one single graphic for all four biotopes in 1981.

2.1 Observations on the damages

The survey on the damages concerning the year 1982 in the biotope no. 21 in the locality of Torrelunga Puleo in the district of Marsala (Trapani) have given the results reported in Tab. IV.

In Fig. 8 are reported the frequency of larva per bunch from 0 to 8 and in Fig. 9 the classes of frequency of the berries hit by moth and/or Botrytis up to the date of September 6th 1982.

Visual observations carried out in other biotopes in the Province of Trapani have not caused damages worthy of mentioning.

The same consideration can be made for the Province of Palermo, with exception of the biotopes of the district of Camporeale where some damages have been noticed, the amount of which not being available. In the same year of 1981 in the Province of Agrigento, in the biotope no. 26

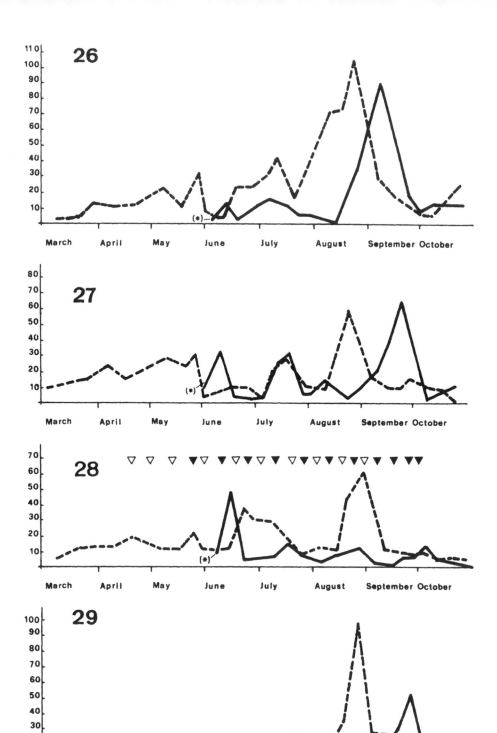

TAB. IV - Percentage of infestation of L. botrana in biotope no. 21.

Date of sampling	Phenological phase	% of bunches infested	% of infested flowers and/or berries
31/05/1982	Flowering - 50% fruit set	22	0.26
21/06/1982	Berries in phase of enlargement	24	0.34
19/07/1982	Half size berries	62	4.77
06/09/1982	Ripe grapes	56	9.23

(locality "Mendola") at harvesting time the 74% of bunches infested was noticed, with an average of 1.4 larva per bunch and with a maximum of 6; in total the 19.6% of the berries were hit by the gray mould.

2.2 Considerations

Several difficulties have contributed to yield results which were not utilizable as previously planned: among them we have the interference of the sprayings in several biotopes and many difficulties in the organization due to the distance from Palermo and the distance between the different biotopes.

From observation carried out on 29 biotopes, even though followed irregularly, we can make the following considerations:

1 - Both species, E. ambiguella and L. botrana, are present in Sicily. The incidence of Eupoecilia is scarce; its capture have been limited. The maximum have been recorded in biotope no. 27 in Castrofilippo (AG) with 30 individuals in one week in the second half of August 1981.
L. botrana have been recorded in all the biotopes, with exception of biotope no. 17 where only one trap has been installed.
The number of captured males have been variable even in the same year and between nearby vineyards. The cause of this evident variability should be investigated having in mind the microclimatic factors; in fact during two different investigations in the same locality, a remarkable concentration of adults of both sexes have been noticed in one vineyard, where it has been easy to capture more than 20

Fig. 6 - Diagram of capturing of males of L. botrana (average/trap) of biotopes 26 to 29. —— : 1981; – – – : 1982; ▽ : fungicide spray; ▼ : fungicide plus insecticide spray; (*): unreported data.

March   April   May   June   July   August   September October

Fig. 7 - Diagram of capturing of males of E. ambiguella (average/trap) of
biotopes 26, 27, 28 and 29 in 1982.

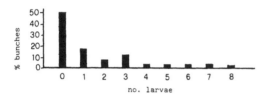

Fig. 8 - Frequency of the larva of moth per bunch of grape

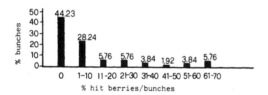

Fig. 9 - Classes of frequency of damaged berries.

individuals on one single plant, while in other nearby zones it was
difficult to find even a single individual.
Actually, we are not able to evidenciate the cause of this noticeable
variability in the capturing of adults and of the same variability in
damage in several biotopes which anyway are not linked only to the
host plant (type of cultivar) but also to other factors which we hope
to study in the future.
In many biotopes in the Province of Palermo capturing have been
limited and only in few of them a higher number of individuals have
been noticed (184 in one month in biotope no. 8).
In the Province of Trapani a capturing highly variable has been
noticed; in fact in 1980 the capturing has been very limited in the 7
biotopes under observation; in 1981 in biotope no. 18 (Messinello)
and in no. 21 (Torrelunga/Puleo) in the first part of August, about
50 individuals have been captured, whereas during the other periods
and in all the other biotopes the capturings have been limited to
just a few individuals (always less than 10).
In any case, from the observation of the graphics, four periods of
flights are clearly noticed. In 1982, although data for all the

80

biotopes are lacking not withstanding the same pheromones were used, the high number of adults (an average of 220 individuals per trap, with a maximum of 334) should be evidenced.

In the area of cultivation of the "Italia" grape in the Province of Agrigento and Caltanissetta the capturing have been higher and more continuous from March to October, unlike to what had been observed in other zones during the same years and in other regions of Italy like Campania (Tranfaglia et al.) and Lazio (Tranfaglia et al. 1980).

2 - It seems interesting that the first captures of both species have been noticed both in 1980 and in 1981 around the end of February and the first decade of March, with more than 2 months of anticipation in comparison to what was also observed in other areas of Southern Italy like the Campania region (Tranfaglia and Viggiani, 1975) and with about 40 days of anticipation in comparison with Sardenia (Deligia et al., 1980) and Puglia (Laccone, 1978).

The larva of L. botrana gathered during the first decade of November in bunches of grapes in the Canicattì zone, by cages under sheds, emerged adults in the same years of 1980 and 1981 between the end of February and the first part of March. These adults survived for a maximum of 10 days.

It still has to be verified if in the field a generation is accomplished on host plants different from grapevines since the first anthophagous on vine starts in May and is completed by the third decade of June - first part of July.

In laboratory the adults have deposited on fuits of ivy (Edera helix) and the larva developed on the same fruits.

3 - Damages caused by L. botrana are very variable in the different years as well as in the diverse biotopes. In all cases the results have been always under the threshold of tolerability in the anthophagous generation.

In the Province of Palermo and Trapani, with the exception of biotope no. 21, no economic damages have been pointed out.

In biotope no. 21 up to the date of July 19th as mentioned before, it was noticed that the 4.77% of the berries were hit, even if a theoretical lost egual to 477 Kg of grapes per ha. (with an average production of 100 qi./ha.) it can be considered that it has been compensated by the enlargement of the berries left, since from such date up till harvesting there have been a fourfold increase in weight.

At harvesting time, the 9.2% of the infested berries could have influenced the quality of the wine. It should be remembered that the population in that biotope reached up to 220 males per trap.

The problem of the moth is important in the cultivation area of the "Italia" grapes where losts come up to 20%. The consequence is higher cost of hand labour needed, only partly returned.

4 - In all the areas under observation, intervention against the moth has proven inefficient except rare cases (see graphs), with serious consequences under the hygienic-sanitary and ecological point of view.

5 - Even though up to today there are no ascertained data on the
    evolution of generations of single pairs, it can be assumed , as
    supposed by Bivona (1955), that in Sicily four generations can be
    completed from May to November.
    After so many decades and although with varied conditions, it still
    seems valid the asumptions by Silvestri (1943), that is that damages
    by moths is much less serious in the Southern zones of Europe than
    those of than the Central-Nothern of Europe.

Acknowledgments

The collaboration of Doctors Mariano Nicolosi, Renato Giuliana,
Eugenio De Vita, the technicians of the Extension Service of the E.S.A.
and of Agricultural School of Canicattì and Mr. Angelo Corsino,
technician of the Institute of Enthomology of Palermo, is kindly
acknowledged.

References

COSTANTINO G. -1939- La tignola dell'uva o verme dell'uva. Boll. Oss.
    Fitopat. Acireale n. 72
DE STEFANI PEREZ T. -1889- Gli animali dannosi alla vite con brevi note
    sul modo di prevenire i loro danni. 27
DELIGIA S.; LOCHE P.; PIRAS S.; FRESU B.; PALMAS M.; PAGLIANI M. -1980-
    Prova di lotta guidata contro Lobesia botrana Schiff.. Atti 2° incon
    tro sulla difesa della vite. Velletri 20-XI-1980: 31-37
MINEO G. -1964- Una grave infestazione di Cetonie pelose alla vite in
    Sicilia (Tropinota hirta Poda e T. squalida Scop.). Boll. Ist. Ent.
    Agr. e Oss. Fitopat. Palermo 5: 155-171
LACCONE G. -1978- Prove di lotta contro L. botrana (Schiff.) (LEPID.-
    TORICIDAE) e determinazione della "soglia economica" sulle uve da
    tavola in Puglia. Annali Fac. Agr. Univ. Bari 30: 717-746
PASTENA B. -1985- Malattie della vite, II vol. Flaccovio, Palermo
STELLWAAG F. -1943- Die Weinbauminsekten der Kulturlander. P. Parey,
    Berlin
TRANFAGLIA A.; VIGGIANI G. -1975- Osservazioni sui voli di L. botrana
    (LEP.-TORTRICIDAE) con trappole a feromone sessuale sintetico e
    prove di lotta. Boll. Ist. Ent. agr. "Filippo Silvestri", Portici
    38: 259-264
TRANFAGLIA A.; WEBER F.; GRANDE C.; PIERETTI M.; AGRESTA M. -1980-
    Relazione sull'attività svolta nel 1980 per il controllo integrato
    degli insetti dannosi alla vite nelle province di Roma e Latina.
    Atti 2° incontro sulla difesa della vite. Velletri 20-XI-1980,
    Regione Lazio Ass. Agr. e Foreste : 3-17
VIVONA A. -1955- La tignoletta dell'uva in Sicilia ed i mezzi di lotta
    più efficaci (Polychrosis botrana Schiff.). Boll. Ist. Ent. agr. e
    Oss. Fitopat. Palermo 1: 205-216

About forecasting damage of *Lobesia botrana* Schiff. carpophagous generation, on the basis of larval mortality of anthophagous generation

C.Grande
Osservatorio per le Malattie delle Piante per il Lazio, Roma, Italy

Summary

It is proposed to discus the possibility of not using chemical controls against larvae of the Lobesia botrana Schiff. (on flowers generation) and a methodologie of mortality (not insecticides) estimation, in order to reduce number of treatments against larvae of the second generation (harmful on fruits).

In some vine areas growes of Latium (Castelli Romani, Aprilia, Cisterna, Latina, Cori) since 1979 no chemical treatments were effectuated against larvae of first generation (antophagus, harmful on flowers) of Lobesia botrana both on table and wine grapes. (1,3,4)

Seven yearly experience 1979-85 led in large areas, whose the wine growing tradition is more or less remote, with various infestation levels of the Lobesia botrana on table grapes cultivars like "Italia", "Cardinal", "A. Lavallée","M. Palieri", "Matilde", "Regina" and wine grapes cultivars "Trebbiano T.", "Malvasia di C.", "Sangiovese", "Merlot", seems to show that chemical treatments adopting direct insecticides against larvae of first generation are inopportune, indipendently from the thresholds of tolerance or damage prestablished. They do not give any utility for the flowers protection, as in that period we have never found any reduction in production or any damage. (2,5)

From the experience, the doubt arises that the treatments against the Lobesia larvae result noxious to the development of the parasites and predators population. The negative effects (diminution of the natural substratum development) fatally appears on the following generation when these important biological control factors are missing.

Consequently the Lobesia generation that causes wastage to the production of wine or table grapes remain the second, or the first carpophagus one, particularly for the lesions of the larvae to the grapes, that are connected to Botrytis cinerea Pers. development at the begenning of the august-september rain. (1,3,6)

It should be considered if the valutation of the Lobesia second generation risk of damage could be done by with the determination of the mortality, checked in the field, not caused by the insecticides, of the larval stages of the first generation.

In case the mortality of the larval stages, of the antophagus generation, were over 50% could we consider a reduction of chemical treatments
against second generation larvae?

As the mortality gradually reaches 95-98% from 50% could we consider
a reduction of the chemical treatments with till their suppresion when it
is near the 98%?

In the current year it has been tried the above mentioned methodology in a viticultural farm near Cisterna of Latina that cultivates, among
other cultures, ten hectares of table grape cultivars "Italia", "A. Laval
lée", and "M. Palieri".They are subdivided in two groups of 5 and 10 year
old grapes, respectively, greffed on Kober 5 BB, distanced 2, 8x2,8 metres.

The test was conducted on few hundreds of "Italia" grape stumps from
the 10 year old plant.

111 inflorescence were marked, with the same number of young larvae.
Ten days later the 111 inflorescence were checked again and the alive larvae, dead larvae and the missing larvae and the chrysalids counted.

The results are reported as follow:

	Alives larvae n.	Dead larvae n.	Missing larvae n.	Chrysalids n.	Grapes n.	Mortality %
1º control	111	–	–	–	111	–
2º control	2	3	105	1	111	95

Because of the great number of missing larvae we could not find in
the same farm or in the other neinghbouring an adeguate number of Lobesia
botrana ripe larvae that should have been object of a particular kind
of observation. Later by the results of the observation, just one prophy-
lactical chemical treatment was made, over the ten hectares of table grape, against larvae of second generation of july.

The result of only this one treatment has been really satisfactory,
as on the 31 st july 1985 almost all the grapes appeared without any erosion produced by Lobesia   larvae.

We think that a reduction of the chemical treatments against first
generation larvae carpophagus should be confirmed with at least a control
of ovideposition or infestation of the grapes, because of the great risk
of damage of second generation larvae.

In the next years the over mentioned methodology should be tried
in some different viticultural areas supported with controls in order to
establish the validity, limits and the adwantage of it, examining in
this way the causes of the dispersion and mortality of the first generation larvae.

All theese questions are submitted for your attention as starting
point of the discussion about the opportunity of the previsione Lobesia
botrana second generation larvae wastage and methodology to use.

We hope the discussion that follows will be wide and profitable.

References

1. Tranfaglia A. et Al.1980 - Relazione sull'attività svolta nel   1980
   per il controllo integrato degli insetti dannosi   alla
   vite nelle province di Roma e Latina. Atti 2° Incontro
   Difesa Vite, Velletri 20 novembre 1980: 3 - 17.

2. Tranfaglia A. et Al. 1981 - Prima esperienza sull'applicazione a livello territoriale della lotta integrata agli insetti
   della vite. Atti 3° incontro sulla difesa integrata del
   la vite. Latina 3-4 dicembre 1981 : 71-90.

3. Grande C.1980 L'attività di divulgazione e di assistenza tecnica del-
   l'Assessorato Agricoltura e Foreste della Regione Lazio
   nel campo della protezione della vite. Atti 2° incontro
   sulla Difesa della vite. Velletri 20 novembre:117-122.

4. Grande C.,Pierretti M. 1981 - L'assistenza tecnica dell'Assessorato
   Agricoltura a favore dei viticoltori laziali. Atti 3°
   incontro su La difesa integrata della vite. Latina 3-4
   dicembre: 157-169.

5. Grande C.Tranfaglia A. 1983 - Les aspects pratiques de la lutte intégrée en viticulture dans la région Latium.
   Atti Riunione gruppo O.I.L.B. aspetti pratici   della
   lotta integrata in viticoltura. Cordoba (Spagna) 22-24
   novembre. In 3° incontro difesa integrata della vite:
   219-230.

6. Grande C. 1985 Supports ultérieurs pour la réalisation pratiques   de
   la lutte intégrée en viticulture dans la région du Latium. VI éme réunion pléniere Lutte intégrée en viticul
   ture. Bernkastel II-14 giugno. In corso di stampa.

Investigations on Auchenorrhyncha accused or suspected to be noxious to vine in Italy*

C.Vidano, A.Arzone & A.Alma
Istituto di Entomologia Agraria e Apicultura dell' Università di Torino, Italy

Summary

The Homoptera Auchenorrhyncha found on vine in various viticolous loc alities of Italy were listed and divided into normally ampelophagous, occasionally ampelophagous, and erratic species. The most important or significant ones from the phytopathological point of view were examined in relation to the nature of their feeding punctures and divided into mesophyll, phloem, and xylem suckers. The phloem sucker Stictocephala bisonia and the xylem sucker Cicadella viridis were considered also for cauline damages caused by their egg-laying wounds. Among the phloem suckers, both the normally ampelophagous Empoasca vitis, which is very common in some parts of Italy, and Jacobiasca libyca, which is well represented and noxious only in some southern loc alities of Sardinia, were responsible of remarkable foliar disorders similar to the ones due to other causes, such as pathogenic agents. The typical phloem sucker Scaphoideus titanus, which is an obligatori ly ampelophagous species spread over northern Italy, was not found responsible of foliar alterations caused by its feeding punctures. The results of the investigations carried out in Italian vineyards, where it was accused to be vector of the Flavescence dorée MLO, induce to increase the researches on pathogenic agents the transmission of which occurs through graft. About vectors of pathogenic agents, such as viruses and mycoplasma, it seems correct not to neglect investig- ations on erratic phloem feeding species, e.g. Euscelidius variegatus, Euscelis incisus, Hyalestes obsoletus, Laodelphax striatellus.

1. Introduction

Various species of Homoptera Auchenorrhyncha can be considered as vit icolous ones for different aspects. Few of them were investigated from the biological and phytopathological points of view. Others were mentioned for

* Research work supported by CNR, Italy. Special grant I.P.R.A. - Sub-
-project 1. Paper N. 621.

Table I - Most significant Auchenorrhyncha found on vine	connection with the vine	involved tissues	generations	overwintering	generic host plants
CIXIIDAE					
Hyalestes obsoletus	o	**	1	-	weeds
DELPHACIDAE					
Laodelphax striatellus	o	**	2	-	grasses
ISSIDAE					
Hysteropterum grylloides	oo	**	1	·	weeds & bushes
FLATIDAE					
Metcalfa pruinosa	oo	**	1	·	weeds - trees
CERCOPIDAE					
Philaenus spumarius	oo	***	1	·	weeds
MEMBRACIDAE					
Stictocephala bisonia	oo	**	1	·	weeds & shoots
CICADELLIDAE					
Aphrodes bicinctus	o	**	1	·	clovers & weeds
Cicadella viridis	oo	***	2	·	weeds & shoots
Empoasca decipiens	oo	**	3	+	weeds
Empoasca vitis	ooo	**	3	+	vine & trees
Jacobiasca libyca	ooo	**	4	+	vine & weeds
Zygina rhamni	ooo	*	3	+	vine & bramble
Macrosteles sexnotatus	o	**	3	·	grasses
Scaphoideus titanus	ooo	**	1	·	vine
Euscelidius variegatus	o	**	3	·	clovers & weeds
Euscelis incisus	o	**	3	-	clovers & weeds

Symbol explanation :

o erratic, oo occasionally ampelophagous, ooo normally ampelophagous species;
* mesophyll, ** phloem, *** xylem;
· egg, - nymph, + adult.

the ampelophily of their adults. Further occasionally viticolous species were neglected. Since these plant sucking insects need to be better known for what concerns their pathological responsibilities in the viticolous field, investigations were carried out with the aim to provide a list of the most significant species classified according to their relationship with the vine and the nature of their feeding punctures.

2. Material and methods

Pluriennial field and laboratory investigations were accomplished in order to divide the Auchenorrhyncha found on vine in various viticolous localities of Italy into normally ampelophagous, occasionally ampelophagous, and erratic species. The most important or significant species from the phytopathological point of view were examined in relation to the nature of their feeding punctures and divided into mesophyll, phloem and xylem suckers. The normally ampelophagous species were able to reproduce themsel ves exclusively or commonly on vine. The occasionally ampelophagous species were rarely found on vine and usually as adults and eggs or as adults and youngs. The erratic species were found on vine only as adults. Foliar and cauline disorders caused by the above differentiated feeding punctures were classified and compared with similar alterations due to other causes. As some auchenorrhynchous species were accused or suspected to be vectors of pathogenic agents to the vine, during the pluriennial field investig-ations particular attention was also devoted to recognize the vines involved by pathogenic agents as a consequence of grafting.

3. Results

Both the data reported in the text and the ones summarized in table I were mainly obtained from 1981 to 1985. They concern vineyards of various provinces in Piedmont (Asti, Alessandria, Cuneo, Torino), Lombardy (Son-drio, Pavia), Venetia (Padova, Treviso, Venezia, Vicenza), Emily and Roma-gna (Bologna, Forlì, Ravenna), Liguria (Imperia), Tuscany (Grosseto, Siena, Livorno), Latium (Roma, Frosinone), Apulia (Bari, Foggia, Lecce), Sardinia (Cagliari), Sicily (Catania, Palermo, Trapani).

The species of Auchenorrhyncha found as adults on vines were more than one hundred, but the most significant or worthy to be mentioned are the following ones : Cixiidae : Cixius cunicularius (Linnaeus), Oliarus cu-spidatus Fieber, O. panzeri Löw, O. quinquecostatus (Dufour), Hyalestes ob-soletus Signoret; Delphacidae : Laodelphax striatellus (Fallén); Dictyopha ridae : Dictyophara europaea (Linnaeus); Issidae : Hysteropterum grylloi-des (Fabricius); Flatidae : Metcalfa pruinosa (Say); Cercopidae : Cercopis sanguinolenta (Scopoli), Aphrophora alni (Fallén), A. salicina (Goeze), Philaenus spumarius (Linnaeus); Membracidae : Centrotus cornutus (Lin-naeus), Stictocephala bisonia Kopp & Yonke; Cicadellidae : Agallia laevis (Ribaut), Penthimia nigra (Goeze), Aphrodes bicinctus (Schrank), Cicadella viridis (Linnaeus), Empoasca alsiosa Ribaut, E. decipiens Paoli, E. solani (Curtis), E. vitis (Göthe), Jacobiasca libyca (Bergevin & Zanon), Zyginidia lineata (Lindberg), Z. pullula (Boheman), Z. ribauti Dworakowska, Z. scutellaris (Herrich-Schäffer), Z. serpentina (Matsumura), Zygina rhamni Ferrari, Macrosteles sexnotatus (Fallén), Scaphoideus titanus Ball, Platymetopius rostratus (Herrich-Schäffer), Anoplotettix fuscovenosus (Ferrari), Allygus mixtus (Fabricius), A. modestus Scott, Euscelidius variegatus (Kirschbaum), Euscelis incisus (Kirschbaum), Conosanus obsoletus (Kirschbaum), Psammotettix alienus (Dahlbom), P. confinis (Dahlbom).

All the Auchenorrhyncha above listed but not included in table 1 revealed to be erratic species. Some of them were common in springtime, such as C. sanguinolenta and C. cornutus, or from June to October, such as the Zyginidia spp. P. nigra, once considered an important viticolous species, was found twice only. The erratic species were usually localized on vines near hedges, woods, headings, ditches, uncultivated areas or in vineyards infested by weeds. Their identification could be useful in view of researches concerning viticolous territories having connection with such ecological conditions.

Among the Auchenorrhyncha reported in table I, only E. vitis, J. libyca, Z. rhamni and S. titanus showed biological characteristics which were retained adequate to include them among normally ampelophagous species. Trees were indicated as host plants of E. vitis because this green leafhopper, well known for its remarkable ampelophily, was usually found in the egg, young and adult stages on various broadleaf trees (Acer spp., Alnus spp., Quercus spp., etc.) and in the adult stage on conifers during wintertime. J. libyca, a green leafhopper notoriously noxious to cotton and vine in Africa, was found in Italian vineyards only near Cagliari, Palermo and Trapani, where uncultivated labiatae, leguminosae, malvaceae were identified as its winter host plants. The three generations a year of the Mediterranean Z. rhamni were observed only on vine, but this white and orange-red leafhopper revealed to need brambles or other evergreen bushes to overwinter. S. titanus, the notorious brownish leafhopper introduced into Europe from North America, was exclusively found on vine in all its life history, which displayed only one generation, with overwintering as egg in the rhytidome of two years old branches.

About the Auchenorrhyncha indicated as occasionally ampelophagous species, an unusual behaviour was noted for E. decipiens; this green leafhopper, common on weeds and infesting various herbaceous cultivations (potato and beet particularly), was able to multiply on vine under experimental conditions; sometimes, it was also found both as nymph and adult on vine in southern Italy. Also H. grylloides, known for its characteristical oothecae fixed on trunks and branches of trees and shrubs, was found here and there on vine as nymph and adult. The flatid M. pruinosa was found on vine both as colonies of youngs and adults in Venetian provinces, but usually only near various infested weeds, bushes, shrubs and tree shoots, mainly along hedges. The spittlebug or froghopper Ph. spumarius was collected as adult on vine everywhere, but always scattered; its symptomatic nymphs were found on bushy vines of nurseries or on vine twigs amid or near involved weeds. The treehopper S. bisonia was represented on vine in particular ecological conditions, but much less than once both as adult and egg-laying. Also the blue-green leafhopper C. viridis was not common on vine; nevertheless it was found very abundant both as adult and egg-laying in nursery vines situated in moist surroundings.

The erratic Auchenorrhyncha reported in table I were considered note-worthy being potential vectors of pathogenic agents. They were rarely collected on vine, except in significant situations, i.e. where the vineyards were infested by their host plants. Interesting combinations were detected for H. obsoletus with Convolvulus arvensis and other weeds, L. striatellus with Cynodon dactylon, Digitaria sanguinalis and other grasses, A. bicinctus with herbaceous leguminosae, M. sexnotatus with C. dactylon, Lolium spp. and other grasses, E. variegatus and E. incisus with wild clovers and various weeds.

The examination of involved tissues showed that only Z. rhamni was a mesophyll feeder and responsible of symptomatic dechlorophyllations. H. obsoletus, L. striatellus, H. grylloides, M. pruinosa, S. bisonia, A. bicinctus, E. decipiens, E. vitis, J. libyca, M. sexnotatus, S. titanus, E. variegatus and E. incisus revealed to be phloem suckers, although in various ways. In any case, none of the above Auchenorrhyncha produced sweet excrements attracting glyciphagous insects like the honeydew of some treehoppers and of the most part of Sternorrhyncha. The examination of thin sections of ribs, stalks and twigs, which were pierced by phloem suckers both in laboratory and field conditions, displayed feeding tracks and consequent tissue alterations related to successive symptomatic foliar and cauline disorders.

As a consequence of radial series of feeding punctures of S. bisonia made by single adults in seasonal shoots and lateral ones, brownish annular stranglings appeared and involved both epidermis and cortex. Then remarkable reactions due to the damaged cambium and phloem were recognized : scarring processes at the level of the trauma, starting from which the seasonal twigs did not mature; thickening, brightness, downward rolling and reddening or yellowing of the leaves placed distally to these cauline alterations. S. bisonia was considered also for its egg-laying wounds, which until few years ago were indicated as very noxious to young fruit trees and also reported as injurious to two years old branches of vine. Cauline and foliar alterations, due to feeding punctures, and branch disorders due to egg-laying wounds of S. bisonia were lately seen very rarely, thanks to a successful case of biological control.

Feeding punctures made by adults and youngs of E. vitis, E. decipiens and J. libyca in ribs of not completely developed leaves were identified as responsible of foliar changes starting from the edges and progressing toward the petiolar sinus. Internal vein browning, due to series of feeding tracks, downward rolling, thickening, brightness, reddening or yellowing, and sometimes marginal burning characterized the involved leaves. In field, vines infested in springtime by more than three-four youngs of E. vitis per leaf showed precocious and persistent foliar symptoms. When eight-ten consecutive leaves were so infested and modified, the involved seasonal shoots appeared shorter and did not mature regularly. Under experimental conditions, responses of Vitis to feeding punctures were easily programmed both with whole vines and parts of them : even single leaves partially reacted according to the infested areas. In field, significant evidences were obtained from Barbera, Cortese and Moscato vines, which showed clear symptomatic foliar disorders only in the infested leaves. Various other cultivars, like Erbaluce, Luglienga, Albana, Trebbiano, Pinot, Sangiovese, Merlot, etc., were found heavily infested by E. vitis, but their foliar disorders seemed due to pathogenic agents too. Second and third generations of E. vitis frequently occurred on leaves already affected by downward rolling, due both to previous attacks of E. vitis and other causes.

The typical phloem sucker S. titanus was not found responsible of foliar alterations caused by its feeding punctures. Feeding tracks of adults and youngs were similar to the ones of Empoasca, but scattered in the ribs, which showed only isolated dark small spots instead of vein browning. Under laboratory conditions, vines infested by about six youngs per leaf during the fifty days of the postembrional development revealed foliar and cauline alterations similar to the ones caused by an analogous infestation of three E. vitis youngs per leaf. In field, while the trophic activity of the tri-voltine E. vitis was begun by adults at the end of April and was continued by youngs and adults with three generations, the one of the univoltine S. titanus began with youngs at the half of May and with adults only at the half of July. S. titanus was found everywhere in northern Italy as an obligatorily ampelophagous species, but never an infesting one being usual ly represented by very few youngs and adults per vine here and there.

The two xylem suckers Ph. spumarius and C. viridis were investigated under laboratory conditions : the first one as adult and nymph, the second one as adult only. Thin sections of pierced twigs revealed feeding tracks always reaching the vessels, often after attempts inside cortical and xylem parenchima. Adults of both species excreted a large quantity of liquid, even 2 milliliters a day from a single individual female, when feeding on succulent shoots. A same quantity of fluid feces dropped from a single froghopper nymph protected by white foam. Vines subjected to relatively intense attacks by these two xylem suckers did not give significant symptoms. In field, the hygrophilous C. viridis was sometimes noxious owing to egg-laying wounds involving seasonal shoots.

For what concerns symptomatological convergency between foliar alterations due to feeding punctures of auchenorrhynchous species and the ones due to other causes, significant cases were checked in various viticolous territories. Everywhere white grape cvs revealed downward rolling of the edges and associated foliar symptoms more frequently than the red ones. About white grape cvs, the most involved ones were : Cortese, Erbaluce (Piedmont); Pinot (Venetia); Albana, Trebbiano (Emily & Romagna). Red grape cvs worthy to be mentioned were : Barbera (Piedmont); Merlot (Venetia); Sangiovese (Emily & Romagna). Less evident cases to be reported concerned the white grape cvs Catarratto and Inzolia (Sicily).

Foliar and cauline symptoms were similar to the ones known for Flavescence dorée and Bois noir, but usually less clear than the ones revealed by Baco 22 A in Armagnac, France. In general, they appeared more marked in not well cultivated vineyards, where frequently weeds known as host plants of above listed auchenorrhynchous species were represented. The findings on vine of H. obsoletus, L. striatellus, M. sexnotatus, A. bicinctus, E. variegatus and E. incisus occurred in such occasions. S. titanus was instead found also in well cultivated vineyards, of course both on healthy and diseased vines. In any case, everywhere vines showing downward leaf rolling, including the ones affected by Leafroll virus, were very attractive to E. vitis, which often appeared quite represented even when well controlled by its natural enemies. The lack of S. titanus and other auchenorrhynchous species indicated that pesticides had been sprayed to control insects or mites.

The pluriennial investigations carried out in Piedmontese vineyards with vines affected or suspected to be affected by Flavescence dorée were not sufficient to prove that the disease was transmitted by the very common S. titanus. On the other hand, affected vine series of Erbaluce, Pinot, and Catarratto, respectively in vineyards of Piedmont, Venetia and Sicily, were recognized as clearly infected through graft. In such cases the patho genic agent could be the Bois noir one.

4. Conclusions

The concentrated and synthetic results, which are exposed above, will be considered separately and analytically in successive reports. They offer an up-to-date panorama of auchenorrhynchous species worthy to be better known for investigations regarding relationships between an important group of plant suckers and involved vines. This research was carried out considering all the publications found on such a subject, but only the basic ones, usually rich in bibliography, can be mentioned. A list of viti colous auchenorrhynchous species was prepared, in a different way, when S. titanus appeared in France (1). Details concerning biological and phyto-pathological aspects of the various auchenorrhynchous species found noxious to vine were published separately for Typhlocybins (2, 3), S. titanus once named S. littoralis (4), S. bisonia once named Ceresa bubalus (5). The almost lack of S. bisonia and related cauline and foliar alterations, both in vineyards and orchards, was predicted with the introduction (6) and diffusion (7) of its specific egg-parasite Polynema striaticorne Girault. Also for the recently introduced M. pruinosa (8) an analogous means of biological control would be welcome.

About S. titanus, Flavescence dorée and Bois noir, various French publications were summarized for the insect (9, 10) and the diseases (11, 12). They were well considered for many aspects, as well as the Italian ones concerning the same subjects (13, 14, 15, 16, 17, 18, 19, 20). Some other phloem sucking auchenorrhynchous species, although found on vine as erratic ones, were listed and examined keeping in mind the kind of their host plants, life history, and transmission of phytopathogenic agents (21, 22, 23, 24). The only two xylem suckers listed, Ph. spumarius and C. viridis, were considered remembering their potential ability in transmitting the pathogenic agent of Pierce's disease (25, 26, 27).

REFERENCES

1. BONFILS, J., SCHVESTER, D. (1960). Les Cicadelles (Homoptera Auchenorrhyncha) dans leurs rapports avec la Vigne dans le Sud-Ouest de la
   France. Annls Epiphyt. 11: 325-336.
2. VIDANO, C. (1963). Alterazioni provocate da Insetti in Vitis osservate, sperimentate e comparate. Annali Fac.Sci.agr.Univ.Torino 1: 513-644.
3. VIDANO, C., ARZONE, A. (1983). Biotaxonomy and epidemiology of Typhlocybinae on Vine. Proceedings, 1st International Workshop on Leafhoppers
   and Planthoppers of economic importance. Commonwealth Institute of
   Entomology, 56 Queen's Gate, London SW7, U.K.: 75-85.
4. VIDANO, C. (1964a). Scoperta in Italia dello Scaphoideus littoralis
   Ball, Cicalina americana collegata alla "Flavescence dorée" della Vite. Italia agric. 76: 1031-1049.
5. VIDANO, C. (1964b). Reperti inediti biologici e fitopatologici della
   Ceresa bubalus Fabricius quale nuovo fitomizo della Vite. Riv.Vitic.
   Enol. 11: 457-482.
6. VIDANO, C. (1966). Introduzione in Italia di Polynema striaticorne Girault, parassita oofago di Ceresa bubalus Fabricius. Boll.Soc.ent.
   Ital. 96: 55-58.
7. VIDANO, C., MEOTTO, F. (1968). Moltiplicazione e disseminazione di
   Polynema striaticorne Girault (Hymenoptera Mymaridae). Annali Fac.
   Sci.agr.Univ.Torino 4: 297-316.
8. ZANGHERI, S., DONADINI, P. (1980). Comparsa nel Veneto di un omottero
   neartico: Metcalfa pruinosa Say (Homoptera, Flatidae). Redia 63: 301-305.
9. SCHVESTER, D. (1973). Insectes vecteurs de maladie a virus et a myco-
   plasmes de la vigne. Riv.Pat.Veg., Supplem., 9: 90-102.
10. MOUTOUS, G., FOS, A., BESSON, J., JOLY, E., BILAND, P. (1977). Résul-
    tats d'essais ovicides contre Scaphoideus littoralis Ball, cicadelle
    vectrice de la flavescence dorée. Rev.Zool.agric.Path.vég. 76: 37-49.
11. CAUDWELL, A., LARRUE, J. (1979). Examen du problème de la Flavescence
    dorée dans le cadre de la sélection sanitaire de bois et plantes de
    vigne. Progrès Agric.Vitic. 6: 128-134.
12. CAUDWELL, A. (1981). La Flavescence dorée de la vigne en France. Phy-
    toma 325: 16-19.
13. BELLI, G., FORTUSINI, A., OSLER, R., AMICI, A. (1973). Presenza di una
    malattia del tipo "Flavescence dorée" in vigneti dell'Oltrepò pavese.
    Riv.Pat.Veg., Supplem., 9: 51-56.
14. OSLER, R., FORTUSINI, A., BELLI, G. (1975). Presenza di Scaphoideus
    littoralis in vigneti dell'Oltrepò pavese affetti da una malattia del
    tipo "Flavescence dorée" della vite. Inftore fitopatol. 25 (6): 13-15.
15. BELLI, G., FORTUSINI, A., OSLER, R. (1978). Present knowledge of dis-
    eases of the type "Flavescence dorée" in vineyards of northern Italy.
    Proc.6th meeting ICVG, Cordoba, 1976, Monografias INIA, 18: 7-13.
16. GRANATA, G. (1982). Deperimenti e giallume in piante di vite. Inftore
    fitopatol. 32 (7-8): 18-20.
17. BELLI, G., FORTUSINI, A., RUI, D., PIZZOLI, L., TORRESIN, G. (1983).
    Gravi danni da Flavescenza dorata in vigneti di Pinot nel Veneto.

Inftore agr. 39: 24431-24433.

18. BELLI, G., RUI, D., FORTUSINI, A., PIZZOLI, L., TORRESIN, G. (1984). Presenza dell'insetto vettore (Scaphoideus titanus) e ulteriore diffusione della Flavescenza dorata nei vigneti del Veneto. Vignevini 11 (9): 23-27.

19. EGGER, E., BORGO, M. (1983). Diffusione di una malattia virus-simile su "Chardonnay" ed altre cultivar nel Veneto. Inftore agr. 39: 25547-25556.

20. CREDI, R., BABINI, A.R. (1984). Casi epidemici di Giallume della vite in Emilia-Romagna. Vignevini 11 (3): 35-39.

21. BRCAK, J. (1979). Leafhopper and planthopper vector of plant disease agents in central and southern Europe. In Maramorosch-Harris: Leafhopper vectors and plant disease agents. Acad.Press. New York. San Francisco. London.

22. FOSTER, J.A. (1982). Plant quarantine problems in preventing the entry into the United States of vector-borne plant pathogens. In Harris-Maramorosch: Pathogens, vectors, and plant diseases : approaches to control. Acad.Press. New York. London. Paris.

23. CATTANEO, E., ARZONE, A. (1983). Ciclo biologico di cicadellidi deltocefalini vettori di MLO. Atti XIII Congr.Naz.It.Ent., Sestriere-Torino: 399-406.

24. SAVIO, C., CONTI, M. (1983). Epidemiologia e trasmissione di micoplasmi dei vegetali. Atti XIII Congr.Naz.It.Ent., Sestriere-Torino: 407-414.

25. SEVERIN, H.H.P. (1950). Spittle-insect vectors of Pierce's disease virus. II. Life history and virus transmission. Hilgardia 19: 357-382.

26. VIDANO, C. (1965). Responses of Vitis to insect vector feeding. Proceedings, International Conference in Virus and Vector of Perennial Hosts, with Special Reference to Vitis. Davis, California, September 6-10, 1965. Univ.California, Div.Agr.Sci., Dept. of Plant Pathology: 73-80.

27. ARZONE, A. (1972). Reperti ecologici, etologici ed epidemiologici su Cicadella viridis (L.) in Piemonte (Hem. Hom. Cicadellidae). Annali Fac.Sci.agr.Univ.Torino 8: 13-38.

Researches on natural enemies of viticolous Auchenorrhyncha*

C.Vidano, A.Arzone & C.Arnò
Istituto di Entomologia Agraria e Apicultura dell'Università di Torino, Italy

Summary

The mesophyll sucker Zygina rhamni, the phloem suckers Empoasca vitis and Stictocephala bisonia, which are all well known Homoptera Auchenorrhyncha owing to typical alterations caused to vine leaves, resulted efficiently controlled by the following natural enemies : the Mymarid Anagrus atomus, an egg-parasite both of Z. rhamni and E. vitis; the Dryinid Aphelopus atratus and the Pipunculid Chalarus sp. prope griseus, both adult-parasites of Z. rhamni; the Pipunculid Chalarus sp. prope spurius, an adult-parasite of E. vitis; the Mymarid Polynema striaticorne, an egg-parasite of S. bisonia. Among the five listed predators of the youngs of Z. rhamni and E. vitis, Chrysoperla carnea was noteworthy. Thanks to the astonishing activity of the egg- -parasites and the additional benefit due to other natural enemies, the use of pesticides to control the three above mentioned Auchenor- rhyncha is judged irrational.

1. Introduction

Among the numerous species of Homoptera Auchenorrhyncha found on vine all over Italy, the Typhlocybins Zygina rhamni Ferrari, Empoasca vitis (Gö the) and the Membracid Stictocephala bisonia Kopp & Yonke, which was once named Ceresa bubalus (Fabricius), are known for symptomatic leaf alterations due to responses to their feeding punctures (1, 2, 3, 4). As a contribution to the biological control of these plant sucking insects, against which the use of insecticides were often recommended without considering the impor- tance of parasitoids and predators, investigations were carried out in or- der to know better the role of their natural enemies. Field and laboratory investigations started in 1968 for S. bisonia, i.e. after introducing (5) and spreading (6) in Italy its specific egg-parasite, the Mymarid Polynema striaticorne Girault, and in 1981 for Z. rhamni and E. vitis.

* Studies of the C.N.R. Working Group for the Integrated Control of Plant pests : 262.

2. Materials and methods

The field surveys took place desultorily in various Italian viticolous territories and every two weeks from May to October in Piedmontese ones.

For what concerns S. bisonia, the frequencies of vine foliar symptoms and the ones of egg-laying wounds both in vine canes and shoots of various broadleaf trees were considered year after year. Then branches showing the characteristical egg-laying wounds were cut and used for laboratory examin ations and rearings, with the aim to check the activity of P. striaticorne.

Z. rhamni and E. vitis were considered both as eggs laid inside vine leaf ribs and as youngs and adults living on the under blade of vine leaves. Their parasitoids were from biological material collected on vines and rea red under laboratory conditions. The activity of their predators was obser ved in field and checked in captivity.

3. Results

Parasitoids and predators of Z. rhamni, E. vitis and S. bisonia found in field or reared under laboratory conditions during our pluriennial in-vestigations are listed in table 1.

The Mymarid Anagrus atomus (Linnaeus) revealed to be a polyvoltine egg-parasite of Z. rhamni and E. vitis, both having three generations a year. Healthy and parasitized eggs of the two leafhoppers were easily found by stereomicroscope from May to October in ribs and less frequently in stalks of vine leaves characterized by feeding puncture alterations. The leaf pierced by the mesophyll feeder Z. rhamni showed symptomatic de-pigmentations. Usually they were not abundant and were more represented in hidden parts of vigorous and bushy vines. The leaf involved by the phloem sucker E. vitis showed at first internal vein browning starting from the edges and then downward rolling, thickening, brightness, reddening or yel-lowing and sometimes marginal burning. The frequency of A. atomus was easier to check in the more abundant eggs of E. vitis. Samplings, which were car-ried out in 1985 during the third generation of this leafhopper in series of twelve basal leaves from single canes, gave the following percentages of parasitized eggs :

cv. Barbera (red), Predosa (Alessandria)	45.6
cv. Erbaluce (white), Caluso (Torino)	50.0
cv. Freisa (red), Bricherasio (Torino)	45.1
cv. Luglienga (white), Torino	46.1

on totals of 79, 72, 71, 63 eggs respectively. The most involved leaves of the above samplings had 11 out of 23 (47.8%), 9 out of 19 (47.4%), 6 out of 11 (54.5%), 7 out of 14 (50.0%) parasitized eggs.

Adults of the three generations of Z. rhamni were parasitized by the Dryinid Aphelopus atratus (Dalman) and the Pipunculid Chalarus sp. prope griseus Coe with various intensity. For example : among 63 adults col-lected on cv. Barbera at Predosa (Alessandria), on 21.9.1981, 30 (47.6%) were parasitized by A. atratus and 7 (11.1%) by C. griseus; among 63 adults collected as above, on 22.7.1984, 40 (63.5%) were parasitized by A. atratus

TABLE I Natural enemies of viticolous Auchenorrhyncha	Parasitoids					Predators				
	Anagrus atomus	Aphelopus atratus	Polynema striaticorne	Chalarus sp. prope spurius	Chalarus sp. prope griseus	Meconema meridionale	Oecanthus pellucens	Malacocoris chlorizans	Chrysoperla carnea	Anystis sp.
Zygina rhamni										
eggs	+									
youngs						+	+	+	+	+
adults		+			+					
Empoasca vitis										
eggs	+									
youngs						+	+	+	+	+
adults				+						
Stictocephala bisonia										
eggs			+							
youngs										
adults										

and 3 (4.8%) by C. griseus. Adults of the three generations of E. vitis were more or less strongly parasitized by the Pipunculid Chalarus sp. prope spurius (Fallén). For example : among 55 adults collected on cv. Barbera at Predosa (Alessandria), on 21.9.1981, 9 (16.4%) were parasitized; among 59 adults collected as above, on 23.9.1984, 30 (50.9%) were parasitized.

Youngs and adults of the Orthopters Oecanthus pellucens (Scopoli) and Meconema meridionale Costa, youngs and adults of the Mirid Malacocoris chlorizans (Panzer), larvae of the Chrysopid Chrysoperla carnea (Stephens) and adults of the Acarus Anystis sp. were predators of youngs of both Z. rhamni and E. vitis. O. pellucens, M. meridionale and Anystis were more represented in vineyards not far from hedges or woods. The most important predator was C. carnea, which has been frequently collected or observed not only as larva but also as adult and egg in vineyards not sprayed with pesticides. Also the activity of M. chlorizans was remarkable, but both its adults and youngs were able to prey upon first and second preimaginal stages of the two leafhoppers.

S. bisonia was considered since cauline and consequent foliar alterations due to its feeding punctures and cane disorders caused by its ovipositor were once frequent on vine. Such symptoms connected with its trophic and egg-laying activities decreased inexorably starting from 1968, when P. striaticorne began to be effective. S. bisonia disappeared slowly, but almost completely from the vineyards of Piedmont and other Italian Regions, where its specific egg-parasite was spread directly or arrived by means of young fruit trees. Very scattered cases of cauline strangling with associated downward leafroll due to S. bisonia were observed lately. Even the shoots of various broadleaf trees, that once were sources of S. bisonia infestations, appeared very poor in egg-laying wounds. Moreover the rare egg-layings of the univoltine S. bisonia were, at the end, always heavily parasitized by the trivoltine P. striaticorne.

4. Conclusions

The parasitoids A. atomus, A. atratus, C. sp. prope spurius, C. sp. prope griseus and the predators O. pellucens, M. meridionale, M. chlorizans, C. carnea and Anystis sp. are reported for the first time as natural enemies of Z. rhamni and E. vitis in Italy. In Europe, where very few investigations have been carried out on natural enemies of viticolous Typhlocybins (7), some specimens of the Trichogrammatid Oligosita tominici Bakkendorf and of the Mymarid A. atomus were reared from eggs of a vine leafhopper erroneously classified as Erythroneura eburnea (Fieber) (8). Actually, in the vineyards near Split, Yugoslavia, where O. tominici and A. atomus were found, only Arboridia dalmatina (Novak & Wagner), E. vitis and Z. rhamni have been collected (Vidano, unpublished data). In North America, where many parasitoids and predators of local grapevine Typhlocybins are known (7), the Mymarid Anagrus epos Girault was emphasized as the most important example of grape pest management and integrated control with regard to the very noxious Erythroneura elegantula Osborne in California (9, 10).

Also in Italy, according to the present preliminary report and unpublished data, A. atomus is to be pointed out for its astonishing importance in the biological control of the Mediterranean vine leafhoppers, as well as for its extraordinary role in controlling typhlocybin pests of maize (11). Finally it is convenient to keep in mind that always in Italy the case of P. striaticorne, well known in the scientific field but ignored in the technical and applied ones, is connected with the disappearance of the phytopathological problem due to S. bisonia. This is a propitious occasion to recommend the safeguard of the very useful above mentioned natural enemies of vine pests from irrational applications of pesticides.

Further investigations are in progress to know better various aspects of natural enemies of viticolous Auchenorrhyncha. Even the leafhoppers Scaphoideus titanus Ball, Jacobiasca libyca (Bergevin & Zanon) and other species are worthy to be considered.

100

5. Acknowledgements

Thanks are devoted to Prof. Gennaro Viggiani, Istituto di Entomologia agraria, Università di Napoli-Portici, for the classification of the Mymarid Anagrus atomus, and to Dr Mark A. Jervis, Department of Zoology, University College, Cardiff, for the classification of Chalarus sp. prope griseus and C. sp. prope spurius, two new species that he is describing.

REFERENCES

1. VIDANO, C. (1958). Le Cicaline italiane della Vite. Boll.Zool.agr.Bachic. (s.II) 1: 61-115.
2. VIDANO, C. (1963a). Alterazioni provocate da Insetti in Vitis osservate, sperimentate e comparate. Annali Fac.Sci.agr.Univ.Torino 1: 513-644.
3. VIDANO, C. (1963b). Eccezionali strozzature anulari caulinari provocate da Ceresa bubalus Fabr. in Vitis. Annali Fac.Sci.agr.Univ.Torino 2: 57-107.
4. VIDANO, C. (1964). Reperti inediti biologici e fitopatologici della Ceresa bubalus Fabricius quale nuovo fitomizo della Vite. Riv.Vitic. Enol. 11: 457-482.
5. VIDANO, C. (1966). Introduzione in Italia di Polynema striaticorne Girault, parassita oofago di Ceresa bubalus Fabricius. Boll.Soc.ent. Ital. 96: 55-58.
6. VIDANO, C., MEOTTO, F. (1968). Moltiplicazione e disseminazione di Polynema striaticorne Girault (Hymenoptera Mymaridae). Annali Fac.Sci. agr.Univ.Torino 4: 297-316.
7. VIDANO, C., ARZONE, A. (1983). Biotaxonomy and epidemiology of Typhlocybinae on Vine. Proceedings, 1st International Workshop on Leafhoppers and Planthoppers of economic importance. Commonwealth Institute of Entomology, 56 Queen's Gate, London SW7, U.K.: 75-85.
8. BAKKENDORF, O. (1971). Description of Oligosita tominici n.sp. (Hym. Trichogrammatidae) and notes on the hosts of Anagrus atomus (L.) and Anaphes autumnalis Foerster (Hym. Mymaridae). Entomophaga 16: 363-366.
9. DOUTT, R.L., NAKATA, J. (1973). The Rubus leafhopper and its egg parasitoid : an endemic biotic system useful in grape-pest management. Environ.Ent. 2: 381-386.
10. FLAHERTY, D.L., PEACOCK, W.L., JENSEN, F.L. (1978). Grape pest management in the San Joaquin Valley. Calif.Agric. 32: 17-18.
11. VIDANO, C., ARZONE, A. (1985). Zyginidia pullula : distribuzione nel territorio e ciclo biologico. Redia (In press).

A new pest of vine in Europe: *Metcalfa pruinosa* Say (Homoptera: Flatidae)

C.Duso

Istituto di Entomologia Agraria dell' Università di Padova, Italy

Summary

Metcalfa pruinosa (Say), an American Flatid planthopper recently found in Venetia, is becoming quickly widespread in North-East Italy, as a new pest of vine. *M. pruinosa* has only one generation per year; the egg, laid under the bark, overwinters and the first nymphs are found on the leaves in May. Population density peaks in mid June in the leaves; in July a large amount of nymphs migrate inside bunches. Adults first appear in July and can be observed as late as October. Dense population of nymphs cause a stunted growth of shoots; nymphs and adults cause a large amount of honeydew and consequently of mould on leaves and inside bunches. The biological control of the species in Italy is unsatisfactorily carried out by unspecific predators. The chemical control of nymphs is difficult due to their mobility and progressive hatching; however adults can migrate from weeds and bushes to vineyards after treatments.

1. Introduction

An American Flatid planthopper *Metcalfa pruinosa* (Say) has appeared in Italy recently (5). In a few years time the area occupied by the insect has become larger and today it can be found in the plains of some provinces of north east Italy (Veneto and Friuli-Venezia Giulia). The fast rate of diffusion of this species is due to its extraordinary polyphagy and ability to colonise herbaceous and tree crops, weeds and bushes. Vine seems to be one of the more infested plants. The *M.pruinosa* life cycle in north east Italy, its distribution and its relation with the numerous host plants are discussed in preliminary notes (2). Following is recent information on this topic.

2. Description of the Species

Like other members of the *Flatidae* family *M.pruinosa* is easily visible; the adult is about 7-8 mm. long (males are slightly smaller than females). The wings, covered by a waxy secretion, are dark greyish-brown and are folded against the sides of the body.

The egg is elongated, oval, translucent white in color, and displays two lateral sinuate grooves.

First instar nymphs are initially white and then become light green; they become covered by a waxy secretion which is particularly visible in 4th and 5th instars nymphs. Description of immature stages is recently carried out (4).

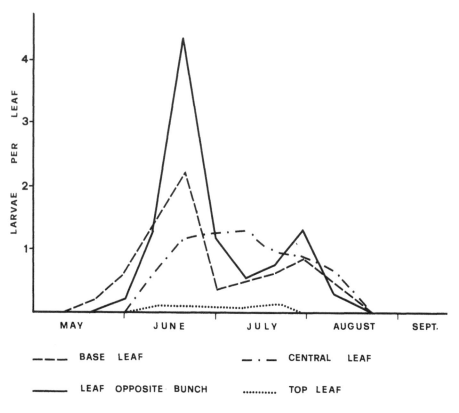

Fig.1: Populations of <u>Metcalfa</u> <u>pruinosa</u> (Say) (immature stages) on vine leaves situated on different parts of shoots in 1984. The highest density of phytophagous mites was observed on leaves opposite bunches. The base leaves and the central leaves was infested at moderate levels during spring and summer.

3. Life history

We report observations carried out in some places of Venetian region (Treviso) in 1983 and 1984.

M.pruinosa overwinters as eggs laid between cracks in the bark or inside the buds; during high infestation more than 30 eggs have been found on a branch of 10 cms; the eggs was sometimes visible on the surface. Hatching begins in the second half of May and continues for over a month. The development of nymphs begins on the underside of leaves. The first instar nymphs which are able to jump considerably are often aggregated on shoots; nymphs initially colonise the leaves near vinestock, moving progressively to the entire shoot. Larvae populations living on leaves peak in the second half of June (Fig.1). The leaves opposite the bunch are infested for a longer time and with more intensity than other leaves so can be considered a good sample for population evaluation (Fig.1). The abundant waxy secretions and the whitish exuviae remain for a long time on the vegetation, therefore the presence of insect is easily

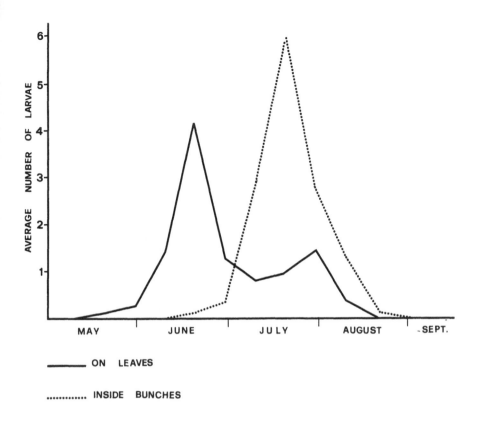

ON LEAVES

.... INSIDE BUNCHES

Fig.2 Populations of Metcalfa   pruinosa (Say) (immature stages) on bunches and
opposite bunches leaves in 1984. Flatid planthoppers peak at mid-June on leaves and at
mid-July inside bunches. This is probably due to a search for a better microclimate.

visible. At the beginning of blooming, it is possible to find the
first nymphs on the rhachis of bunches. During fruit-setting, density
of nymphs decreases on leaves and progressively increases inside
bunches (Fig.2).   This migration is probably due to the search for
better microclimate. Many fifth instar nymphs mature inside bunches.
In some of these, more than 10 specimens of M.pruinosa have been
found with abundant honey-dew on rhachis.
        The first adults appear in the second half of July both on
leaves and inside bunches. During high infestations, adults spread to
shoots and branches and in other cases inside the vegetation.
        Towards the end of August some nymphs can still be found,
sometimes, on the surface of two year old branches. Egg-laying
begins in the first half of September; adults can survive until the
first frost.
        The life history of M.pruinosa in Italy seems similar to the
one in Ohio (Jubb, pers. comm.) and in Texas (1); in this area the
presence of the Flatid is reported two months earlier.

4. Economic importance

When M.pruinosa first appeared in Venetian region no important damage was observed. The presence of nymphs is generally linked to "esthetic" damages (persistence of exuviae and waxy secretions). However, in some cases the presence of a large aggregation of nymphs can reduce the growth of vegetative apex, a large production of honey-dew and development of mould, especially inside bunches. This phenomenon has been observed on bunches containing at least 3-4 specimens. Honey-dew production increases with the progressive appearance of adults; in some cases this phenomenom caused a slight attack of wasp. The presence of many adults at harvest time disturbs the workers because the insects stick to their skin, sometimes causing irritation.

The heavy infestations found on vines are nearly always associated to the presence of natural vegetation (Acer campestre L., Hedge Maple; Robinia pseudo-acacia L., Locust; Rubus ulmifolius Schott, European Blackberry) bordering the fields; these and other species represent pockets of infestation and allow the migration of planthoppers to vines. Many infested weeds such as Stinging Nettle (Urtica dioica L.) or European Glorybind (Convolvulus arvensis L.) favour dispersion of M.pruinosa in the environment.

5. Possibility of biological control

In its original territories, M.pruinosa has many natural enemies. Among these, Dryinids (Psilodryinus typhlocybae (Ashmead) and Neodryinus sp. (Ashmead) are considered important (1, 4). The adults can be attacked by the moth Epipyrops barberiana Dyar (4). Larval mites (Leptus sp.) has been found attached to the thorax and abdomen of nymphs (4).

In Italy during our research no parasites of this species were found. In some cases, unspecific predators like Lacewing flies (Neuroptera : Chrysopidae) were observed; adults are sometimes prey of spiders.

6. Conclusions

Damages caused by M.pruinosa are nearly always "esthetic"; but high infestations can cause large production of honey-dew and development of mould inside bunches and on leaves.
On pruning, a large number of eggs is eliminated with positive results for population control of Flatid planthoppers in successive years (3). In some cases, the elimination of natural vegetation infested by M.pruinosa caused a decrease in damages in vineyards.

Sprays against nymphs can be justified only in few cases (more than 5-6 nymphs per bunch). The success of application of insecticides is limited by prolonged hatching period and mobility of planthoppers which migrate from natural vegetation to vine some days after treatment.

REFERENCES

1. DEAN, H.A. and BAILEY, J.C.(1961). A Flatid Planthopper , Metcalfa pruinosa (Say). J.Econ.Ent.,54: 1104-1106.

2. DUSO, C. (1984). Infestazioni di Metcalfa pruinosa nel Veneto. Inf.Fit.,5: 11-14.

3. WALDEN, B.H. (1922). The mealy flatas, Oremensis pruinosa Say. Connecticut Agric.Expt.Sta.Bull. 235: 189-90.

4. WILSON, S.W. and MCPHERSON, J.E. (1981). Life history of Anormenis septentrionalis , Metcalfa pruinosa and Ormenoides venusta with descriptions of immature stages. Ann.Ent.Soc.of America, 74: 299-311.

5. ZANGHERI, S. and DONADINI, P. (1980). Comparsa nel Veneto di un omottero neartico: Metcalfa pruinosa Say (Homoptera, Flatidae). Redia, 63:301-305.

A survey of the grape phylloxera (*Viteus vitifoliae* (Fitch)) problem a century after its introduction

A.Crovetti & E.Rossi
Istituto di Entomologia Agraria dell' Università di Pisa, Italy

Summary
After a brief report of some notices about the diffusion and about the present distribution of the grape phylloxera (Viteus vitifoliae (Fitch)), the Authors give a survey of some of the research subjects presently going on phylloxera in Italy. These themes are. schematized in 9 points: the first four describe studies on different aspects of phylloxera on european grapevines on their own roots, others (points n. 5 and 7) indicate the eco-ethological observations carried on phylloxera in field and in laboratory. Another branch of studies in which many Entomologists are investigating, is the one regarding the causes determining the appearance of foliar galls of phylloxera on the leaves of grafted european grapevines. Also some researches on the relations between phylloxera and other organisms (bacteria, fungi and, especially, viruses) are reported.

As few other insects the grape phylloxera was ill-famed among a public much larger than the narrow circle of the Entomologists. Known today to the experts as Viteus vitifoliae (Fitch), it is probably more famous with the old name of Phylloxera vastatrix Planch., given to it by a French Entomologist just to underline its destructive power . The story of its diffusion is today well known: it was introduced into Europe from the American Continent at the half of the last century as a consequence of a large importation of American grapevines necessary to face a severe reduction of European vineyards attacked by oidium (Uncinula necator) and peronospora (Plasmopara viticola).

Almost immediately the infestation spread out all over the Old Continent. To give an idea of the quick diffusion of the pest, in tab.1 the dates of the first signalized appearance of the phylloxera in some Countries are reported: it is easy to see that between 1863 and 1883 in all the European Countries the presence of the Aphid was observed. As concern the problem in Italy, in tab.2 we report the chronological list of the first finding of the grape phylloxera in several Italian localities. In fig.1 the present world distribution of the pest is shown in relation to the presence of wild and cultivated grapevines.

Since the beginning of its diffusion, many experts tried to find a solution to control the grape phylloxera. Surprisingly, in 1869 already, during a Viticulture Congress held at Beaune (FR), Laliman suggested for the fist time the possibility of grafting the European grapevines varieties on American stocks. With this method it was possible to take advantage of the incapacity of the pest to form galls on the leaves of Vitis vinifera utilizing, at the same time, the good resistance showed by the roots of American grapevines.

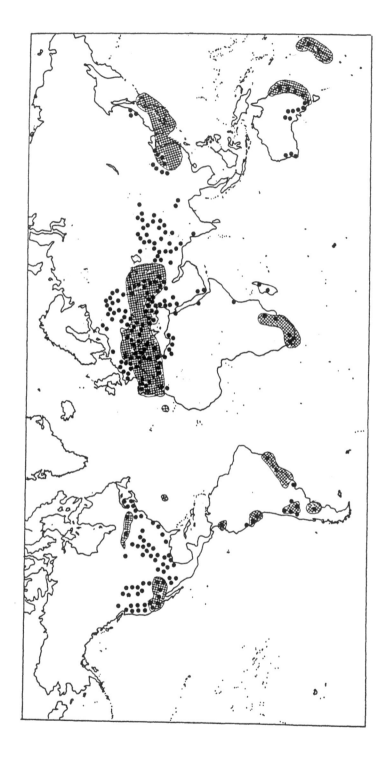

Fig. 1 – Present geographic distribution of the grape phylloxera (_Viteus vitifoliae_ (Fitch)) (grey areas) in relation to the presence of grapevine (_Vitis_ spp., rapresented by the black points) in the world.

Tab. 1 – Dates of the first detection of the grape phylloxera in several world Countries.

YEAR	COUNTRY	YEAR	COUNTRY	YEAR	COUNTRY
1863	U.K.	1879	Ungary	1883	Switzerland
1863	France	1879	Australia	1883	Algeria
1865	Portugal	1880	Austria	1885	Turkey
1872	Madeira	1880	Crimea	1885	Russia
1875	Germany	1880	Yugoslavia	1885	New Zealand
1876	Spain	1880	Romania	1890	Peru
1879	Italy	1882	California	1890	Argentina

Tab. 2 – Dates of the first findings of the Aphid in Italy.

YEAR	REGION or LOCALITY	YEAR	REGION or LOCALITY	YEAR	REGION or LOCALITY
1879	Milan	1886	Novara	1892	Rome
	Como	1888	Elba Island	1894	Cuneo
1880	Sicily	1888	Grosseto	1894	Pise
1883	Sardinia	1888	Siena	1896	Turin
1885	Bergamo	1888	Catanzaro	1897	La Spezia

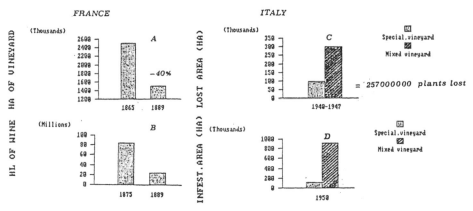

Fig. 2 – Figures of the losses caused by the grape phylloxera in France (A and B) and in Italy (C and D).

As Barbagallo (2) has recently remarked, we can use the term "tolerance" rather than "resistance" to define the reaction of the association of stock and scion to the phylloxeric attack.

In spite of this early suggested solution to the phylloxeric problem, the Aphid was able to produce enormous damages. In fig.2 the reported figures give an idea of the dimensions of the disaster in terms of decreasing of production and attacked areas in France and in Italy. As regards Italy, it is shown the data concerning the damages caused by the grape phylloxera and in specialized vineyards and in the ones in a mixed planting system: 110,000 ha and 928,000 ha respectively were destroyed according to a census taken in 1950 (29 cited in 13).

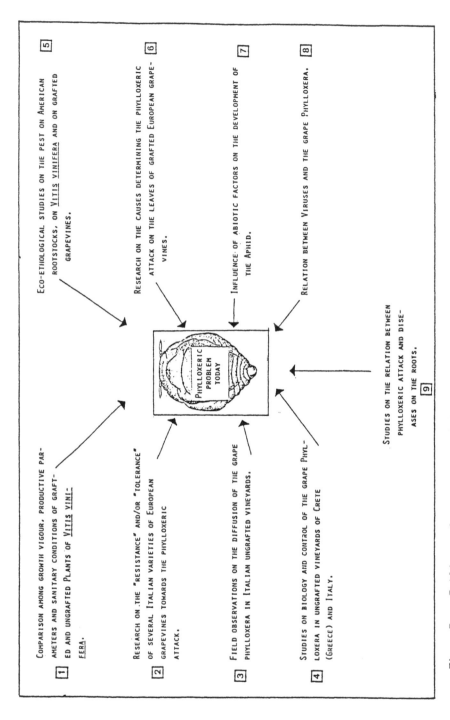

COMPARISON AMONG GROWTH VIGOUR, PRODUCTIVE PAR-
AMETERS AND SANITARY CONDITIONS OF GRAFT-
ED AND UNGRAFTED PLANTS OF VITIS VINI-
FERA. [1]

RESEARCH ON THE "RESISTANCE" AND/OR "TOLERANCE"
OF SEVERAL ITALIAN VARIETIES OF EUROPEAN
GRAPEVINES TOWARDS THE PHYLLOXERIC
ATTACK. [2]

FIELD OBSERVATIONS ON THE DIFFUSION OF THE GRAPE
PHYLLOXERA IN ITALIAN UNGRAFTED VINEYARDS. [3]

STUDIES ON BIOLOGY AND CONTROL OF THE GRAPE PHYL-
LOXERA IN UNGRAFTED VINEYARDS OF CRETE
(GREECE) AND ITALY. [4]

ECO-ETHOLOGICAL STUDIES ON THE PEST ON AMERICAN
ROOTSTOCKS, ON VITIS VINIFERA AND ON GRAFTED
GRAPEVINES. [5]

RESEARCH ON THE CAUSES DETERMINING THE PHYLLOXERIC
ATTACK ON THE LEAVES OF GRAFTED EUROPEAN GRAPE-
VINES. [6]

INFLUENCE OF ABIOTIC FACTORS ON THE DEVELOPMENT OF
THE APHID. [7]

RELATION BETWEEN VIRUSES AND THE GRAPE PHYLLOXERA. [8]

STUDIES ON THE RELATION BETWEEN
PHYLLOXERIC ATTACK AND DISE-
ASES ON THE ROOTS. [9]

PHYLLOXERIC
PROBLEM
TODAY

Fig. 3 – Outline of some research topics studied today in the ambit of the
project on "Possibility of growing ungrafted European grapevines",
sponsored by the Italian "Ministero della Pubblica Istruzione".

112

Tab. 3 - Research projects on the grape phylloxera in Italy: subjects, Sponsors and Partecipants.

RESEARCH PROGRAM	SPONSORED BY	PARTECIPANTS
Possibility of growing ungraft-ed European grapevines	Italian "Ministero della Pubblica I-struzione".	Ist.di Coltivazioni Arboree-Univ.di Firenze Ist.di Coltivazioni Arboree-Univ.di Pisa Ist.di Coltivazioni Arboree-Univ.di Catania Ist.di Patologia Vegetale-Univ.di Pisa Ist.di Patologia Vegetale-Univ.di Catania Ist.di Entomologia Agraria-Univ.di Pisa Ist.di Entomologia Agraria-Univ.di Catania Ist.di Zoologia-Univ.di Modena Dip.di Botanica-Univ.di Pisa
Improvement in the integrated control in the vineyards	CCE and the Italian Ministero della A-gricoltura e Fore-ste.	Ist.Sperimentale per la Zool. Agr.-Firenze Ist.di Entomologia Agraria-Univ.S.C.Piacenza Ist.di Entomologia Agraria-Univ.di Milano Ist.di Entomologia Agraria-Univ.di Padova Ist.di Entomologia Agraria-Univ.di Pisa S.Sperim.Lotta Fitosan.Reg.Piemonte-Torino

The enormous economic importance that the grape phylloxera assumed in few years and its peculiar life cycle interested many Entomologists, especially at the beginning of the present century. The contribution given by the Italian School (Grassi and Coll., 1912-1927) , by the German one (Borner and Others, 1910-1924) and by many other Scientists from other Countries (i.e., 27, 16,36,21,17) allowed an almost complete knowledge of the biology of this Aphid and on the American grapevines and on the European ones.

Neverthless, recent events have shown that there are still irresolute questions about the life history of the grape phylloxera on grafted grapevines. The events we are referring to are the proposed return to the planting of ungrafted European grapevines (26 and 6) and, on the other hand, the more and more frequent manifestation of phylloxeric galls on the leaves of grafted European vines (3) and the supposed discovery of resistant strain of the Aphid (15 and 8).

As a consequence of the renewed interest toward the grape phylloxera, the hypothesis about the existence of "ecotypes", "biotypes", "strains" and "races" into the phylloxeric population is emerged again. Some examples of this are represented by the papers of Italian and foreign Authors (9,31,32,33 and 30).

The importance of a complete explanation of the obscure points still existing in the knowledge of this pest, is testified also by the interest taken in it by some National and International Organizations sponsoring the Research. For example, since 1984 the CEE promoted researches to give a contribution to viticulture problems: in particular, among them there is a project titled "Improvement of the Integrated Pest Control in the Vineyards" in which studies on grape phylloxera are included (see tab.3). Also the Italian "Ministero della Pubblica Istruzione" is, sponsoring a research program on the "Possibility of growing ungrafted European vines" in which several University Institutes are collaborating, as shown in tab.3.

An outline of the themes of the modern research carried out in the ambit of this program is shown schematically in fig.3.

Tab. 4 - Salient characteristics and results of the observation carried out
in the ungrafted vineyard of the Gorella farm at Follonica
(Grosseto, Italy).

DESCRIPTION	OBSERVATIONS		
	1982	1983	1984
Area: 18,000 sm	Only 3	No incre-	No incre-
Year of planting: 1978	vines we-	ase in	ase in
Cultivars grown: Sangiovese,Treb-	re found	the inf-	the inf-
biano toscano,Canaiolo,Malvasia	infested	estation	estation
bianca, Malvasia nera.			
Characteristics: ungrafted vines.			
Soil type: clay.			

As it is possible to see, a first argument (indicated as point n.1) is
the one regarding the evaluation of the yield of ungrafted European
grapevines in terms of productivity and resistance or tolerance to pests
and diseases. The "Istituto Sperimentale per la Zoologia Agraria" of
Florence and the "Istituto di Coltivazioni Arboree" of Pise are working on
this subject. In particular, the researchers of the second Institute are
working on an experimental vineyard composed by 500 ungrafted European
grapevines of the cultivars Sangiovese and Trebbiano . These plants are
periodically controlled with the aim to study the cited characteristics in
comparison with a grafted vineyard.

The second argument listed in fig.3 (point n.2) deals with the
investigation on varieties of Vitis vinifera resistant or tolerant to the
grape phylloxera. For this purpose (22) in 50 cement containers, 10 of
which filled by sandy soil and 40 by loam soil, 400 ungrafted European
grapevines of 20 different varieties were planted and infested with grape
phylloxera to study the reaction to the pest of the different cultivars in
the two different soil conditions.

The third point of fig.3 (point n.3), is referred to an investigation
carried out for several years with the aim to take a census of some of the
ungrafted grapevines existing in Italy (5). The results of this
investigation seem to confirm a clear relation between type of soil and
occurrence of the phylloxeric attack, relation already observed by many
other Authors in literature (18).All these vineyards were visited at least
once to investigate on the eventual presence of grape phylloxera on the
leaves and/or on the roots, judging at the same time a more general
sanitary condition of the plants, to evaluate a possible relation between
phylloxeric infestation and the occurrence of fungal, bacterial and viral
diseases . In particular, in an experimental vineyard planted in 1978 at
Follonica (Grosseto, Italy) by the "Istituto di Coltivazioni arboree" of
Florence, this kind of observations were repeated annually, starting from
the year of the planting. The results of these pluriennal research are
synthetized in tab.4 : it shows that the phylloxeric infestation spread
very slowly and, appearently, up to this moment it has not caused any
damage to the vines.

Another branch of the studies carried out on the grape phylloxera,
(point n.4 in fig.3), regards the investigations on the biology and on the
control of the Aphid in a particular environment, the one of Crete (Greece)
where almost all the vineyards are ungrafted. The island was considered
uninfested until few years ago, but since the very beginning of our
observations started in 1982 with the collaboration of Specialists of the
"Istituto di Coltivazioni Arboree" and of "Patologia Vegetale" of Pise, we

noticed that the infestation is largely diffused in the most part of the vineyards. With the effective assistance of several members of the Sub-Tropical Plants and Olive Tree Institute of Chania (Crete)(*) , trials on the possibility of a chemical control of the pest were carried out. According to us, in fact, considering several factors as the diffusion of the ungrafted grapevines in the island, the determinant importance of viticulture in the local economy and the heavy production losses observed, the chemical control of the grape phylloxera could be the best countermeasure to apply immediately, in view of a long term solution of the problem: the replacement of the grapevines on their own roots with the grafted ones. However, it is necessary to remember that also the chemical approach to the phylloxeric problem showes several difficulties and because the practical and economic difficulties to control the root forms of the Aphid, and because the hazard to the biological equilibrium of the environment involved in those treatments. Trials on the chemical control of the root forms of phylloxera are going on also in a mother plantation situated in Emilia Romagna (Italy).
Numerous other researches have been started on American grapevines too.
In some vine nurseries of the district of Pise, for example (point 5 in fig.3), eco-ethological studies on the grape phylloxera were carried out (23,24 and 25). Using infested material coming from the same fields (point n. 7), a laboratory research on the influence of abiotic factors on the rate of development and on the mortality of phylloxera was begun : up today, the results of the studies carried out on the termic influence on the rate of the embrionic development are available (4) while similar trials on the larval instars and on the adults are almost concluded.
As many others Aphids, the grape phylloxera is studied also by Plant Pathologists because the relations between the phylloxeric attack and the manifestation of bacterial, fungal and especially viral diseases of the grapevines (19).
As shown in the points 8 and 9 of fig.3, also in Italy these studies are going on and the first results has been already published (35).
     Last but not less important research argument we want to talk about, is the one described in the point n.6, regarding the investigations carried out on the causes of the manifestation of foliar galls of phylloxera on Vitis vinifera. More and more frequent are in our Country the notices about the presence of leaf infestations on grafted European grapevines (31,32,33,12,3). This phenomena is not new: in literature we can find many papers dealing with similar manifestations, and in Italy (14,37) and abroad (i.e. 10,11,34). All these Authors describe the phenomena as an occasional event but the causes indicated to explain it are different: many of them (i.e. 37) considered the presence of foliar galls on Vitis vinifera as a consequence of heavy attacks on American grapevines situated in the vicinity , while the hypothesis pointed out in the recent studies of Strapazzon and Girolami (31,32,33) is a deep change in the biology of the Aphid, as a consequence of the existence of new ecotypes, so that the foliar infestation derives directly from females hatched from the winter egg. Other Authors have put forward also the proposal of change in the host (3 and 12) even if with some reservations. The genetic selection operated on the European varieties to improve the productivity , the organoleptic characteristics of the grape and the affinity between scion and rootstock are invoked as the main causes of the host change. Besides these

---

(*) We thank in particular Dr. N. Psyllakis, Dr. S. Michelakis, Dr. V. Alexandrakis and Dr. V. Vardakis for their collaboration.

115

Tab. 5 - Listing of recent detections of foliar phylloxeric galls on on grafted grapevines in Italy.

LOCALITY	AGE (Years)	VARIETY and STOCK	CITED AUTHORS	OBSERVATIONS
Lonigo	7	Tokai rosso on Kober 5BB	Strapazzon and Girolami (1983), (1985a), (1985b).	
Latina	5	Italia on Kober 5BB	Grande, 1984	
"	"	Palieri on Kober 5BB		
"	"	Lavalle' on Kober 5BB		
Padova	6	Moscato on Kober 5BB	Setti, 1984	
Pisa	?	Kober 5BB (V. nursery)	* *	
Udine	?	Merlot on Kober 5BB	Setti, 1984	Uninfested
		Pinot on Kober 5BB		Scarcely inf.
		Tokai on Kober 5BB		Highly inf.
Udine	?	Cabernet on Kober 5BB	Barbattini, Pravisani and Zandigiacomo 1985	Only one plant infested: a stock Kober 5BB
Udine	12	Verduzzo on Kober 5BB	Conti, Quaglia and Rossi, 1985	Highly inf.
Udine	16	Moscato rosa on Kober 5BB	*	* *
		Tokai on Kober 5BB		Scarcely inf.
		Pinot grigio on Kober 5BB		* *
		Pinot bianco on Kober 5BB		* *
		Schiava freisa on Kober 5BB		Uninfested
		Merano on Kober 5BB		*
		Merlot on Kober 5BB		
		Moscato bianco on Kober 5BB		*
		Silvaner on Kober 5BB		*

Tab. 5 cont.

LOCALITY	AGE (Years)	VARIETY and STOCK	CITED AUTHORS	OBSERVATIONS
Udine	7	Tokai on K.5BB and 420 A	Conti, Quaglia and Rossi, 1985	Highly infested
		Ribolla on K.5BB and 420 A		" "
		Picolit on K.5BB and 420 A		" "
		Sauvignon on K.5BB and 420 A		Scarcely inf.
		Malvasia on K.5BB and 420 A		" "
		Verduzzo on K.5BB and 420 A		" "
		Pinot bianco on K.5BB and 420 A		" "
		Pinot grigio on K.5BB and 420 A		" "
		Riesling on K.5BB and 420 A		" "
		Refosco on K.5BB and 420 A		" "
Udine	11	Picolit on Kober 5BB	Conti, Quaglia and Rossi, 1985	Highly infested
		Verduzzo on Kober 5BB		Scarcely inf.
		Pinot on Kober 5BB		" "
Ravenna	9	Sangiovese ungrafted	Conti, Quaglia and Rossi, 1985	Scarcely inf.
		Trebbiano ungrafted		" "
		Albana ungrafted		Uninfested

considerations, according to our observations (see tab.5) also the variety of the rootstock seems to have a determinant importance in the sensitivity of the European scion. Many of the grapevines showing foliar infestation were grafted on the same rootstock, Kober 5BB that is one of the most common stock used in Italy. This diffusion is a consequence of its good adaptability to different soil conditions, even if the growers know very well its sensitivity to the phylloxeric attack demonstrated also by its medial position in the scale of sensitivity suggested by Pastena (20).

In conclusion of this brief survey, we wish to express the hope that this renewed interest in the grape phylloxera could give a contribution not only from a scientific point of view, but also from an applicative one, since the numerous difficulties nowadays the Viticulture meets with.

REFERENCES

1. ALDRICH J.M., 1923 - The grape phylloxera. 17th Bienn.Rept.Oregon Satte Bd.Hortic. : 191-192.
2. BARBAGALLO S., 1980 - La resistenza delle piante nella lotta contro i fitofagi. Prospettive di controllo biologico degli insetti in agricoltura, Collana del Progetto Finalizzato "Promozione della qualità dell'ambiente", Consiglio Nazionale delle Ricerche, AQ/1/51-56: 155-164.
3. BARBATTINI R., PRAVISANI L., ZANDIGIACOMO P., 1985 - Presenza di fillossera nel Pordenonese. L'Informatore Agrario, XLI, 18: 91-96.
4. BELCARI A., COGNETTI G.C., 1983 - Influenza della temperatura sullo sviluppo degli stadi preimaginali di Viteus vitifoliae (Fitch) (Rhyncota, Phylloxeridae). 1 - Durata dello sviluppo dell'uovo in generazioni epigee a temperature costanti. Frustula Entomologica, n.s. VI (XIX): 413-420.
5. CONTI B., QUAGLIA F., ROSSI E., 1985 - Preliminary results of the investigations on the diffusion of the Grape Phylloxera (Viteus vitifoliae (Fitch)) in some plantings of European grapevines on their own roots in Italy. Proceedings of the Expert's Group Meeting on "Integrated Pest Control in Viticulture", Portoferraio, Italy, 26-28 September 1985 (in press).
6. CROVETTI A., 1980 - La difesa dai fitofagi della vite. From: "Difesa antiparassitaria e diserbo chimico per le regioni Umbria Liguria Toscana", Proceedings of the Course held in Pisa, November, 6-11, 1980: 85-94.
7. DALMASSO G., 1967 - Ritornare alle viti franche di piede anche in terreni fillosserati?. Il Coltivatore , 113: 121-125.
8. GLUSHKOVA S.A., 1979 - Resitance of the vine louse to hexachlorobutadiene as a result of frequent soil fumigation and way of preventing it. (in russian) Ustoichivost' vreditelei k khimicheskim sredstvam zashchnity rastenii. Ed. by Zil'bermints, I.V.; Smirnova A.A., Matov, G.N., Moscow, USSR: 54-58.
9. GOIDANICH A., 1960 - Term "Fillossera della vite" in: "Enciclopedia Agraria", R.E.D.A., Rome, 1960: 682-698.
10. GOLLMICK F., SCHILDER F.A., 1941 - Histologie und morphologie der rebenblatter in ihren beziehungen zum reblausbefall. Mitt. biol. Reichsanst., 65: 57-59.
11. GORKAVENKO E.B., SYREL'SHCHIKOVA L.P., 1979 - Control of the leaf form. (in russian) Zashchita Rastenii, 12: 38-39.
12. GRANDE C., 1984 - Dopo più di un secolo la Fillossera, il piccolo "pidocchio" della vite, ritorna alla ribalta. L'Informatore Agrario, XL, 30: 44-46.
13. GRANDI G., 1951 - Istituzioni di Entomologia. Edagricole, Bologna, Vol.I: 868-865.
14. GRASSI B., FOA A., GRANDORI R., BONFIGLI B., TOPI M., 1912 - Contributo alla conoscenza delle fillosserine e in particolare della fillossera della vite. Tip. Nazionale G. Bertera, Roma, 456 pp.
15. KHMELEVSKAYA M.A., 1979 - The development of resistance by phylloxera to hexaclorobutadiene in the Crimea.(in russian) Ustoichvost' vreditelei k khimicheskim sredstvam zashchity rastenii. Ed. by Zil'bermints I.V., Smirnova A.A., Matov, G.N., Moscow, USSR: 51-54.
16. KOZHANCHIKOV I.V., 1930 - Data for the study of pests and diseases of the vine. Results of a survey of the plantations of european vine in Uzbekistan in 1929. (in russian). Uzbekist. oputn. Sta. Zashch. Rast., 17, 80 pp.
17. MAILLET P., 1958 - La Phylloxéra de la vigne. Quelques faits biologiques et les problémes qu'ils soulévent. Proceedings 10th International Congress of Entomology, Montreal, August 17-25, 1956, I: 75-84.
18. MARCOVITCH S., 1934 - The woolly aphids in Tennessee. Journal of

Economic Entomology, 27 (4): 779-784.
19. OCHS G., 1960 - Ubertragungsversuche von drei Rebviren durch Milben und Insekten. Z. angew. Zool., 47: 485-491.
20. PASTENA B., 1972 - Trattato di Viticoltura Italiana. Tip. Sirte, Roma, 1049 pp.
21. PRINTZ Y.I., 1937 - Contribution to the question of the changes in the virulence of Phylloxera of different biotypes. (in russian). Plant Protection, 12: 137-142.
22. QUAGLIA F., ROSSI E., 1985 - Susceptibility of some cultivars of European grapevines to the foliar attack of the grape phylloxera (Viteus vitifoliae (Fitch)): observations of the triennium 1982-1984. Proccedings of the Expert's Group Meeting on "Integrated Pest Control in Viticulture", Portoferraio, Italy, 26-28 September 1985 (in press).
23. RASPI A., ANTONELLI R., 1985 - Grape Phylloxera (Viteus vitifoliae (Fitch)) infestation on American vine leaves in a nursery at S.Piero a Grado (Pisa) during the years 1982-1983. Ibid. (in press).
24. RASPI A., ANTONELLI R., CONTI B., 1985 - Preliminary observations on the composition of the Grape Phylloxera (Viteus vitifoliae (Fitch)) population in American vine leaves. Ibid.(in press).
25. RASPI A., BELCARI A., ANTONELLI R., CROVETTI A., 1985 - Epigean development of Grape Phylloxera Viteus vitifoliae (Fitch) in Tuscan nurseries of American nurseries of American vines, during the years 1982-1983. Ibid. (in press).
26. SCARAMUZZI F., 1979 - Viticoltura moderna e fillossera. Giornale di Agricoltura, 89 (41): 40-44.
27. SCHNEIDER-ORELLI O., LEUZINGER H., 1924 - Vergleichende Untersuchungen zur Reblausfrage. Vierteljahrsschr. Naturforsch. Ges. Zuerich, LXIX: 1-50.
28. SETTI M., 1984 - Fillossera della vite: un problema che si ripropone. Proceedings of "Giornate Fitopatologiche", Sorrento, 26-29 March 1984: 307-314.
29. SPAGNOLI A., 1948 - I danni della fillossera. Istituto Centrale di Statistica, Roma.
30. STEVENSON A.B., 1970 - Strain of the grape phylloxera in Ontario with different effects on the foliage of certain grape cultivars. Journal of Economic Entomology, 63 (1): 135-138.
31. STRAPAZZON A., GIROLAMI V., 1983 - Infestazioni fogliari di fillossera (Viteus vitifoliae (Fitch)) con completamento dell'olociclo su Vitis vinifera (L.) innestata. Redia, LXVI: 179-194.
32. STRAPAZZON A., GIROLAMI V., 1985a - Aspetti dell'infestazione di fillossera (Viteus vitifoliae (Fitch)) su vite europea. Proc. of the 14th National Congress of Entomology, 5.28.1985/6.1.1985, Palermo (Italy): 633-641.
33. STRAPAZZON A., GIROLAMI V., 1985b - La fillossera su viti europee. L'Informatore Agrario, XLI, 20: 73-76.
34. TUPIKOV V.K., 1932 - Ecological characteristics of the habitat of Phylloxera in Tuapse.(in russian) Bull.Plant. Prot., 4: 45-70.
35. VANNACCI G., LORENZINI G., TRIOLO E., 1983 - First report on the microflora associated to grape vine roots infested by "Phylloxera". Rivista di Ortoflorofrutticoltura italiana, 68: 21-23.
36. VASIL'EV I.L., 1929 - On the race of Ukrainian Phylloxera. (in russian). Vestn. Vinodel. Ukrainui, XXX (1): 13-14.
37. ZANARDI D., 1962 - La fillossera "gallecola" su foglie di vite nostrana. Il Coltivatore, 12: 371-374.

Leaf and root infestation of *Viteus vitifoliae* (Fitch) on *Vitis vinifera* (L.) grafted on American roots and ungrafted

A.Strapazzon

Istituto di Entomologia Agraria dell' Università di Padova, Italy

Summary

Viteus vitifoliae (Fitch) populations have been sampled in 1984 and 1985 in North-West Italy on roots and leaves of cv. Tocai rosso grafted on Kober 5BB . The results show:

1- Leaf infestation is linked to the amount of fundatrix galls and to the regular formation of new leaves.

2- The flights of sexuparae have been extensive and continued in summer and part of autumn.

3- The increase of root nodosities in spring is not linked to the falling of newly hatched larvae from leaf galls onto the soil; root infestations are linked to the overwintering root forms.

On both grafted (Kober 5BB) and ungrafted European vines (cv. Tocai Rosso) a similar root infestation of Viteus vitifoliae was found in 1985.

1. INTRODUCTION

Viteus vitifoliae (Fitch) foliar infestations on Vitis vinifera (L.) have been reported in some Italian regions (15, 7, 1). Injury to European cultivars by the leaf population of Grape phylloxera seems however limited (17).

Foliar infestations can increase root infestations on grafted Vitis vinifera . A larger number of root gall-like nodosities seem present in biotopes with leaf infested vineyards (15).

Basic information on Grape phylloxera behaviour and damages on ungrafted Vitis vinifera is needed. In fact, a normal yield in ungrafted vineyards seems possible (3, 4, 12).

Two years observation on the relation between leaf and root infestations of Grape phylloxera are reported.

2. MATERIALS AND METHODS

Research was done on cv. Tocai Rosso in two hillside commercial vineyards in the Venetian region: 2.1 Lonigo (Vicenza) 1984 and 1985 on cv. Tocai Rosso grafted on Kober 5BB; age: 7 years; "pergola" training system; 2.2 Lovolo (Vicenza) 1985 on cv. Tocai Rosso grafted on Kober 5BB and ungrafted; age: 8 years; "Sylvoz" training system.

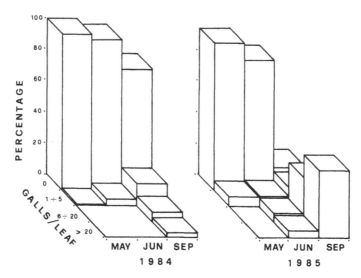

Figure 1 - Percentage of new leaves infested by _Viteus vitifoliae_ (Fitch) observed on cv. Tocai Rosso in mid-May, end of June and end of September, 1984 and 1985 (class of infestation: leaves with 0; 1-5; 6-20 and over 20 galls per leaf) (Lonigo - VI). In 1985, leaf infestation has been higher than in 1984. The presence of a large number of fundatrix galls (in mid-May) caused a rapid spring increase of leaf infestation.

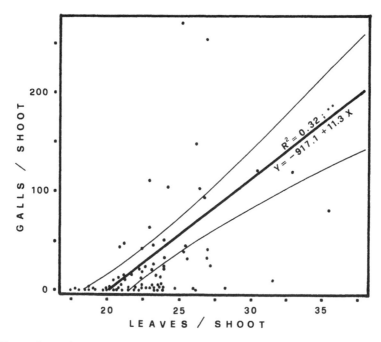

Figure 2 - Regression line and relative confidence belt (P < 0.05) between the number of leaves and galls per shoot of _Viteus vitifoliae_ (Fitch) at the end of the summer. (Lonigo VI -1984) (R2 = 0.32). The relation between growth of shoots and phylloxera leaf infestation is significant. In summer the amount of shoot growth can regulate leaf infestation.

122

2.1. Lonigo (Vicenza)

2.1.1. Leaf Infestation

The number of leaves per shoot and the number of leaf galls were counted periodically (Fig. 1, 3 and 4) on 1 shoot on the central part of the vine, on 18 vines in three different rows. In autumn, 1984, samples were taken from 76 vines (4 shoots per vine) with different infestations.

Trapping of newly hatched larvae which fell from leaves was done with horizontal sticky traps placed on 9 vines (under foliage). Larval count was done on 100 cm2 per each trap (4 areas of 25 cm2 per trap) from June to September 1985 (Fig.4).

2.1.2. Root Infestation

Samples of roots were collected on 12 vines in 1984 and 9 in 1985. Each sample was taken below the first 10 cm of soil. Vine roots were collected surrounded by soil. Samples were taken from both midway between adjacent rows and near the vinestock. In the laboratory, from each sample 3 groups of 10 root tips were observed under a stereoscopic microscope. On each root tip group the number of nodosities and the number of different stages of phylloxeras were counted. Nodosities with or without the presence of root forms were believed caused by phylloxera according to Cox et al. (2). It cannot be excluded that some galls are caused by root vine Roundworms.

2.1.3. Activity of Sexuparae

In summer and part of autumn in 1984 and 1985 the population density of winged sexuparae (alatae) was monitored by the use of sticky yellow "Polionda" traps measuring 17 x 20 cm. "Temo" glue (Kollant S.p.A. - Vigonovo VE - Italy) was applied to both surfaces. 6 traps in 1984 and 9 in 1985 were placed 2 m above the soil under the vine foliage. The traps were periodically changed (Fig. 3 and 4); alatae were counted by a stereoscopic microscope on both surfaces.

2.2. Lovolo (Vicenza)

Root infestation (5 samplings in 1985) and leaf infestation (1 sampling at the end of summer) were taken from 8 vines situated in 2 adjacent rows of cv.Tocai Rosso grafted on Kober 5BB and ungrafted.

3. RESULTS AND DISCUSSION

3.1.1. Leaf Infestation

In both years, the beginning of leaf infestation coincided with vine budding. Fundatrix galls in 1984 on 0.7% of leaves (average of 7.7 leaves / shoot) and in 1985 on 5.9% of leaves (average of 5.7 leaves / shoot) were found in mid May sampling (Fig. 1). The percentage of leaves infested by phylloxera at the end of June was higher in 1985 than in 1984 (Fig.1). A great population increase was observed in summer; at the end of September the percentage of leaves infested was 16% in 1984 and 32% in 1985; on leaves formed in the summer (after 3rd decade of June) the percentage was 22.7% (1984) and 83.5% (1985) (Fig.1).

In 1985, the widespread presence of fundatrix galls caused an abrupt increase of infested leaves (Fig.1). On American vines similar observations were made by Stevenson (13). On the other hand, new leaves are necessary for the formation of new galls. In 1984, a statistically significant relationship of regression between the number of leaves and galls per shoot was found (Fig.2); in summer, both the growth of shoots and the sequence of leaf generations appear to be influenced by the rainfall, frequent and heavy in 1984 (234mm). In 1985 both leaf generations and lengthening of shoots were blocked by the long period of drought. The few leaves which formed afterwards had no galls (Fig.4). Therefore, according to other workers (9, 10) climatic factors regulate the amount of the summer leaf infestation.

In spring, 1985, the falling of newly hatched larvae from leaf galls onto the ground was low (0.7 larvae per 100 cm2) although galls on leaves were numerous (Fig.4). It has been reported that rare neogallecolae - radicicolae (newly hatched larvae which migrate to roots) can be found in the early leaf generations (9).

In summer, the number of fallen larvae found was very high (even 179.4 newly hatched larvae on 100 cm2) (Fig.4). The falling of larvae some weeks after the formation of new leaf galls was observed (Fig.4).

3.1.2. Root Infestation
In 1984 and 1985, root nodosities of <u>Viteus vitifoliae</u> in samples collected from midway between the adjacent rows and from near the vinestock, showed no statistical difference. Some date has been reported in figures 3 and 4.

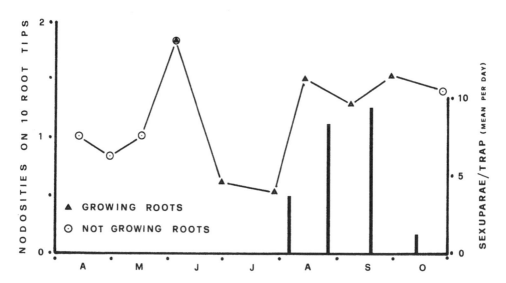

Figure 3 - Number of nodosities on 10 root tips and catches of <u>Viteus vitifoliae</u> winged sexuparae per trap a day (Lonigo VI)(1984). The presence of new roots was observed from June to September (▲ = growing roots; ○ = roots not growing). A significant increase of nodosities at the beginning of June and August was found. A high number of sexuparae was collected for over 3 months and until autumn - from Strapazzon and Girolami (15).

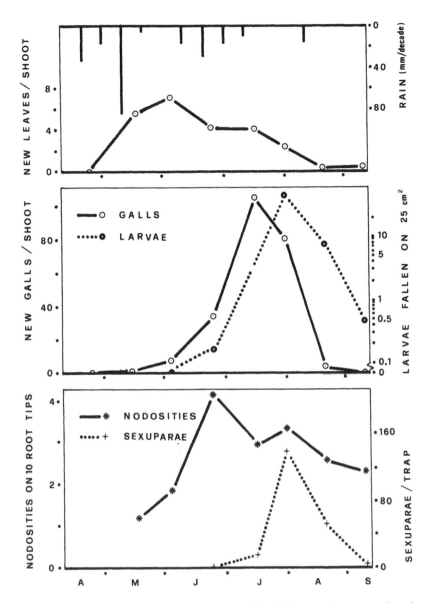

Figure 4 – Growth of shoots in different periods of the season (average number of new leaves/shoot) and rainfall (top figure). New phylloxera leaf infestation (average number of new galls per shoot) and newly hatched larvae fallen down from leaves (catches on horizontal, sticky traps) (middle figure). Root infestation (new nodosities formed at root tips) and catches of sexuparae (grown on roots) on vertical, sticky, yellow traps.

When new leaves no longer form, the development of phylloxera (new galls) is blocked and, therefore, the fall of larvae from leaves to the ground also stops. The nodosities start from overwintering root forms independent from the fall of newly hatched larvae. The summer peak may be linked to the fall of larvae, which, however, has a limited connection to nodosity formation. The flight of alatae quickly decrease after the high mid-summer peak (Lonigo, VI).

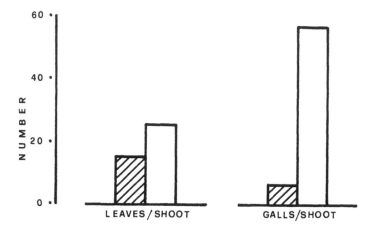

<u>Figure 5</u> - The number of leaves per shoot and leaf galls per shoot of <u>Viteus</u> <u>vitifoliae</u> on cv. Tocai rosso grafted on Kober 5BB and ungrafted (Lovolo VI - September 1985). Both a significantly higher infestation and number of leaves per shoot was found on grafted vines.

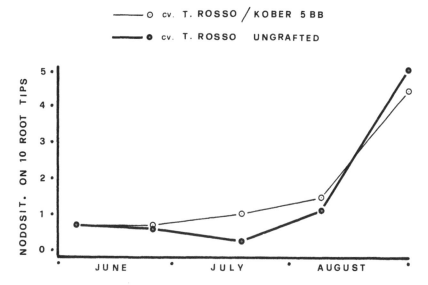

<u>Figure 6</u> -Phylloxera root infestation (nodosities/10 root tips) on vines of cv. Tocai Rosso grafted on Kober 5BB and ungrafted (Lovolo VI - 1985). The number of root nodosities on both rootstocks was similar.

In the laboratory the number of phylloxera counted on root tips was low. A rapid migration of radicicolae larvae from nodosities was observed. The significantly different number of root forms in the samples taken near the vinestock and midway between rows in 1984 (16) was not found in 1985.

In spring of both 1984 and 1985 a significatively abrupt increase of nodosities (P < 0.01) was observed (Fig.3 and 4). The beginning of root infestation coincided with the growth of spring vine roots. At the end of June 1985, the percentage of root tips damaged was very high (41.5%) (Fig.4). In June 1984, unfortunately, nodosities on growing roots (yellowish) or on ones which were not growing (brown), were not separately counted.

The beginning of root infestation seems to occur independent from leaf infestations. In 1984, when an increase of nodisities in spring took place, leaf infestation was very low (Fig.1 and 3); in 1985 a leaf gall increase and the beginning of root infestation occurred at the same time (Fig.4). Furthermore, in 1985 the falling of newly hatched larvae from leaves to the ground has been slight (0.7 per 100 cm2); this number does not appear high enough to cause the abrupt increase of nodosities recorded on 24 June 1985 (from 1.85 to 4.15 every 10 root tips) (Fig.4). Therefore, the spring increase of nodosities on grafted European vines starts independently from leaf generations.

In August 1984, a second, significant, even though smaller peak of nodosities (P < 0.05) was observed (Fig.3). In 1985 summer samples, the presence of nodisities was lower than the spring peak (P < 0.05) (Fig.4). The high number of larvae which fall down seems to increase root infestation in summer (Fig.4); however, falling larvae are not strictly linked to root population densities.

3.1.3. Activity of Sexuparae

The catches of _viteus_ _vitifoliae_ winged sexuparae (alatae) in both years are reported in figures 3 and 4. In 1984 and 1985, the length of the flight period is about 3 months and finishes only in Autumn. In 1984, the catches of alatae resulted high and constant; in 1985, they quickly decreased in the late, dry summer. According to other works (9, 5, 6) dry soil conditions cause a low number of grape phylloxera sexuparae to emerge.

However, the variety of vine influences the emergence of alatae; in some hybrid varieties, different amounts of alatae were found (14). Furthermore, the alatae show a preference for certain European vine varieties (15).

3.2. Grape Phylloxera on cv. Tocai Rosso grafted and ungrafted

Leaf and root infestation of vines (cv. Tocai Rosso) grafted on Kober 5BB and ungrafted are compared in Figures 5 and 6. In grafted vines both the number of leaves and galls per shoot were significantly higher than in ungrafted vines (P < 0.01).

The number of damaged root tips on both rootstocks (Kober 5BB and ungrafted) was similar; at the end of August, a significative increase of nodosities (P < 0.01) was observed (Fig 6). A significantly higher amount of nodosities (P>0.01) was noted on Kober 5BB in the only sampling on 18 July 1985 (Fig.6). However, the number of radicicolae phylloxera collected on old roots (n.6 x 10 cm in length and 2 - 7 mm in diam) in both rootstocks was similar (n.43

on Kober 5BB; n.47 on ungrafted). The expected higher root infestation on ungrafted European vines by the wintering root forms (9, 10) and higher fecundity of phylloxera (8) was not confirmed.

Root infestation damage also differs according to the root vines susceptibility to phylloxera punctures (11). However, radicicolae activity does not block the growth of thriving roots; these were deformed but still continued growing. This suggests that in thriving rootstocks (e.g. Kober 5BB), a high root phylloxera infestation can be supported without evident injuries to vines.

Acknowledgment
    Thanks to C. Guarnieri for her assistance in the collection of data, and to Dr. Moretti (Ist.Sper.Viticol., Conegliano V. - Treviso) for permitting the use of the ungrafted vineyard. Thanks are expressed to Prof. V. Girolami (Univ. di Padova) for the comments and suggestions which improved both the content and presentation of this paper.

REFERENCES

    1. BARBATTINI R., PRAVISANI L., ZANDIGIACOMO P., 1985 - Presenza di fillossera nel Pordenonese. L'Informatore Agrario, XLI (18): 91-96.
    2. COX J.A., VAN GELUWE J., LAWATSCH D., 1960 - Hexachlorocyclopentadiene, A Promising New Insecticide for the Control of the Root Form of the Grape Phylloxera. J. Econ. Entomol., 53 (5): 788-791.
    3. DALMASSO G., 1961 - Ritorna in discussione il tema della ricostruzione antifillosserica. Il Coltivatore e G.V.I., 107 (7-8): 222 - 225.
    4. DALMASSO G., 1967 - Ritornare alle viti franche di piede anche in terreni fillosserati?. Il Coltivatore e G.V.I., 113 (5-6): 121 - 125.
    5. DAVIDSON W.H., NOUGARET R.L., 1921 - The grape Phylloxera in California. Tech. Bull. U.S. Dep. Agric.: 903 (see: De Klerk, 1974).
    6. DE KLERK C.A., 1974 - Biology of _Phylloxera vitifoliae_ (Fitch)(_Homoptera : Phylloxeridae_) in Sauth Africa. Phytophylactica, 6: 109-118.
    7. GRANDE C., 1984 - Dopo piu' di un secolo la Fillossera, il piccolo "pidocchio" della vite, ritorna alla ribalta. L'Informatore Agrario, XL (30): 44-46.
    8. GRANETT J., BISABRI-ERSHADI B., CAREY J., 1983 - Life tables of phylloxera on resistant and susceptible grape rootstocks. Ent. exp. et appl., 34: 13-19.
    9. GRASSI B., FOA' A., GRANDORI R., BONFIGLI B., TOPI M., 1912 - Contributo alla conoscenza delle fillosserine ed in particolare della fillossera della vite, con riassunto - teorico pratico della biologia della fillossera della vite. Tip. Naz. G. Bertero, Roma, pp. 456.
    10. MAILLET P., 1957 - Contribution a l'étude de la biologie du phylloxéra de la vigne. Annales de Scien. Nat. Zool., XIX: 283-410.
    11. PETRI L., 1910 - Ricerche istologiche su diversi vitigni in rapporto al grado di resistenza alla fillossera. Rend. R. Accad. Lincei, XIX (1' sem.): 578-585.
    12. SCARAMUZZI F., 1979 - Viticoltura moderna e fillossera. Gion. di Agricol., 89 (41): 40 - 44.

13. STEVENSON A.B., 1970 - Endosulfan and Other Insecticides for Control of the Leaf Form of the Grape Phylloxera in Ontario. J. Econ. Entomol., 63 (1): 125-128.

14. STEVENSON A.B., JUBB G.L. Jr., 1976 - Grape Phylloxera: Seasonal Activity of Alates in Ontario and Pennsylvania Vineyards. Env. Entom., 5 (3): 549-552.

15. STRAPAZZON A., GIROLAMI V., 1983 - Infestazioni fogliari di fillossera (_Viteus vitifoliae_ (Fitch)) con completamento dell'olociclo su _Vitis vinifera_ (L.) innestata. REDIA, LXVI: 179-194.

16. STRAPAZZON A., GIROLAMI V., 1985 - Aspetti dell'infestazione di fillossera (_Viteus vitifoliae_ (Fitch)) su vite europea. Atti XIV Congr. Naz. di Entomol., Palermo - Erice - Bagheria, 28/5 - 1/6/1985: 633-641.

17. STRAPAZZON A., GIROLAMI V., 1985 - La fillossera su viti europee. L'Informatore Agrario, XLI (20): 73-76.

Preliminary results of the investigations on the diffusion of the grape phylloxera (*Viteus vitifoliae* (Fitch)) in some plantings of ungrafted European grape-vines on their own roots in Italy

B.Conti, F.Quaglia & E.Rossi
Istituto di Entomologia Agraria dell' Università Pisa, Italy

Summary

A survey of the observations carried out on ungrafted grapevines grown in several italian regions is reported. Those plantings were visited and the eventual presence of the grape phylloxera (*Viteus vitifoliae* (Fitch)) on the roots and/or on the leaves was detected. Some considerations on the sanitary conditions of the vines were expressed and also the eventual relation between soil type and the occurence of phylloxeric attack is investigated.

1. INTRODUCTION

The idea of a return to the growing of ungrafted *Vitis vinifera* is recurrent in literature since the beginning of Sixties (2, 5) and more recently (1, 4).
Neverthless this hypothesis needs a verification in the light of the economic incidence of the insect that was the "key species" for the European ungrafted grapevines: the grape phylloxera (*Viteus vitifoliae* (Fitch)).

It is easy to understand the practical importance that such a topic could have and from an economic and from an agronomical point of view. The interest in this theme is testified also by the support that some European and Italian Organizations are giving to several research programs in this field. In particular, for a quinquennium, the Institutes of Agricultural Entomology of Pisa and Catania, the ones of Plant Pathology of Pisa and Catania, the Institute of Zoology of Modena and the Fruits Science Institutes of Pisa, Firenze and Catania are collaborating in a research program sponsored by the Italian Ministry of Public Education with the aim to evaluate the possibility of the growing of ungrafted European grapevines.

In this paper are reported the first results of a preliminary research carried out to take a census of some ungrafted vineyards in several Italian Regions looking at the same time, into the phytopathological and agronomical problems associated to them.

2. REPORT OF THE INVESTIGATION

During the Winter 1982, through direct interviews to Vine-Growers Associations, Research Institutes and Regional Agricultural Organizations, it was possible to find out some ungrafted vineyards in several italian regions. As it is shown in fig.1, the localities found for the research were situated in 5 Regions: Emilia Romagna, Liguria, Tuscany, Umbria and Sicily. Beginning from the Spring 1983 until the Summer 1985, the indicated ungrafted vineyards were visited repeatedly to survey the eventual presence

Tab. 1 - Data collected in the ungrafted vineyards regarding their general
characteristics (col. 1,2,3 and 4), agronomical and soil features
(col.5 and 6) and considerations on the general sanitary
conditions of the plantings (col.7) with particular reference to
the eventual detection of the grape phylloxera on the roots
and/or on the leaves (col.8 and 9). The vineyard marked with *
situated at Savio (Ravenna) was uprooted in 1984.

REGION	LOCALITY (District)	AREA (sqm)	YEAR OF PLANT.	CULTIVAR	SOIL TYPE	SANITARY CONDITION	PRESENCE OF G.PHYLLOXERA	
							ON THE ROOTS	ON LEAVES
		2.000	1976	S.GIOVESE,TREBBIANO, ALBANA	CLAY	GOOD	YES	YES
	TEBANO (RA)	8.000	1979	MERLOT,BARBERA,MO- SCATO,ALBANA,S.GIO- VESE,RIESLING,CABER- NET,ANCELLOTTA,TOKAI PINOT B.,CROATINA, MALVASIA DI CANDIA	DISOMOG- ENEUS	GOOD	YES	NOT
E M I L I A R O M A G N A	PERGOLA (RA)	5.000	1978	ALBANA	CLAY	GOOD	YES	NOT
	BONCELLINO (RA)	5.000	1933	TREBBIANO,UVA D'ORO	SANDY	GOOD	NOT	NOT
		8.000	1976	TREBBIANO	LOAM	Viroses + Armill.	NOT	NOT
	SAVARNA (RA)	27.000	1976	TREBBIANO	LOAM	VERY GOOD	NOT	NOT
	SAVIO (RA)	26.500	1945*	FORTANA	LOAM	VERY GOOD	NOT	NOT
		20.000	1980	S.GIOVESE	LOAM	VERY GOOD	NOT	NOT
	CERVIA (RA)	72.000	1979	S.GIOVESE,ANCELLOTTA	SANDY	VERY GOOD	NOT	NOT
		45.000	1982	S.GIOVESE	SANDY	VERY GOOD	NOT	NOT
	VOLANIA (FE)	24.000	1980	ANCELLOTTA,MERLOT, S.GIOVESE,TREBBIANO	SANDY	GOOD	NOT	NOT
LIGURIA	RIOMAGGIORE (SP)	3.000	1930	VERDELLO	LOAM	GOOD	YES	NOT

Tab.1 (cont.)

REGION	LOCALITY (District)	AREA (sqm)	YEAR OF PLANT.	CULTIVAR	SOIL TYPE	SANITARY CONDITION	PRESENCE OF G. PHYLLOXERA	
							ON THE ROOTS	ON LEAVES
T U S C A N Y	FOLLONICA (GR)	18.000	1978	S.GIOVESE,CANAIOLO, TREBBIANO T.,MALVA- SIA BIANCA,M. NERA	CLAY	VERY GOOD	YES	NOT
	PIAN D'ALMA (GR)	5.000	1980	TREBBIANO,MALVASIA BIANCA,M. NERA, S. GIOVESE	CLAY	VERY GOOD	NOT	NOT
	MONTECARLO (LU)	5.000	1969	TREBBIANO	LOAM	GOOD	NOT	NOT
U M B R I A	AIUCCIA (TR)	30.000	1977	RUBECCIO, TREBBIANO	CLAY	Viroses + Armill.	YES	NOT
		100.000	1980	VERDELLO	LOAM	Viroses + Armill.	NOT	NOT
	LE PRESE (TR)	4.000	1980	GREGHETTO	SANDY	Viroses + Armill.	NOT	NOT
SICILY	KHAMMA (TP)	5.000	BEG.OF CENTURY	ZIBIBBO	LOAM	GOOD	YES	NOT

of a phylloxeric attack on the roots and/or on the leaves. At the same
time, it was asked the owners to fill up a questionnaire with the aim of
collecting all the possible informations on the those vineyards: varieties
present and correspondent age,agronomical practices utilized, control
strategies applied and yield. During every visit to the plantings, also
samples of the soil were collected and analysed for the texture. In these
views researchers in Plant Pathology and Fruits Science participated:
their opinions respectively about the phytopathological conditions, the
growth vigour and the yield of the vineyards were used to express a
qualitative evaluation of the general sanitary status of the fields.

Near the described observations, on the occasion of the occurrence of
heavy phylloxeric attacks on the leaves of European grafted grapevines,
several vineyards placed in Friuli Venezia Giulia were visited (**). Also
in these cases, similar questionnaires were filled up, enriched by notices
about the association stocks/grafts since this union seems to have an
influence on the susceptibility of the plant .

The results of the observations carried out for 4 years on roots and on
the leaves of european ungrafted grapevines are summarized in tab.1. In the
first two columns the names of the regions and of the visited localities
are reported. The third column indicates the areas of the ungrafted
vineyards followed, in the fourth column, by the year of the planting and
the names of the varieties grown.

(**) We thank Dr. P.L. Carniel of the Osservatorio per le Malattie delle
Piante of Gorizia for the collaboration.

Fig. 1 - Map showing the distribution of the ungrafted vineyards visited.
The different symbols used correspond to the different soil
types (▲ = loam; □ = sandy; △ = clay).

In the last three columns the most important informations collected are
shown: in fact, the soil types are listed next to an evaluation of the
general conditions of the vineyard (col.7) with particular specifications
in the cases of detection of viroses and/or mycelium of Armillariella
mellea and finally (col.8) the results of the samplings executed on roots
and leaves are reported.
The first comment deriving from the data is the influence of the soil
texture on the possibility of a phylloxeric attack: the observations
carried out confirmed the absence of the Aphid in all the vineyards lieing
on sandy soils. On the other hand, the difficulties of the development of
the grape phylloxera in such a soil type is well known in literature
(i.e.,3). Moreover, in the cases in which the infestation was heavy and
diffused on all the roots, no external symptoms were visible with the
exception of the vineyard visited in Umbria. Neverthless in this last case,
the unhealthy look was probably caused also by other concomitant factors
like the presence of the mycelium of Armillariella mellea and the poor
drainage of the soil. A particular mention is necessary for the ungrafted
vineyard situated at Follonica (Grosseto). Formed (4) by ungrafted
grapevines of five varieties (Sangiovese, Trebbiano, Canaiolo, Malvasia
bianca, Malvasia nera), it was planted in 1978 with the aim to evaluate its
productivity in comparison to the one of similar grafted vineyards. During
four years of observations, it was possible to follow the development of
the phylloxeric infestation: in fact, the sampling executed in 1981 and

Tab. 2 - Results of the observations carried out on the foliar attack occurred in 1984 on grafted European grapevines in Friuli Venezia Giulia. The first 6 columns regard general, agronomical and soil characteristics of the vineyards; in col.7 an evaluation of the entity of the foliar attack is reported: ***= heavy (more than 50% of the leaves phylloxerated with more than 50% of the leaf area attacked); **=medium (25% of the foliage with galls occupying about the 50% of the leaf area); *= scarce (less than 25% of the leaves attacked and less than 25% of the leaf area damaged). The results of the investigation on the presence of root forms of the grape phylloxera are shown in col.8. In col.9 the stocks are listed.

REGION	LOCALITY (District)	AREA (sqm)	YEAR OF PLANT.	SOIL TYPE	CULTIVAR	FOLIAR GALLS	ROOT FORMS	STOCK
F	CERVIGNANO	20.000	1973	LOAM	VERDUZZO	***	YES	Ko. 5BB
R	(UD)	20.000	1969	LOAM	MOSCATO R.	***	YES	Ko. 5BB?
I					TOKAI	**	YES	id.
U					PINOT G.	**	YES	id.
L					PINOT B.	**	YES	id.
I					SCHIAVA F.		YES	id.
					MERANO		YES	id.
V					MERLOT		YES	id.
E					MOSCATO B.		YES	id.
N					SILVANER		YES	id.
E								
Z		100.000	1978	CLAY	TOKAI	***	YES	420A/5BB
I					RIBOLLA	***	YES	id.
A	BUTTRIO				PICOLIT	***	YES	id.
	(UD)				SAUVIGNON	*	YES	id.
G					MALVASIA	*	YES	id.
I					VERDUZZO	*	YES	id.
U					PINOT B.	*	YES	id.
L					PINOT G.	*	YES	id.
I					RIESLING	*	YES	id.
A					REFOSCO	*	YES	id.
		2.000	1974	CLAY	PICOLIT	***	YES	Ko. 5BB
	BUTTRIO				VERDUZZO	*	YES	id.
	(UD)				PINOT	*	YES	id.

1982 didn't show any presence of infestation even if several American grapevines around the vineyard showed heavy foliar infestation. On the contrary, in 1983 the examination of the roots evinced the beginning of the attack. In that occasion, only two grapevines resulted infested. Next year any progress in the infestation was observed . Also in this case, no symptoms of the attack were observed on the leaves and the general look of the phylloxerated plants was the same of the other ones. As regards the presence of phylloxera on the leaves, it was a very rare event on the ungrafted plants observed: from tab.1 it is possible to see that only an Emilian vineyard resulted with scarce foliar galls.

On the contrary, on the foliage of the grafted vineyards observed in Friuli Venezia Giulia a heavy phylloxeric infestation was observed (see tab. 2). This phenomena has suggested a possible influence of the stock on the susceptibility of the scion. This argument is actually studied and it seems to have a serious foundation.

REFERENCES
1. CROVETTI A., 1980 - La difesa dai fitofagi della vite. From "Difesa antiparassitaria e diserbo chimico per le regioni Umbria Liguria e Toscana", Proceedings of the Course held in Pise, 6-11 November 1980: 85-94.
2. DALMASSO G., 1967 - Ritornare alle viti franche di piede anche in terreni fillosserati?. Il Coltivatore, 113: 121-125.
3. NOUGARET R.L., LAPHAM M.H., 1928 - A study of the phylloxera infestation in California as related to types of soils. Technical Bulletin n.20, USDA, Washington D.C.: 1-38.
4. SCARAMUZZI F., 1979 - Viticoltura moderna e fillossera. Giornale di Agricoltura, 89 (41) : 40-44.
5. ZANARDI D., 1962 - La fillossera "gallecola" su foglie di vite nostrana. Il Coltivatore, 12: 371-374.

Susceptibility of some cultivars of ungrafted European grape-vines to the foliar attack of the grape phylloxera (*Viteus vitifoliae* (Fitch)): Observations of the triennium 1982-1984

F.Quaglia & E.Rossi
Istituto di Entomologia Agraria dell' Università di Pisa, Italy

Summary
Twenty varieties of european wine grapes were tested for their susceptibility to the foliar attack of the grape phylloxera (Viteus vitifoliae (Fitch)). The plants tested were all ungrafted. The trial was carried out growing the young plants in containers. The results showed an erratic trend of the phenomena difficult to interpretate up to the present point of the experiment.

1. INTRODUCTION

The foliage injury of the grape phylloxera Viteus vitifoliae (Fitch) on european grapevines is well known and its occurrence caused sometimes problems to the vine-growers. Recently in Italy heavy attacks on the leaves of european grapevines were observed (1,2,3,4,5).

In this paper observations on the response of several cultivars of ungrafted european vines to the grape phylloxera foliar infestation are reported.

2. MATERIALS AND METHODS

The original aim of the research was to individuate the susceptibility or the tolerance of several ungrafted varieties of european grapevines to the phylloxeric attack on the roots. The cultivars chosen for the test are listed in the explanation of fig. 1, in which the pattern of the experiment is shown. As it is possible to see, 12 vines of the varieties "Albana" (E), "Grignolino" (O), "Nebbiolo" (Q), "Malvasia istriana" (U) and "Raboso di Piave" (V) and 16 of the remaining cultivars were used in the trial. In fifty cement containers (ten of which containing sandy soil, the other forty filled with loam soil), 6 ungrafted vines of different cultivars chosen at random and 2 american grapevines (Vitis rupestris, type Rupestris du Lot) european cultivars were planted and were grown appling the current agronomical practices and the usual fungicide treatments.

In spring 1981, shoots and leaves with phylloxeric galls coming from a nursery situated at S.Piero a Grado (Pisa, Italy) were used to infest the american grapes. Since may 1982, weekly controls were executed on the vines foliage until its fall and these observations were repeated during the next two year. The attacked leaves were countersigned since the very beginning of the infestation. The number of galls for each leaf was counted and the diameter of the galls was measured during all its development, from its appearance to the maximum size. The trial is still going on.

Fig. 1 — Pattern of the trial carried out on 300 grapevines of 20
different varieties planted in 50 cement containers: A=
Sangiovese toscano, B= Mammolo, C= Schiava freisa, D= Dolcetto,
E= Albana, F= Merlot, G= Cabernet Sauvignon, H= Bonamico, I=
Tokai friulano, L= Corvina, M= Pinot grigio, N= Canaiolo nero,
O= Grignolino, P= Prosecco, Q= Nebbiolo, R= Trebbiano toscano,
S= Barbera, T= Cortese, U= Malvasia istriana, V= Raboso di
Piave, ✱= Rupestris du Lot.

3. RESULTS AND DISCUSSION

During 1982, the 55 % of european varieties, for a total of 14 plants
(correspondent to the 4.7 %), showed presence of foliar galls; in 1983 the
two values increased to the 75% (15/20 varieties) and to the 10.7% (32/300
plants) respectively, while the next year the attacked plants were 5%
(15/300) in total and the infested varieties were 12/20 (60%) (see tab.1
and figs.2 and 3). So it is possible to see how to an hight number of
affected varieties corresponds a very low value of injured plants. As it
is shown in tab.1, in which data concerning each cultivar are reported,

Tab. 1 - Summary of the data of the phylloxeric foliar infestation collected during three years (1982/84) correspondent to the different varieties of ungrafted European wine grape.

CULTIVAR	YEAR	TOT.Nr. OF INF. PLANTS (%)	TOT.Nr. OF INF. LEAVES	AVER.Nr. OF GALLS/ INF. LEAF	AVER. DIAM. OF GALLS (mm) (MIN.;MAX)
	1982	1 (6.2)	1	1.0	4.0
A	1983	2 (12.5)	3	3.3 + 0.6 -	3.4 + 0.1 (3.0; 3.5)
	1984	1 (6.2)	2	2.5 + 0.7 -	3.3 + 0.3 (3.0; 3.7)
	1982	1 (6.2)	2	2.0 + 1.4 -	1.3 + 0.6 (0.7; 2.0)
B	1983	3 (18.7)	4	1.5 + 1.0 -	1.0 + 0.2 (0.8; 1.5)
	1984	1 (6.2)	1	2.0	1.2 + 0.3 (1.0; 1.5)
	1982	1 (6.2)	2	1.0 + 0.0 -	3.7 + 0.3 (3.5; 4,0)
C	1983	1 (6.2)	3	1.3 + 0.6 -	3.4 + 0.6 (2.5; 4.0)
	1984	0	-	-	-
	1982	1 (6.2)	2	1.5 + 0.7 -	4.3 + 0.5 (4.0; 4.5)
D	1983	1 (6.2)	3	2.3 + 1.5 -	3.9 + 0.3 (3.5; 4.5)
	1984	1 (6.2)	2	3.0 + 0.0 -	4.0 + 0.3 (3.8; 4.5)
	1982	1 (8.3)	1	1	5.0
E	1983	1 (8.3)	2	1.5 + 0.7 -	4.5 + 0.5 (4.0; 5.0)
	1984	0	-	-	-

Tab. 1 (cont.)

CULTIVAR	YEAR	TOT.Nr. OF INF. PLANTS (%)	TOT.Nr. OF INF. LEAVES	AVER.Nr. OF GALLS/ INF. LEAF	AVER. DIAM. OF GALLS (mm) (MIN.;MAX)
F	1982	1 (6.2)	1	1.0	3.0
	1983	1 (6.2)	2	2.0 + 1.4 −	3.2 + 0.3 (3.0; 3.8)
	1984	1 (6.2)	1	2.0	3.2 + 0.3 (3.0; 3.5)
G	1982	3 (18.7)	7	2.4 + 1.3 −	4.2 + 0.6 (3.0; 5.0)
	1983	1 (6.2)	2	2.5 + 0.7 −	4.3 + 0.5 (3.8; 5.0)
	1984	0	−	−	−
H	1982	1 (6.2)	8	3.7 + 1.7 −	3.4 + 0.4 (2.7; 4.0)
	1983	2 (12.5)	11	3.3 + 2.0 −	3.6 + 0.6 (2.8; 4.5)
	1984	2 (12.5)	2	3.0 + 1.7 −	3.2 + 0.4 (2.5; 4.0)
I	1982	1 (6.2)	1	1.0	4.0
	1983	4 (25)	9	2.9 + 1.4 −	3.7 + 0.4 (3.0; 4.3)
	1984	2 (12.5)	7	2.7 + 1.7 −	3.5 + 0.4 (2.8; 4.0)
L	1982	0	−	−	−
	1983	3 (18.7)	9	8.0 + 3.4 −	2.1 + 0.5 (1.5; 3.0)
	1984	1 (6.2)	4	1.7 + 0.9 −	1.7 + 0.6 (1.0; 2.5)

140

Tab. 1 (cont.)

CULTIVAR	YEAR	TOT.Nr. OF INF. PLANTS (%)	TOT.Nr. OF INF. LEAVES	AVER.Nr. OF GALLS/ INF. LEAF	AVER. DIAM. OF GALLS (mm) (MIN.;MAX)
	1982	2 (12.5)	15	5.4 + 2.5 -	1.2 + 0.4 (0.8; 2.4)
M	1983	3 (18.7)	26	8.1 + 3.5 -	1.8 + 0.4 (0.8; 2.6)
	1984	1 (6.2)	11	8.2 + 3.8 -	1.6 + 0.3 (0.8; 2.2)
	1982	0	-	-	-
N	1983	2 (12.5)	13	5.5 + 2.7 -	2.1 + 0.6 (1.5; 2.8)
	1984	1 (6.2)	5	4.2 + 1.9 -	1.8 + 0.2 (0.8; 2.3)
	1982	0	-	-	-
O	1983	1 (8.3)	6	3.7 + 1.2 -	1.7 + 0.4 (1.0; 2.5)
	1984	1 (8.3)	4	2.7 + 1.7 -	1.4 + 0.3 (1.0; 2.0)
	1982	1 (6.2)	2	1.0 + 0.0 -	3.7 + 0.3 (3.5; 4.0)
P	1983	4 (25.0)	25	2.2 + 1.3 -	1.9 + 0.5 (1.0; 2.8)
	1984	2 (12.5)	8	2.7 + 1.3 -	1.7 + 0.5 (1.0; 2.5)
	1982	0	-	-	-
Q	1983	3 (25.0)	7	1.8 + 0.8 -	1.8 + 0.7 (1.0; 3.0)
	1984	1 (8.3)	2	3.0 + 1.4 -	1.7 + 0.6 (1.0; 2.5)

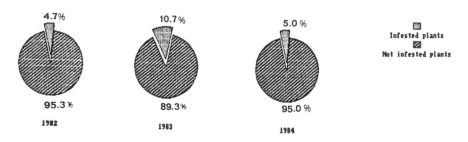

Fig. 2 — Percent of vines showing foliar phylloxeric attack in 1982, 1983 and 1984, respectively.

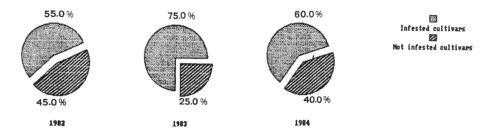

Fig. 3 — Percent of infested varieties in the years 1982, 1983 and 1984, respectively.

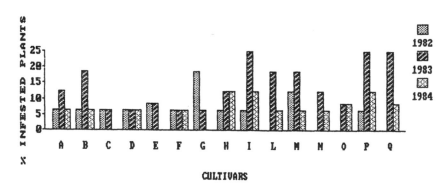

Fig. 4 — Bars of the percentage of vines of each cultivars infested in the years 1982, 1983 and 1984 (for the legend of the letters see the explanation of fig.1)

also the amount of the leaves with galls and the mean number of these ones/infested leaf are likewise low.

As concern the susceptibility of the different varieties, from the triennal observations it is possible to point out an erratic trend of the phenomena. "Pinot grigio" (M), "Bonamico" (H),"Tokai friulano" (I), and "Prosecco" (P) appeared the most attacked considering and the percent of the plants and the number of leaves and the average number of galls/infested leaf (see tab.1 and figs. 4-6).

142

Fig. 5 - Bars of the number of infested leaves on each cultivars during 3
years of observations (1982,1983 and 1984)(for the legend of the
varieties, see the explanation of fig.1).

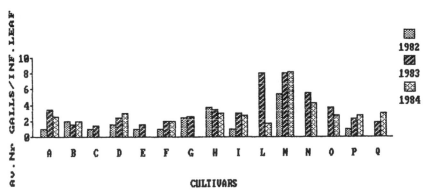

Fig. 6 - Bars showing the average number of galls/infested leaf for each
attacked variety in the triennium 1982-1984 (for the legend of
the cultivars, see the explanation of fig.1).

Fig. 7 - Bars showing the average size (mm) of the galls detected on the
leaves of the infested cultivars (for the legend of the
varieties see the explanation of fig.1).

On the ground of these first results the phylloxeric foliar infestation on european grapevines varieties seems to have a very modest influence on the plants. This is confirmed also by the general small final size of the galls (see fig.7) which indicates the uncompleted development of the phylloxeric population inside them. More, it was observed that the galls were produced only during the early vegetative period (max. five to six leaves unfolded) while on the following foliage just very rare and open galls occurred. On the contrary, during the trial the 100% of Rupestris du Lot plants were infested and almost all of their leaves showed galls which took up an average more than 50% of the foliar surface.

REFERENCES

1.   BARBATTINI R., PRAVISANI L., ZANDIGIACOMO P., 1985 - Presenza di fillossera nel Pordenonese. L'Informatore Agrario, XLI, 18: 91-96.
2.   GRANDE C., 1984 - Dopo più di un secolo la Fillossera, il piccolo "pidocchio" della vite, torna alla ribalta. L'Informatore Agrario, XL, 30: 44-46.
3.   STRAPAZZON A., GIROLAMI V., 1983 - Infestazioni fogliari di fillossera (Viteus vitifoliae (Fitch)) con completamento dell'olociclo su Vitis vinifera (L.) innestata. Redia, LXVI: 179-194.
4.   STRAPAZZON A., GIROLAMI V., 1985 - Aspetti dell'infestazione di fillossera (Viteus vitifoliae (Fitch)) su vite europea. Proc. XVI National Congress of Entomology, Palermo, 5.28.85-6.1.85:
5.   STRAPAZZON A., GIROLAMI V., 1985 - La fillossera su viti europee. L'Informatore Agrario,LXI, 20: 73-76.

Preliminary observations on the composition of the grape phylloxera (*Viteus vitifoliae* (Fitch)) population in American vine leaves

A.Raspi, R.Antonelli & B.Conti
Istituto di Entomologia Agraria dell' Università di Pisa, Italy

Summary

The Authors investigated into the composition of a Viteus vitifoliae Fitch population existing from May to October on leaves of Kober 5BB vine rootstock. The grape phylloxera stages taken into consideration were the first instar larvae (L1), the 2nd-4th instar larvae (L2-4) and the females of gallicolae found during the various test periods inside leaf-galls of four different diameters: less than 1 mm, between 1 and 2 mm, between 2 and 3 mm and greater than 3 mm, on young leaves (apical), intermediate-age leaves (middle) and on those completely developed (basal) on the shoot. An analysis of the data obtained shows that it is not useful to test leaves at random, but that it is sufficient to examine a fair number of galls with diameters between 2 and 3 mm and over 3 mm on middle and basal leaves of the shoot. In fact, inside these leaf galls one may find egg-laying females at the same time as eggs, and newly hatched larvae who have not yet left the mother gall and the number of which, differing from period to period, supplies valuable information on the dynamics of the grape phylloxera population.

1. INTRODUCTION

In the course of an interdisciplinary programme of "Possibility of growing ungrafted European vines" the Istituto di Entomologia Agraria dell'Universita' di Pisa carried out tests on Viteus vitifoliae (Fitch) infestation on European vines, and at the same time investigations were performed on American vine rootstocks for further eco-ethological and morphological examination of phylloxera.

The approach to the subject was conducted through field studies aimed at investigating the dynamics of the population, and with laboratory tests which were directed towards limiting the thermal limits and the optimal ranges of development of aphid.

In the case of the field studies first place was given to a detailed study of the composition of the phylloxera population on American vine leaves. The purpose of this was to set up a simplified method of sampling for studies on the dynamics of the population which, because of the high number of phylloxera generations and their overlapping (1) (2), hence the presence of all stages of aphid in every period, is a very difficult problem to resolve.

2. MATERIALS AND METHODS

In a nursery of American vines situated at San Piero a Grado (Pisa), a

LEAVES OF THE SHOOT

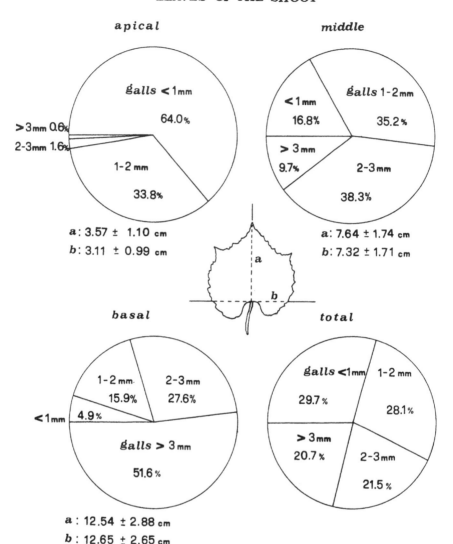

Fig. 1 – Kober 5BB vine rootstock leaves. The percentage relationship among the four categories of galls chosen (under 1 mm, between 1 and 2 mm, between 2 and 3 mm, and over 3 mm) on apical, middle and basal leaves of the shoot, characterised by different average measurements for the arbitrary axes a and b.

plot of about 5000 m² was chosen, containing Kober 5BB hybrids which, at first examination showed signs of more infestation than other common vine rootstocks.

The experimental project foresaw sampling every ten days, from May to October, taking the vine leaves according to an appropriate method. Sample were taken from a constant number of plants (six) chosen at random each time; from each one of these samples 3 young leaves were taken from the apical part of the shoot, 3 of intermediate age were taken from about halfway along the shoot and 3 completely developed leaves from the base, for a total number of 54 leaves (18 apical, 18 middle and 18 basal). These were measured along two arbitrary axes (a and b in Fig.1).

The total number of galls on each leaf were counted; moreover, on a sub-sample of 6 apical, 6 middle and 6 basal leaves in each period, the galls were first of all divided into four arbitrary categories according to diameter (less than 1 mm, between 1 and 2 mm, between 2 and 3 mm and greater than 3 mm) and were thereafter opened to examine the different gallicola stages.

First instar larvae (L1), 2nd–4th instar larvae (L2–4) and females of gallicolae (1) could be distinguished in the gall leaves. In addition, the eggs and newly-hatched larvae still inside the galls containing egg-laying females were also counted.

3. RESULTS AND DISCUSSION

3.1 Percentage relationship between the different categories of diameter of the galls found on apical, middle and basal leaves of the vine shoot.

All results obtained refer to the three categories of apical, middle and basal leaves taken from the shoot and the average length (b axis) was, respectively: 3.11 cm ± 0.99, 7.32 cm ± 1.71, 12.65 cm ± 2.65 (Fig.1).

In the apical leaves the most numerous galls are those with a diameter below 1 mm (64.0 %), followed by the 1–2 mm ones (33.8 %) (Fig.1). In the middle leaves the most numerous galls are those of the 2–3 mm and 1–2 mm (73.5 % in all). In the case of the basal leaves 51.6 % of the galls were greater then 3 mm, followed by the 2–3 mm ones (27.6 %) (Fig.1).

By sampling an equal number of apical, middle and basal leaves'we had a sample embracing the 4 categories of galls in a fairly proportional manner (between 21 % and 30 %) (Fig.1).

3.2 Percentage composition of the phylloxera population on apical, middle and basal leaves of the vine shoot.

If we take into account the percentage composition of the forms found (L1, L2–4, egg-laying females (*)) during the course of sampling (14/5/1985 – 20/10/1985) we can see all the stages on the apical and middle leaves, even if in the apical ones, where the population is mainly L2–4 (51.9 %) and L1 (42.0 %), the egg-laying females are very few (6.1 %) and are not to be found in all of the periods (Fig.2).

In the middle leaves the most numerous are the egg-laying females (52.7 %) and the N2–4 (42.8 %), whereas the L1 (4.5 %) are very few and sometimes non-existent (Fig.2).

Lastly, on the basal leaves there are no L1, the L2–4 are very few (5.7 %) but the egg-laying females are numerous (94.3 %) (Fig.2).

3.3 Percentage relationship between the stages found in the different categories of diameter of the galls.

If we leave aside the position of the leaves for a moment and consider the population inside the four categories of galls we can see that in the galls below 1 mm 92.7 % is made up of larval stages, of these the L1 prevail (56.7 %) followed by the L2–4 (36.0 %) (Fig.3). In galls between 1

(*) The females lay their eggs only a few hours after the last moult, so it is there fore difficult to find females without eggs in the galls.

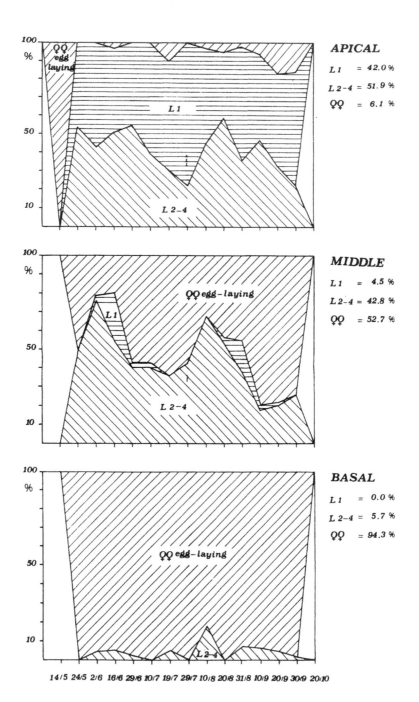

Fig. 2 – <u>Viteus vitifoliae</u> (Fitch) – Percentage composition of the stages found during the different periods of the test in apical, middle and basal leaves of Kober 5BB shoot. On the right is the percentage ratio of the stages.

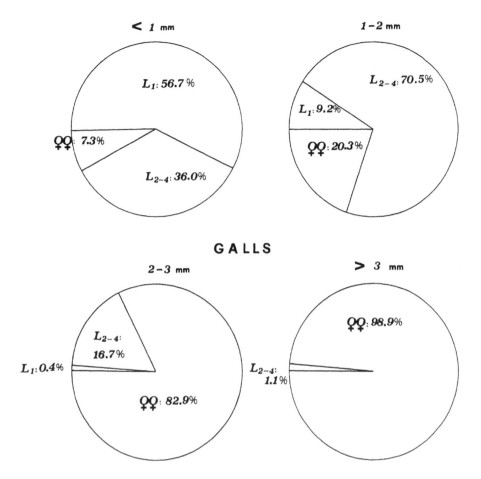

Fig. 3 - <u>Viteus vitifoliae</u> (Fitch) - Percentage ratio among the different stages found in galls smaller than 1 mm, between 1 and 2 mm, between 2 and 3 mm and larger than 3 mm on Kober 5BB leaves.

and 2 mm the larval stages represent 79.7 % , 70.5 % of which are L2-4. The 2-3 mm diameter range galls and the ones over 3 mm show a prevalence of egg-laying females (82.9 % and 98.9 % respectively) (Fig.3).

3.4 <u>Percentage composition of the phylloxera population in different categories of galls on apical, middle and basal leaves of the vine shoot.</u>

If we examine the population in galls of different diameters on apical, middle and basal leaves, we achieve even more detailed results.
In the apical, middle and basal galls smaller than 1 mm (Fig.4) all three stages taken into account are present (except for the L1 in basal galls). In the apical leaves the L1 prevail (65.8 %), in the middle the L2-4 prevail (56.0 %) and in the basal ones the egg-laying females prevail (98.0 %) (Fig.4). It is important to note, however, that the females are present in galls smaller than 1 mm even on the apical and middle leaves but only, as happens in the case of the basal ones, during the last stage of the cycle (mid-August to mid-October) (Fig.4).

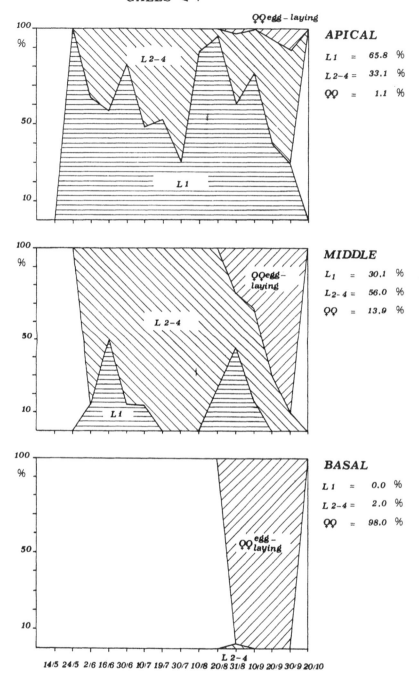

GALLS < 1

APICAL

L1   =   65.8 %
L2-4 =   33.1 %
♀♀   =   1.1 %

MIDDLE

L1   =   30.1 %
L2-4 =   56.0 %
♀♀   =   13.9 %

BASAL

L1   =   0.0 %
L2-4 =   2.0 %
♀♀   =   98.0 %

14/5 24/5 2/6 16/6 30/6 10/7 19/7 30/7 10/8 20/8 31/8 10/9 20/9 30/9 20/10

Fig. 4 - Viteus vitifoliae (Fitch) - Percentage composition of the stages found during the different periods of the test inside galls smaller than 1 mm, on apical, middle and basal leaves of Kober 5BB shoot. On the right is the percentage ratio of the stages.

In the galls in the 1-2 mm range (Fig.5) the L2-4 are definitely the most numerous both in the apical and in the middle ones (77.6 % and 72.3 %, respectively). The egg-laying females are always present in the middle and basal galls, but only during two periods in the apical ones. In the apical leaves the L1 (16.1 %) prevail over the egg-laying females (6.3 %) whereas the latter are more (24.6 %) than the L1 in the middle ones, and in the case of the basal leaves, where the L1 are absent, they are decidedly more numerous (81.4 %) than the L2-4 (Fig.5).

There are no L1 in the galls in the 2-3 mm range (Fig.6). In the apical leaves the L2-4 prevail (63.0 %) whereas, in the middle and basal ones the egg-laying females represent the majority of the population (77.6 % and 90.3 %, respectively) and are always present from May to mid-October.

In the galls larger than 3 mm diameter (Fig.7) there are no L1 and only a few L2-4 (6.86 %), to be found only on the middle leaves and concentrated between mid-May to mid-June. The egg-laying females represent the prevalent stage in this category of galls and they are therefore always present on middle and basal leaves (Fig.7). It must be noted that these are only to be found in the apical ones around mid-May; these are therefore the fundatrices.

4. CONCLUSIONS

With the sampling method described we were able to see that the categories of galls chosen (less than 1 mm, between 1 and 2 mm, between 2 and 3 mm and larger than 3 mm) are not evenly distributed on the leaves. In fact those under 1 mm are mainly to be seen on the apical leaves of the shoot. The 1-2 mm and the 2-3 mm galls are much more abundant in the middle leaves, whereas in the basal ones the galls over 3 mm prevail (Fig.1).

In spite of the fact that in every one of the test periods all stages of Aphid are present (Fig.2) the Viteus vitifoliae population inside the galls results to be considerably different from one to another of the diameter categories chosen (Fig.3). In fact, the L1 prevail in the galls under 1 mm (56.7 %). Though the L2-4 are well represented in the under 1 mm galls they prevail in those between 1 and 2 mm (70.5 %). The egg-laying females are the 82.9 % in the 2-3 mm galls and then represent almost the entire population in those larger than 3 mm (98.9 %) (Fig.3).

If we examine the population in galls of different diameters on apical, middle and basal leaves of the shoot we achieve even more detailed results (Figs. 4,5,6 and 7).

By using this method our aim was to classify the different compositions of phylloxera population in galls of different diameters and on leaves with different ages (apical, middle and basal of the shoot) on Kober 5BB vine rootstock, the most commonly used in Italy. This study was particularly important for setting up simplified sampling methods for achieving specific results, for example to be able to follow the dynamics of phylloxera population and hence draw up an appropriate programme of chemical control in vine nurseries.

We realize from the results obtained that it is useless to sample the leaves at random and it is better to take samples of a fair number of galls with diameters ranging from 2 to 3 mm, and over, on middle and basal leaves of the shoot during every period (except May, when the egg-laying fundatrices can be found even on apical leaves). In fact, inside these galls egg-laying females are to be found at the same time as eggs and newly-hatched larvae the number of which, being different from period to period, will supply us with important information on the dynamics of the phylloxera population.

GALLS 1-2

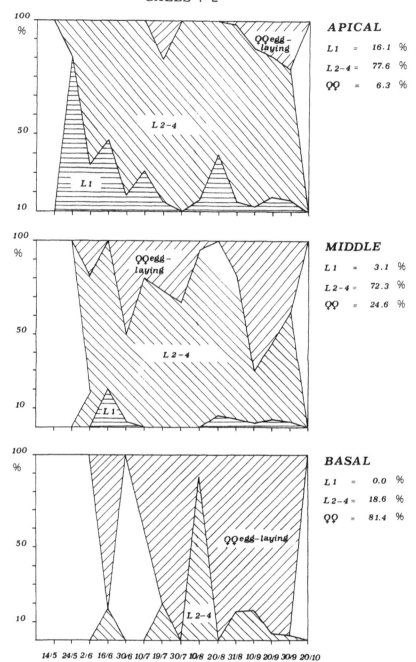

Fig. 5 – <u>Viteus</u> <u>vitifoliae</u> (Fitch) – Percentage composition of the stages found during the different periods of the test inside galls sized 1 to 2 mm, on apical, middle and basal leaves of Kober 5BB shoot. On the right is the percentage ratio of the stages.

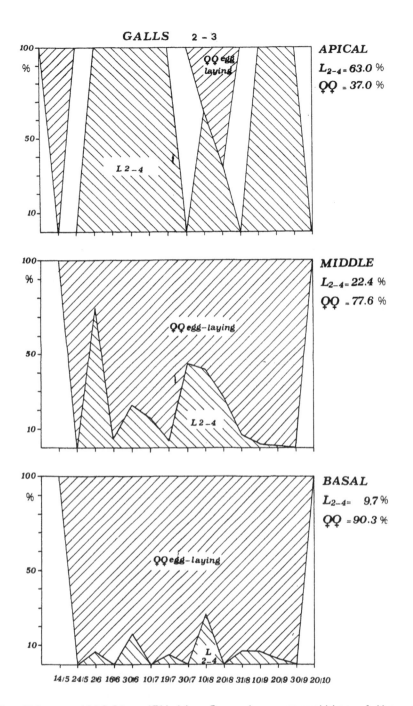

Fig. 6 – <u>Viteus</u> <u>vitifoliae</u> (Fitch) – Percentage composition of the stages found during the different periods of the test inside galls sized 2 to 3 mm, on apical, middle and basal leaves of Kober 5BB shoot. On the right is the percentage ratio of the stages.

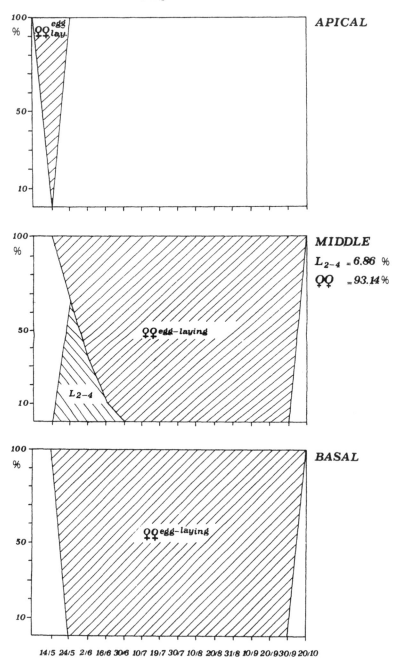

Fig. 7 — <u>Viteus</u> <u>vitifoliae</u> (Fitch) — Percentage composition of the stages
found during the different periods of the test inside galls larger
than 3 mm, on apical, middle and basal leaves of Kober 5BB shoot.
On the right is the percentage ratio of the stages.

REFERENCES

1. GRASSI B., FOA A., GRANDORI R., BONFIGLI B., TOPI M. (1912). Contributo
   alla conoscenza delle fillosserine e in particolare della fillossera
   della vite. Tip. Naz. G. Bertera, Roma, pp. 1-456 + XIX tav.
2. STEVENSON A. B. (1966). Seasonal development of foliage infestations of
   grape in Ontario by Phylloxera vitifoliae (Fitch) (Homoptera:
   Phylloxeridae). Can. Ent. 98 no. 12 pp. 1299-1305.

Grape phylloxera (*Viteus vitifoliae* (Fitch)) infestation on American vine leaves in a nursery at San Pietro a Grado (Pisa) during the years 1982-1983

A.Raspi & R.Antonelli

Istituto di Entomologia Agraria dell' Università di Pisa, Italy

Summary

The Authors illustrate grape phylloxera infestation, from May to October in the years 1982 and 1983, on Kober 5BB vine rootstock leaves in a nursery at San Piero a Grado (Pisa). They examine in detail and discuss the results obtained, taking into account the increase in the average number of galls per leaf, the percentage composition of the population present during the various periods, the egg-laying females, the eggs and newly-hatched larvae inside the mother galls.

1. INTRODUCTION

During the years 1982 and 1983, in a vine nursery at San Piero a Grado (Pisa), together with studies on the composition of the phylloxera population on American vine leaves, research was also conducted on the course of the infestation and on the dynamics of phylloxera population. These field-studies were carried out in collaboration with laboratory research aimed at defining the thermal limits and the optimal development range of aphid (1) (8).

2. MATERIALS AND METHODS

The experimental project was the same as that used for studying the composition of the phylloxera population from May to October on leaves of Kober 5BB vine rootstocks (4).

In fact, every ten days samples were taken of the epigean part of six plants chosen at random each time. From a shoot of each of these 3 apical, 3 middle and 3 basal leaves were taken, reaching a total of 54 leaves (18 apical, 18 middle and 18 basal) (4). The number of galls upon these were then counted and the stages of the phylloxera gallicolae (2) inside them was recorded.

The experimental field was a plot of about 5000 m², with Kober 5BB hybrids, in a nursery of American vine rootstocks located at San Piero a Grado (Pisa).

3. RESULTS AND DISCUSSION

The phylloxera infestation in the vine nursery which we studied showed to behave slightly different in 1982 as compared to 1983.

In fact, in 1982 the phylloxera attack showed up with leaf galls containing the egg-laying fundatrices as the first ten days of May. At the beginning these only involved a few of the plants, often next to each

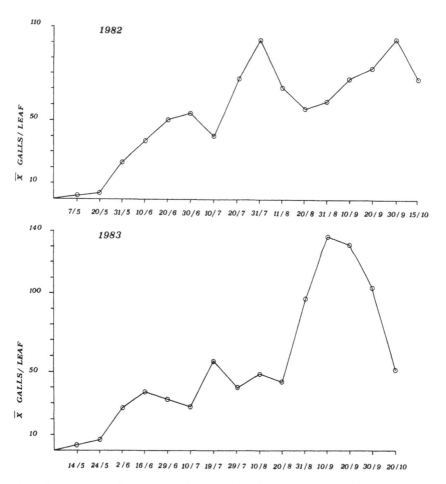

Fig. 1 – Average number of galls per leaf found during the various test periods on Kober 5BB in 1982 (top), and in 1983 (bottom).

other, in different rows of the experimental field. Infestation then spread rapidly involving all plants by the end of June.

In fact, if we take into account the average number of galls found on the leaves (Fig.1) we can see that, in 1982, this remained low until mid-May when it increased continuously, reaching 55 at the beginning of July, 103 (the maximum) at the end of the month and then, after various fluctuations, at the end of September' it went back to the maximum number before dropping.

In 1983 on the other hand, infestation showed up with the presence of egg-laying fundatrices in mid-May. Infestation, though involving all the plants in the plot from the end of June onwards, was less than during the year before. This was probably due to the temperature being on average lower than in 1982, particularly during the last ten days of May and the entire month of June. In fact, the average number of galls per leaf was below 50 before the second ten-day period of July, it varied until mid-August when it rose considerably, then maximum values of 137 (Fig.1) were reached during the first ten days of September.

The fluctuation in the average number of galls per leaf recorded in

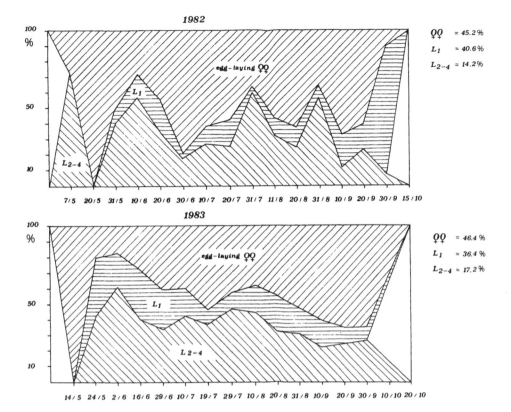

Fig. 2 – Percentage composition of the different stages of <u>Viteus</u>
<u>vitifoliae</u> (Fitch) on Kober 5BB leaves during various periods in
1982 (top), and in 1983 (bottom).
The percentage relationship of the stages found is also given.

different periods both in 1982 and in 1983 is probably linked to different
factors (e.g., generations overlapping) but it is mainly due to great
variability of the number of galls found on each single leaf.

On analysis of the percentage composition of grape phylloxera
populations throughout various periods of 1982 and 1983 it can be seen that
from May to October egg-laying females, L2-4 and L1 are always present
inside the galls (Fig.2). In particular, it may be seen that the percentage
relationship of the different stages found between May and October does not
differ much from one year to the other. In fact, in 1982 the egg-laying
females found were 45.2 % of the population, the L2-4 were 40.6 % and the
L1 on the apical leaves were 14.2 %, whereas in 1983 these values were 46.4
%, 36.4 % and 17.2 %, respectively (Fig.2).

However, on analysis of the composition of the population during the
various periods in the two-year study (Fig.2), we can see that because of
premature overlapping of the generations and the consequent presence of all
stages of the phylloxera, distinguishing the different generations after
that is rather difficult.

If, when performing the test one takes into account, during the
various periods, not only the percentage of the egg-laying females, but
also that of the eggs layed and the newly-hatched larvae which have not yet
left the mother gall, then more information is obtained Figs.3 and 4).

159

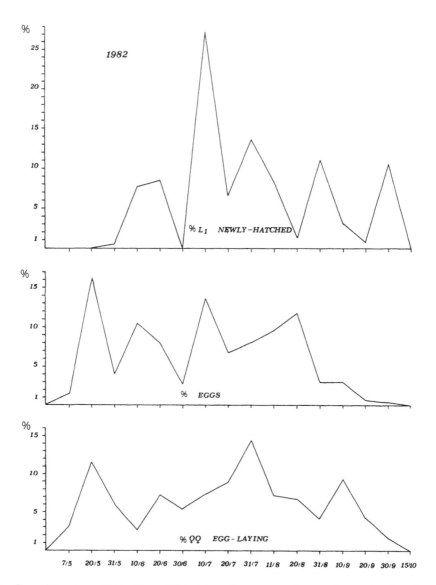

Fig. 3 - <u>Viteus vitifoliae</u> (Fitch) - The percentage of egg-laying females,
eggs and newly-hatched larvae found inside mother-galls during the
various test-periods in 1982.

For the May-October period of 1982 (Fig.3) there was a series of peak
periods of the presence of newly-hatched larvae, eggs and egg-laying
females all distinguishable one from another, whereas during the May-July
period of 1983 (Fig.4) only three peak periods are easily picked out. From
this moment onwards generation overlapping, probably due to the rise in
temperature during the last part of July and the beginning of August and
also aided by favourable rainfall conditions during August and September,
did not allow the distinction of individual peaks. All the same, the
percentages concerning newly-hatched larvae, eggs and egg-laying females in
both 1982 and in 1983, when considered all together, do not permit a

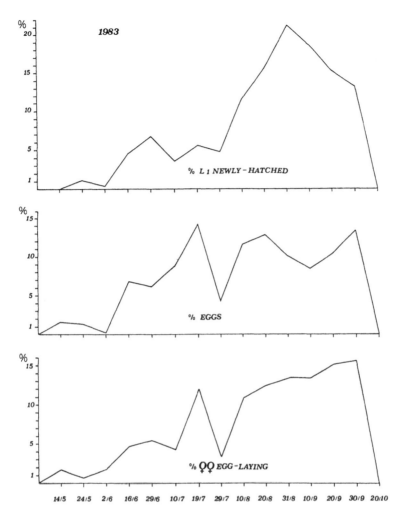

Fig. 4 - <u>Viteus</u> <u>vitifoliae</u> (Fitch) - The percentage of egg-laying females, eggs and newly-hatched larvae found inside mother-galls during the various test-periods in 1983.

definite distinction of the different, subsequent generations in each of the two years under study.

4. CONCLUSIONS

    Phylloxera infestation in our study vine nursery showed up in the first and second ten-day period of 1982 and 1983, respectively, with the presence of the egg-laying fundatrices in galls on the leaves. Few plants were infested at the beginning but by the end of June all the plants in the experimental plot were involved.
    However, due to the high number of gallicola generations and to their overlapping, it is very difficult to perform a field-study of the infestation course.

In fact, the increase in the average number of galls per leaf is an indication of the increase in infestation (Fig.1), but there are fluctuations in these values throughout the various periods probably due to different factors (e.g. generation overlapping) though mainly due the high variability in the number of galls on each single leaf.

An analysis of the percentage composition of the population inside the galls in the various periods (Fig.1) shows that from May to October all stages of the phylloxera are present and therefore neither random sampling nor sampling equal numbers of different aged leaves supplies information on the dynamics of the population or on the number of generations which have taken place. More information may be obtained by sampling a fair number of galls measuring 2-3 mm or over on the middle and basal leaves of the shoots (4). In fact, inside these galls egg-laying females, eggs and newly-hatched larvae who have not yet left the mother gall for the apical leaves where they will form new galls, may all be found. The percentage of egg-laying females, eggs and newly-hatched larvae in the various periods, though not leading to a sure distinction of the subsequent generations, does give information on the dynamics of the phylloxera population (Figs.3 and 4) and besides giving the average number of galls per leaf, supplies important information on the best moment for chemical control in the vine nursery.

REFERENCES

1. BELCARI A., COGNETTI G. C. (1983). Influenza della temperatura sullo sviluppo degli stadi preimmaginali di Viteus vitifoliae (Fitch) (Rhynchota, Phylloxeridae). 1 - Durata dello sviluppo dell'uovo in generazioni epigee a temperature costanti. Frust. Entom., Pisa, (n.s.) VI (XIX): 413-420.
2. GRASSI B., FOA A., GRANDORI R., BONFIGLI B., TOPI M. (1912). Contributo alla conoscenza delle fillosserine e in particolare della fillossera della vite. Tip. Naz. G. Bertera, Roma, pp. 1-456 + XIX tav.
3. JUBB G. L. Jr. (1977). Estimating grape phylloxera (Homoptera: Phylloxeridae) gall numbers on single grape leaves. Entomological News, 88 (3/4): 77-80.
4. RASPI A., ANTONELLI R., CONTI B. (1985). Preliminary observations on the composition of the grape phylloxera (Viteus vitifoliae Fitch) population in american vine leaves. Integrated pest control in viticulture. Expert's Group Meeting, Portoferraio, Italy, 26-28 September 1985.
5. RASPI A., BELCARI A., ANTONELLI R., CROVETTI A. (1985). Epigean development of grape phylloxera in Tuscan nurseries of american vines during the years 1982-1983. Integrated pest control in viticulture. Expert's Group Meeting, Portoferraio, Italy, 26-28 September 1985.
6. STEVENSON A. B. (1966). Seasonal development of foliage infestations of grape in Ontario by Phylloxera vitifoliae (Fitch) (Homoptera: Phylloxeridae). Can. Ent. 98 no. 12: 1299-1305.
7. STEVENSON A. B. (1970). Endosulfan and other insecticides for control of the leaf form of the grape phylloxera in Ontario. J. Econ. Entomol. 63 (1): 125-128.
8. STRAPAZZON A., GIROLAMI V. (1983). Infestazioni fogliari di fillossera (Viteus vitifoliae (Fitch)) con completamento dell'olociclo su Vitis vinifera (L.) innestata. Redia, Firenze, 56: 179-194.

Grape phylloxera, *Viteus vitifoliae* (Fitch): Heavy foliar infestation in a vineyard near Florence

G.Del Bene

Istituto Sperimentale per la Zoologia Agraria, Firenze, Italy

Summary

After some considerations about the present phylloxera risks for Italian viticulture,the author gives an account of a remarkable infestation
on the leaves of ungrafted vines (commonly called "French vines") in a
vineyard near Florence. Data concerning 4 years old plants were recorded
in 1984 and 1985. The first galls appeared at the end of May only in a
very small percentage (1%) of plants, but a strong attack (72%) occurred at the end of July. In 1984, despite of the heavy foliage damage,
only a weak infestation in the roots was noted and no winged migrant
female was captured. On the contrary, in 1985, a good number of winged
migrants was present from the middle of august.
Although further investigations are in progress, the preliminary results
obtained seem to indicate that infestation does not appreciably affect
production.

In recent years the grape phylloxera, *Viteus vitifoliae* (Fitch), has
again become the subject of attention in Italy for two reasons:
1. Agronomic policy (7) of using ungrafted vines with the intention of
establishing simple vineyards, with production characteristics of high
quality and quantity.
 The using of ungrafted vines, on the one hand, has its advantages espe
cially on account of low management costs and of its fruiting from the second year, but, on the other, it again raises the problem of phylloxera,
which had previously been solved with American rootstocks.
2. Constantly more frequent leaf infestation on European vines grafted on
resistant rootstocks.
 So far in Italy the following wine cultivars have been indicated as
receptive: Merlot, Tocai rosso, Moscato bianco, Garganega, Malvasia istria
na, verduzzo friulano, Sangiovese (9), Pinot grigio (1) ; and the following table cultivars: Italia, Alfonso Lavallee and Michele Palieri (4).
 Leaf infestation does not appear to be responsible for remarkable dam
age; however, it must not be underestimated as it demonstrates that a new
relationship is being created between the insect and the host plant, with
the completion of the olocycle even on grafted European vines. The reason
for this may be a gradual adaptation on the part of the aphid to the Euro-

pean vine, due to both the now accepted existence of a high number of biotypes of phylloxera (2, 3, 6, 8) and to adjustments on the part of the
host plant, which have come about through selection, growing practices, and
the influence of the rootstock.

In 1982 the Istituto Sperimentale per la Zoologia Agraria began research to determine the presence and the attack of *Viteus vitifoliae* on
grafted and ungrafted vines in various areas of Italy, and to evaluate by
means of artificial infestation the tolerance of selfrooted plants of the
Corvina cultivar, belonging to three clones.
This paper limits itself to referring to a severe infestation found
on vines commonly known as "French vines" in a vineyard close to Florence,
and to figures relevant to 1984 and 1985 (until August).

The vineyard which is composed of about 1500 selfrooted plants of two
and four years, has been affected particularly in the aerial part of the
older plants. On 1% of these the first galls appeared at the beginning of
May. In 1985 the percentage of infested plants rose to 25% in a week; it
reached 59% during June and 72% by the end of July.
In this last month, the plants, which were divided into four categories according to the number of damaged leaves, showed that 27% belonged to
the 3rd (from 10 to 20 leaves with galls) and 20% to the 4th (more than 20
leaves with galls, with a maximum of 489 observed). Moreover, on the more
damaged plants the infestation was not only on the leaves, but also on the
shoots and tendrils,where distorsions were found. These were due to *gallicolae* , which were present together with eggs and *larvae;*on the younger
leaves the galls were no longer recognisable individually, but ran into
each other in elongated wrinkles containing a very large number of females,
eggs and *larvae* . In 1984 the number of galls per leaf, which was generally
between 20 and 50, in some cases exceeded 200; in 1985 the average was 46,
with maximum values of 96. During the summer season the presence of eggs
and *larvae* in the gall appeared constant (table I).
With regard to the root system, in 1984 it was only possible to ana-
lyse three plants, which showed only very little nodosities (0 - 13) and
rare *larvae* (0 - 6).
The absence of winged migrant females corresponded to this low root
infestation. Throughout the whole season they were never captured with
sticky yellow chromotropic traps. In 1985, however, the winged migrants,
which appeared in the middle of August, were present in a very high number,
even 68 per trap per week. This fact leads to the assumption that this year
there is a heavy presence of *radicicolae* : relevant controls are still in
progress as are those to ascertain the presence of sexual forms and overwintering eggs. However, on the basis of the large number of winged migrant
females captures, it may be concluded that on the cultivar under observation, phylloxera is able to complete its olocycle.
So far, production does not appear to be affected even in the most
infested vines, and it is within the average values for undamaged plants,
6.4 Kg per plant.

Table I

Average number of eggs and *larvae* of *Viteus vitifoliae* in leaf galls (minimum and maximum in brackets).

		1984	1985
May	eggs	65 (20 - 120)	
	larvae	2 (0 - 11)	
June	eggs	135 (35 - 280)	66 (0 - 135)
	larvae	4 (0 - 20)	8 (0 - 36)
July	eggs	58 (7 - 212)	52 (0 - 140)
	larvae	14 (0 - 54)	9 (0 - 40)
August	eggs	54 (12 - 112)	35 (0 - 100)
	larvae	9 (0 - 35)	4 (0 - 21)

With regard to natural enemies, at the beginning of June, in galls produced by the *fundatrix* and occupied by eggs and *larvae* , a predatory mite was found. The individuals (1 or 2 larval forms per gall) were identified by M. Castagnoli as *Allothrombium* sp.

It is of importance to continue research in this sector, which apart from the finding of some entomophagous species, such as *Scymnus haemorroidalis* , *Coccinella decempunctata* , *Chrysopa* spp., *Trombidium holosericeum* (5) has not been particularly developed, since the problem of phylloxera had previously been resolved by using resistant rootstocks.

REFERENCES

1. BARBATTINI, R. et al.(1985). Presenza di fillossera nel Pordenonese. Inform. Agr., A. 41, n.18: 91-96.
2. BÖRNER, C., HEINZE, R. (1957). Aphidina Aphidoidea. In: Sorauer P. and Bluck H., "Handbuck der Pflanzenkrankheiten". P. Parey, Berlin u. Hamburg, 5, 402 pp.
3. GOIDANICH, A. (1960). Fillossera della vite. In: Encicl. Agr. It., Reda, Roma, 4: 682-698.
4. GRANDE, C. (1984). Dopo più di un secolo la Fillossera, il piccolo "pidocchio" della vite, ritorna alla ribalta. Inform. Agr., A. 40, n. 30: 44-46.

5. GRASSI, B. (1912). Contributo alla conoscenza delle Fillosserine ed in particolare della fillossera della vite. Tip. Nazionale Bertero, Roma: 456 + LXXV pp.

6. JUBB, G.L. (1976). Grape Phylloxera: incidence of foliage damage to wine grapes in Pennsylvania. J. Econ. Entomol. 69(6): 763-766.

7. SCARAMUZZI, F. (1979). Viticoltura moderna e fillossera. Giornale di Agricoltura, A. 89, n. 411: 40-44.

8. STEVENSON, A.B. (1970). Strains of the grape phylloxera in Ontario with different effects on the foliage of certain grape cultivars. J. Econ. Entomol. 63(1): 135-138.

9. STRAPAZZON, A., GIROLAMI, V. (1983). Infestazioni fogliari di fillossera (*Viteus vitifoliae* (Fitch)) con completamento dell'olociclo su *Vitis vinifera* (L) innestata.

Epigean development of grape phylloxera, *Viteus vitifoliae* (Fitch), in Tuscan nurseries of American vines, during the years 1982-1983

A.Raspi, A.Belcari, R.Antonelli & A.Crovetti

Istituto di Entomologia Agraria dell' Università di Pisa, Italy

Summary

In the two-year period of 1982-1983, during sampling for studies on *Viteus vitifoliae* (Fitch) population dynamics on American vine rootstocks, there were various fluctuations in the population which hindered identification of the number of the possible generations produced by the insect. With the aid of theoretic constants obtained through laboratory data and concerning the length of the aphid generation, it was possible to identify the number of actual generations performed by the gallicolae form.

1. INTRODUCTION

During the two-year period of 1982-1983, in a series of observations aimed at defining sampling methods for studies on the dynamics of *Viteus vitifoliae* populations on American vine leaves, we came up against several difficulties, especially in calculating the exact number of aphid generations with these sampling methods.

In fact, analysis of the curves pertaining to the fluctuations seen in the course of sampling did not permit an exact calculation of the number of generations produced by the insect because of generation overlapping and hence the simultaneous presence of all stages on the leaves.

It was therefore considered opportune to compare field data (4) (5) with certain preliminary results obtained in the laboratory and referring to the influence of temperature constants on the development rate of the gallicolae forms of phylloxera with the thermal constants obtained from these data a theoretic calculation of the generation was therefore achieved.

2. MATERIALS AND METHODS

The data concerning the rate of development from first instar larvae to egg-laying females were obtained from observations performed at three constant temperatures: 20 °C, 25 °C and 30 °C. The tests were carried out by deliberately infesting young Kober 5BB hybrids with larvae born at least 24 hours previously.

The average development rates hence obtained were subsequently added to those already known for the eggs (1) obtaining, therefore, the entire development.

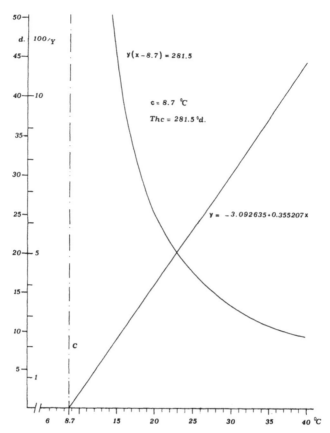

Fig. 1 – Equilateral hyperbole and regression line indicating, respectively, the duration (days) and the rate of development (100/y) of the entire development of <u>Viteus</u> <u>vitifoliae</u> (Fitch) at constant temperatures. Also indicated are the equations and the thermal constants of the species and their developmental zero (dotted line).

The data were then used for calculating the theoretic constants by means of the thermal summation (*).
The observations on the dynamics of the populations were performed on samplings in a vine nursery located in the province of Pisa during 1982 and 1983 (4) (5).

3. <u>RESULTS AND CONCLUSIONS</u>

3.1 <u>Thermal summation</u>
The data relative to the duration at the three temperature were used

----

(*) A further study reports more details on the influence of the variable and constant temperatures on the entire development and on the single stages.

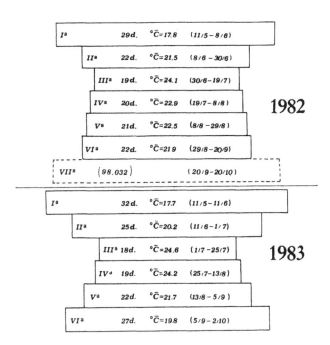

I^a	29d.	$°\bar{C}=17.8$	(11/5 – 8/6)
II^a	22d.	$°\bar{C}=21.5$	(8/6 – 30/6)
III^a	19d.	$°\bar{C}=24.1$	(30/6 – 19/7)
IV^a	20d.	$°\bar{C}=22.9$	(19/7 – 8/8)
V^a	21d.	$°\bar{C}=22.5$	(8/8 – 29/8)
VI^a	22d.	$°\bar{C}=21.9$	(29/8 – 20/9)
VII^a	(98.032)		(20/9 – 20/10)

1982

I^a	32d.	$°\bar{C}=17.7$	(11/5 – 11/6)
II^a	25d.	$°\bar{C}=20.2$	(11/6 – 1/7)
III^a	18d.	$°\bar{C}=24.6$	(1/7 – 25/7)
IV^a	19d.	$°\bar{C}=24.2$	(25/7 – 13/8)
V^a	22d.	$°\bar{C}=21.7$	(13/8 – 5/9)
VI^a	27d.	$°\bar{C}=19.8$	(5/9 – 2/10)

1983

Fig. 2 – Viteus vitifoliae (Fitch). – Schematic representation of the theoretic number of generations produced by gallicolae forms during 1982 and 1983. Each square shows, respectively, the number of days required to complete the generation, the average temperature and the period during which the generation took place.

to calculate the "thermal summation", the behaviour of which is shown in the hyperbole in fig.1.

For the developmental zero the embryo stage was chosen (c = 8.7°C) since the egg stage is the most sensible one. In fact, the developmental zero for the post-embryo stage showed a lower value (c = 7.8°C).

The thermal constant (Thc = 281.5°g) was obtained from the sum of the egg Thc and that of the post-embryo development. The graph was then drawn up with the regression line, obtained from the reciprocal values of the hyperbole, which expresses the development rate of aphid and the relative equation (y = -3.092635 + 0.355207x).

3.2 Theoretic calculation of the generations

With the thermal constants obtained in the laboratory and using the ten-day thermic averages recorded in 1982 and 1983 during the test-periods when the gallicola forms were present, a theoretic calculation of the number of possible generations produced by the gallicolae over the two-year period was then elaborated. As starting point for the calculation the second ten-day period of May was chosen since field studies showed that this was the period when the fundatrices layed their eggs (**).

(**) The generation produced from overwintering egg and which ends when the fundatrices lay their eggs, has therefore been excluded from the calculation.

169

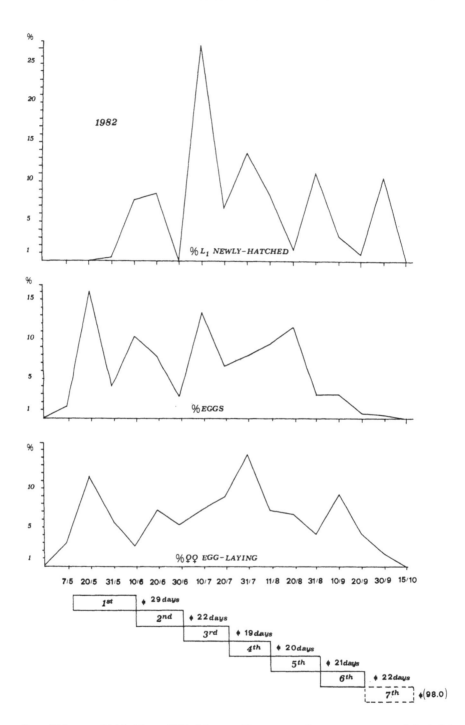

Fig. 3 - <u>Viteus</u> <u>vitifoliae</u> (Fitch). - The percentage presence of hatching larvae, eggs and egg-laying females found during the 16 ten-days periods of the experiment in 1982. Below, the theoretic calculation of the generations.

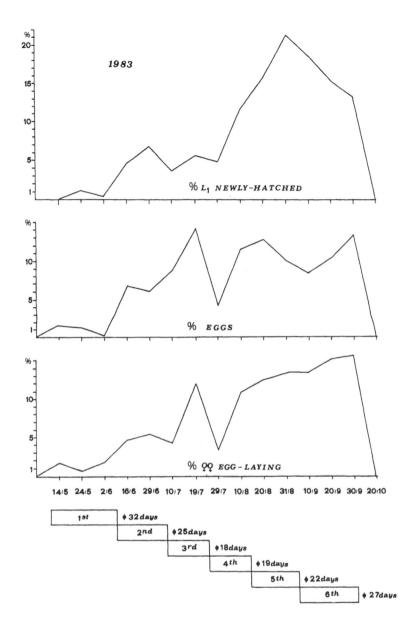

Fig. 4 – <u>Viteus</u> <u>vitifoliae</u> (Fitch). – The percentage presence of hatching
larvae, eggs and egg-laying females found during the 15 ten-days
periods of the experiment in 1983. Below, the theoretic
calculation of the generations.

Fig.2 schematically shows the number of possible generations
produced by the aphid during the two years of observation.
In 1982 the insect produced 6 complete generations plus one partial
one (98.032) which could have been completed under certain climate
conditions. In 1983 6 complete generations were calculated. Worthy of note
is the wide variability of the number of days necessary for completing a

generation: from a minimum of 18 days at an average temperature of 24.6 ºC to a maximum of 32 days with an average of 17.7 ºC.

Figs.3 and 4, concerning the sampling for the dynamics of the gallicolae form of the population, show the theoretic duration of generations for appropriate checking against peaks relative to the percentage of newly-hatched larvae, eggs and egg-laying females seen with sampling.

In 1982 (Fig.3) the points indicating the lowest percentage of presence of the three stages coincides to a great extent with the end of one generations and the beginning of another.

In 1983 (Fig.4) only the first three generations can be detected (from the first ten days of May till the third ten-day period of July) as having counterproof in the theoretic calculation.

From the first ten-day period of August onwards, because of enhanced generation overlapping (4) (5) the number of generations can no longer be detected using the field data alone, whereas the theoretic calculation confirms that even in this second year the aphid produced 6 generations.

To conclude, it may be stated that the number of generations is strictly influenced by climate conditions (in Germany, aphid produce 4-5 generations (2) while in Sicily they can produce 8 or more (3)) even from the plant itself. In fact, the thermic averages in October and part of November could allow the insect to carry on its development, but the plant, from the first part of October onwards loses its receptibility (***) and hinders the completion of further colonies of newly formed gallicolae.

REFERENCES

1. BELCARI A., COGNETTI G. C. (1983). Influenza della temperatura sullo sviluppo degli stadi preimmaginali di Viteus vitifoliae (Fitch) (Rhynchota, Phylloxeridae). 1 - Durata dello sviluppo dell'uovo in generazioni epigee a temperature costanti. Frust. Entom., Pisa, (n.s.) VI (XIX): 413-420.
2. BREIDER H. (1952). Beitrage zur Morphologie und Biologie der Reblaus. Dactylosphera vitifolii Shim. Z. ang. Ent. BD 33 (4): 517-543.
3. GRASSI B., FOA A., GRANDORI R., BONFIGLI B., TOPI M. (1912). Contributo alla conoscenza delle fillosserine e in particolare della fillossera della vite. Tip. Naz. G. Bertera, Roma, pp. 1-456 + XIX tav.
4. RASPI A., ANTONELLI R., CONTI B. (1985). Preliminary observations on the composition of the grape phylloxera (Viteus vitifoliae (Fitch)) population in American vine leaves. Integrated pest control in viticulture. Expert's Group Meeting, Portoferraio, Italy, 26-28 September 1985.
5. RASPI A., ANTONELLI R. (1985). Grape phylloxera (Viteus vitifoliae (Fitch)) infestation on American vine leaves in a nursery at San Piero a Grado (Pisa), during the years 1982-1983. Integrated pest control in viticulture. Expert's Group Meeting, Portoferraio, Italy, 26-28 September 1985.

(***) In effect, the newly-hatched larvae, from the first part of October onwards can no longer find foliage substrate - made up of young reddish apical leaves - which reacts to their attack. This has been discovered many times in the laboratory when, during the course of observations, several plants remained free of infestation because of the absence of this type of leaf.

Studies on the correlation between the presence of some *Drosophila* species and the appearance of grape-vine sour-rot in Lombardy (Italy)

G.C.Lozzia & A.Cantoni
Istituto di Entomologia Agraria dell'Università di Milano, Italy
A.Vercesi & M.Bisiach
Istituto di Patologia Vegetale dell'Università di Milano, Italy

Summary

The constant presence of various species of Diptera Drosophilidae, and in particular of Drosophila fasciata Meigen, on grapes affected with 'grapevine sour-rot', leads to believe that the insect plays an important role in the diffusion and development of the disease, present in many vineyards of Lombardy.
Grapevine sour-rot is an alteration produced by yeasts and bacteria that penetrate into the grape through wounds; they are constantly joined by some Drosophilidae attracted by the strong smell of acetic acid and of ethyl acetate. Researches carried out on various viticultural areas of Lombardy have led to point out how, together with Drosophila fasciata Meigen, also Drosophila funebris Fabr. and Drosophila fenestrarum Fall. are present.
In order to verify the role of these insects in the transport of the grapevine sour-rot agents, the yeast attractiveness towards the vinegar flies has been measured by tests through a proper olfactometre. Laboratory data confirm that a group of yeasts separated from bunches affected with grapevine sour-rot prove to be particularly attractive towards Drosophila adults. Moreover, tests on transmission by adults and the finding of yeast in the feeding channel of all the development phases of vinegar fly, demonstrate how the insect is particularly important for the spreading of the disease.

1. Introduction

Connection phenomena between insects and microorganisms are very common in nature and the specific relations sometimes reach very high evolution levels. These relations assume a remarkable importance as far as the cultivated plants are concerned, when the associated organisms are pathogenous agents of the plants. This is the case of the connection between some Drosophila spp. and grapevine sour-rot that takes place in the vineyards of Northern Italy in the period foregoing the vintage.

Grapevine sour-rot is a lytic disease that reveals itself on compact bunch species with thin skin grapes. Grape-stones assume a brownish colour at the beginning, afterwards the skin becomes thin and splits, the juice drops on the grape-stones below, letting out a sour smell of acetic acid (2); around and on the rotting bunch it has been noted the appearance of a big number of grown-up Diptera Drosophilidae; inside the infected berries we can observe larvae and insect pupae together with several already empty puparia.

The aim of this research is to investigate on some aspects of this connection that causes remarkable damages to the production of wine grapes both for the reduction of product and for the increase of flying acidity that is determinant for must.

173

2. Materials and methods

In order to determine which was the distribution of some Drosophila species in several viticultural areas of Lombardy, in September 1983 it has been carried out a collection of samples in some vineyards placed in the di stricts of Brescia, Bergamo, Pavia, Milano and Sondrio. Data concerning ly- ing surface, growing-up system and species cultivated in several areas are quoted in table I. In each of the examined vineyards fifteen bunches infe- sted with Diptera have been drawn. In case of infestation was not macrosco- pically relevant inside the vineyard, it was anyway drown the same number of bunches chosen by chance in the area in order to verify if there were eggs of Drosophilidae on them.

Each bunch was put in a glass container; in laboratory the containers were put in climatic cells at a temperature of 25°C and UR 70%, and were da ily examined to follow the eventual development of Diptera. Insects appeared on bunches have been examined under a microscope and classified according to the key of Duda (5).

A similar collection of samples has taken place in September 1984.

3. Found species and geographic distribution

The results of the research carried out in 1983-84 show that, in the vi ticultural areas which have been studied, three species of Drosophilidae are present, and precisely: Drosophila fasciata Meigen, Drosophila funebris Fabr. and Drosophila fenestrarum Fall.. Clearly prevailing both as number of pre- sent beings and as width of distribution is D. fasciata Meigen. This latter has been constantly drown in every area. Sometimes, as in Ponte Valtellina and S. Colombano, the whole population belonged to this species; in the other cases, beings of D. fasciata constituted a relevant part as much as to reach in Cellatica 80% of the population.

Less significant is the presence of the other two Drosophilae: it is anyway remarkable the wide distribution of D. funebris. As far as D. fene- strarum is concerned, as it has been drown only in one case, it seems that is should be considered as only sporadically present in the vineyards of Lom bardy.

During various investigations effected in the before mentioned vineyar- ds, it has been remarked a noteworthy and constant correlation between the presence of these Diptera and the grapevine sour-rot. The insect has resul- ted absent in those vineyards that, because of cultivation process, climate conditions and phytoiatric treatments, did not present any rot in the bunch. This has taken place in 1983 in Cigognola (Pavia) and in Rodengo Saiano (Bre scia) where the bunchs of grapes were sound. The maximum intensity of infe- station has been remarked in a vineyard in Cellatica (Brescia) in which the lytic phenomena due to grapevine sour-rot were very evident.

Intermediate intensity has been reached in other vineyards in which it has been again remarked that Diptera gather in preference around bunches af- fected with grapevine sour-rot.

In the vineyards investigated during 1984, only one species, Drosophila fasciata, has been reported, with a diffusion higher than that of the previ- ous year. This event is partially explainable considering the unfavourable climate conditions of September 1984, characterized by drizzling and fre- quent rain that, besides increasing the incidence of the lytic phenomena to- wards the bunch, has exalted the characteristics of higher adaptability to the meteoric agents of this species.

TABLE I - Place, surface, cultivar, growing-up system and lying of the experimental particles.

Place	Surface (ha)	Cultivar	Growing-up system	Lying
Rodengo S. (Brescia)	5,5	Pinot nero Pinot chardonet	Silvoz	level
Rodengo S. (Brescia)	2	Merlot, Cabernet Nebbiolo	Silvoz, Pergola trentina	level
Cellatica (Brescia)	4	Schiava, Marzemino Barbera	Silvoz	level, hilly
Padenghe (Brescia)	1,5	Trebbiano di Lugana	Guyot	hilly
Torre de' Roveri (Bergamo)	3	Merlot, Cabernet, Pinot bianco	Silvoz	level, hilly
Cigognola (Pavia)	1	Pinot nero, Bonarda Barbera	Guyot	hilly
Bosco Casella (Pavia)	1,5	Pinot nero, grigio Riesling, Bonarda	Guyot	hilly
S. Colombano (Milano)	1	Croatina, Malvasia Barbera, Bonarda	Guyot	hilly
S. Colombano (Milano)	4	Croatina, Malvasia Barbera, Bonarda	Pergola trentina	hilly
Ponte Valtellina (Sondrio)	1,5	Nebbiolo, Pignola	Pergola	hilly
Villa Tirano (Sondrio)	1	Nebbiolo, Pignola	Guyot	hilly

Also in those areas that in the previous year presented grapes substantially sound, were evident, although in quite low percentage, bunches with the characteristic symptoms of grapevine sour-rot. Even in this case, therefore, it has been remarked a correlation between the presence of lytic processes towards the bunch, and the appearance of Drosophilidae in the areas.

4. Laboratory tests

4.1 Olfactometric tests

In order to verify if the different strains of yeast isolated from bunches affected with grapevine sour-rot exerted an attraction on Drosophila fasciata, specific tests by olfactometre have been carried out.

This is formed by a plexiglas container, that has a cubic structure of 55 cm for each side, on which front face are inserted two square section tubes, of 30 cm lenght, where are placed the Petri plates containing the examined strains. Each tube is separated from the body of the olfactometre by a removable shutter.

Inside the container 100 adults of D. fasciata have been introduced for each test. The insects were coming from breeding on artificial substrate pre pared in order to avoid any preliminary contact between Diptera and microorganisms similar to those to test.

Each of 140 strain of yeast under examination, only 46 of which are identified, has been inoculated on plates containing agar-malt and incubated for three days at 28°C.

At the beginning of the experiment, on the extremity of each tube, a Petri plate has been applied by adhesive tape, placed in order to show the agarized surface toward the inner part of the olfactometre. For each test it has been utilized a plate inoculated with one of 140 strains of yeast and a control represented by a plate containing only agar-malt. After having placed the plates on the extremity of the tubes, the removable shutters placed at the extremity of the same has been opened, giving a start to the attraction test. At intervals of an hour it has been made the computation of the beings attracted into each tube; each test lasted four hours.

5. Results

All the tested yeasts turned out to be able to attract vinegar flies, even if according to different standards; they range from a minimum standard of 1-2 to a maximum of 50-52 attracted beings.

Data concerning the control have been excluded from the statistical table, as its inexistant attractiveness towards the tested strains was evident.

From the obtained data it comes out that the general tendency is that of an increase of the number of the attracted insects as time of observation goes on. This fact is due to the increase of the concentration of flying attractive compounds that takes place inside the olfactometre tubes. The obser ved standards have been statistically analyzed, using a UNIVAC computer. Utilizing a program of automatic calculation STAT-JOB, it has been, first of all, carried out the variation analysis. In this analysis survey times and 140 strains of yeast have been considered as independent variables, while the computation of the insects attracted into the olfactometre tubes during the stated times has been considered as a dependent variable.

Results of variation analysis are listed in table II.

TABLE II - Variation analysis

	Deflection	F.D.	Variation	Test F	Significance
Tot. deflection	63286,000	575			
Yeast	54437,000	143	380,67832	20,18	oo
Times	754,625	3	251,54166	13,33	oo
Error	8094,375	429	18,86801		

From the obtained results it turns out that the data concerning the com putation of attracted beings of vinegar fly show, with 99% of probabilities, meaningful differences both as regards the different phases of the experiment, and as regards the different strains of tested yeast.

On the ground of these results, it has been provided, therefore, by the same program, to calculate the means of attraction data both as regards the different strains of yeast and as regards the different phases. Data concerning the average attraction standard for each yeast have been subsequently utilized to carry out a multiple comparison among means (Duncan's multi-ple range test). Aim of this test is to gather the yeasts that do not pre-sent significant differences among their average standard of attraction. It has been taken into account only one standard of significance, calculating the means that differred from each other with 99% of probability. Neverthe-less, many tested microorganims belong, at the same time, to many groups; this tendency is more and more evident as we move to intermediate standards of attraction (see plate 1). The remarkable number of considered standards prevents from a visualization of the differences existing among strains by the use of letters, usual practice in the exposition of the results of Dun-can test.

Therefore, it has been decided to integrate the results of the multiple comparison with the standards of the arithmetic mean of attraction obtai-ned for each yeast, in order to establish arbitrary groups of microorganisms in order of their capacity of attraction towards Drosophila . These groups of yeasts have been formed in a completely arbitrary way, taking care, never theless, to make the beginning to the single group clash with that of some groups expressed in Duncan test for a significance level equal to 99%.

The groups obtained according to the mentioned method are listed in ta-ble III and visualized in plate 1.

TABLE III - Yeasts subdivision according to their attractiveness towards
             D. fasciata Meig.

Group 1: highly attractive yeasts   more than 30 Drosophila

Group 2: attractive yeasts   from 21 to 30

Group 3: yeasts with a medium attractiveness   from 14.76 to 20

Group 4: scarcely attractive yeasts   from 11.25 to 14.75

Group 5: weakly attractive yeasts   from 1 to 11.24

The comparative examination of table III and IV points out that spe-cies belonging to most of genera took into consideration, Hanseniaspora, Candida, Hansenula, Zygosaccharomyces, Endomyces and Issatchenkia are inclu-ded in the groups turned out to be particularly attractive towards D. fascia-ta. An exception are the genera Kluyveromyces included in group 5, Pichia, of which the only representative belongs to group 4, and Torulaspora which, nevertheless, is present among the yeasts with a mean attraction.

Observing the distribution of the several species in the mentioned grou ps, it is evident the remarkable difference existing among the attractive-ness exerted by the single strains. Such phenomenon is particularly marked as regards Hanseniaspora uvarum, which representatives are included both in the group of highly attractive beings, and among yeasts provided with such a characteristic. This result is confirmed by the contemporaneous presence of strain 75 C. steatolytica in group 1 and of strains 64 and 108 belonging to the same species in the last group. The same remark concerns I. terricola, whose strain 104 is classified as attractive, while strain 97 is able to e-xert only a weak attraction towards vinegar flies.

This activity seems, therefore, to be linked more than to one or more genera of yeasts, to the specific strain, probably utilized in relation to different phisiological characteristics of the being.

YEASTS' ATTRACTIVENESS ON _Drosophila fasciata_ Meig.

DUNCAN'S MULTIPLE RANGE TEST (SIGNIFICANCE LEVEL 0.1)

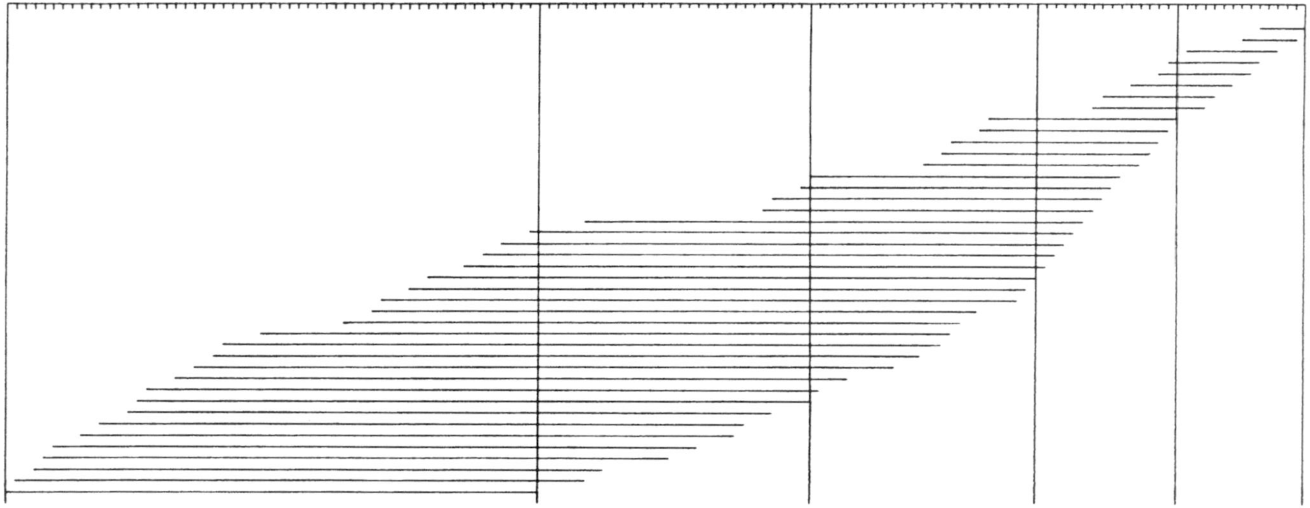

GROUP 1	GROUP 2	GROUP 3	GROUP 4	GROUP 5
71 145 116 111	115 61 13 52	15 53 77 14 93 102	83 5 98 99 135 19 11 24	144 47 85 146 69 82 92 128 3 125 64 78 88 113 101 95 57 137
70 75 9 141	112 30 60 36	119 129 45 86 140 65	79 55 87 118 42 68 132 22	54 89 12 114 121 1 20 25 100 39 44 34 66 94 2 4 10 120 6
59 138 32 29	7 104 103 134	142 63 105 17 27 73	28 40 74 109 16 49 117 8	31 97 46 106 107 108 38 58 21 26 33 84 91 131 48 124 35 18
139 110	62 123 126	76 96 136 80 122 43	72 143 37 41	51 50 23

PLATE I

178

TABLE IV - Identificated yeast species and corresponding strains.

Species	Strains
Candida boidinii	117
deformans	121
intermedia	132
kefyr	116
krusei	129,134
pseudotropicalis	122,128
steatolytica	64,75,108
utilis	112,115
valida	39
vini	124
Endomyces fibuliger	126
Hanseniaspora uvarum	3,5,6,8,19,25,32,33,83,86,87,90,91,143,146
Hansenula jadinii	111,118
petersonii	119,123
Issatchenkia terricola	97,104
Kloeckera apiculata	125
Kluyveromyces fragilis	114
Pichia mucosa	135
Saccharomycopsis lypolytica	113,120
Torulaspora delbrueckii	136
Zygosaccharomyces bailii	41
ronscii	131

6. Correlation between vinegar flies and grapevine sour-rot

During the period previous to the gather of grapes, the connection between grapevine sour-rot and vinegar flies seems to be quite always prejudi-
cial to the production of wine grapes. In fact, to the fermentative action
produced by yeasts and bacteria agents of grapevine sour-rot, it is summed
the berry emptying action produced by larvae in active growing-up.

Experimental researches carried out on the past demonstrate that the
yeasts are present as microsymbions in the intestinal canal of the insect,
where they carry out the important function of vitamin synthesis that the
insect is not able to produce. The yeasts also constitute a nutriment with
high protein contents for the larvae; these latter, if grown-up in an environment not very rich of proteins, as precisely the tissues of the sound ber
ry, remain filiform and are seldom able to reach maturity (4).

Associated microorganisms that live in the feeding channel of the insect, are provided with a constant source of food, and are protected from

some unfavorable environment factors, such as the extreme heat, the light and the drying up.

As far as diseases that manifest themselves on vegetables are concerned, it is known that many pathogenous microorganisms are able to produce a big quantity of inoculation; nevertheless, a necessary condition to produce an infection is that the inoculation itself is brought into contact with liable beings. Wind is an efficacious dissemination agent, but the inoculation of several pathogenous is not suitable for this kind of dispersion (3). In particular the cells of the colonies of yeasts and bacteria are often surrounded by mucilaginous exudates, which make impossible the detachment of propagulae by the wind. Moreover, in environments with a scarse relative humidity, these latter dry up very quickly, precluding quite completely their vitality. In these cases, insects, and in particular vinegar flies with their hairy body and their legs covered with bristels, are fit for the transport of propagules.

In order to verify if some strains of yeast isolated from grapes affected with grapevine sour-rot could be transported on the body of the insects maintaining untouched their vitality characteristics, some laboratory tests have been carried out, utilizing the same olfactometre used in attractiveness tests. In one of the two tubes it has been placed a plate containing sterile agar-malt, while in the other one it has been placed a plate of agar-malt inoculated three days before with L 85 yeast. L 85 strain, which is not yet identified, has been chosen because its colonies of pink colour are particularly evident. Afterwards, 30 adults of Diptera, grown up on a sterile artificial substrate, have been introduced into the tube containing the inoculated plate. After half an hour, the shutters of the two tubes have been removed in order to allow the free circulation of the insects from one plate to the other. Thereafter, the plate that was not inoculated has been removed and incubated at 30°C; after three days we could establish the appearance on this plate of some colonies of pink yeast that, examined under a microscope, showed cells identical to those of the original L 85 strain. In this case, the transport of propagulae had been evidently encrusted to adults, that infecting one another in several ways on the inoculated plate, consequently infected the sterile plate.

Nevertheless, infection can take place only if yeasts, which are not able to penetrate actively through the skin of grapes-stones, are brought into contact with the tissues of the berry.

The grown-up Diptera Drosophilidae are provided with a sucking and lapping mouth apparatus that allows them to feed themselves on a liquid substrate. When food is represented by a solid substrate, the insects lays its labella on food and regurgitates the contents of the ingluvies together with salivary secretions rich of enzyms. These latter are able to melt many substrates that afterwards are sucked again into the oesophagus. Experimental researches have ascertained that the liquid present in the ingluvies of Drosophilidae contains remarkable quantities of vital microorganisms (6).

The capacity of vinegar flies of functioning as vectors of the agents of grapevine sour-rot, and more in general of various microorganisms, is linked, nevertheless, not only to the passive transport of propagulae effected by adults, but also to the opportunity that pathogenous are transmitted by inner way during the several stages of development of the insect.

Results of the researches carried out on Drosophila melanogaster (1) demonstrate the persistence of bacteria till adults of the second generation. The method suggested by Ark and Thomas (1), with slight modification, has been adopted to examine the possibility of transmission of yeasts isolated from bunches affected with grapevine sour-rot in the following stages of development of Drosophila fasciata. On this purpose some females of the Dipteron have been placed to lay on a sterile substrate. The new-born larvae have

been maintained for 9 days on the substrate. At the same time, many plates
of agar-malt have been inoculated with L85 yeast and incubated for three
days at 28°C. Afterwards, the larvae have been transferred sterily into
the plates containing L 85 and left there to feed themselves with yeast.
The superficial sterilization of larvae has been obtained by immersion into
a solution of mercuric chloride.

At this point, each larva has been rolled on a plate of agar-malt in
order to check the absence of microorganism on the tegument. The same larva
has been sectioned transversely and each of the to obtained portions has
been distributed on two sterile plates containing agar-malt. All the three
plates have been incubated in order to remark the development of any kind
of microorganism. After three days since the inoculation the results listed in table 5 have been obtained.

TABLE V - Results concerning the experiments on the inner contamination
of larvae.

	Plate 1	Plate 2	Control
Drosophila n. 1	+	+	−
Drosophila n. 2	+	+	−
Drosophila n. 3	+	+	−
Drosophila n. 4	+	+	−
Drosophila n. 5	+	+	−
Drosophila n. 6	+	+	−
Drosophila n. 7	+	+	−
Drosophila n. 8	+	+	−

(+) presence ⏋f yeast colonies
(−) absence of yeast colonies.

It is interesting to remark how the control has always resulted negati
ve as a confirmation of the absence of external contamination. Instead, the
colonies grown-up on the other plates should have originated, therefore,
from an inner source of inoculation.

A similar test has been carried out to establish whether also the pupae
presented inner contaminations. A part of the larvae, after having been fed
with L 85 yeast, have been left on the plates in order to let them get into
pupae. Afterwards, a superficial sterilisation of puparia has been carried
out by immersion into a solution of mercuric chloride. Then, each pupa has
been rolled on a Petri plate containing agar-malt in order to control the
superficial sterility. The same pupa, after having been sectioned transversely, has been inoculated into a Petri plate containing agar-malt. The three
plates have been incubated at 28°C and daily checked to examine the development of eventual present microorganism. Three days after the inoculation
we have obtained the results listed in table VI.

Even in this case the external contamination proved by the control has
resulted negative, while the plates inoculated with the residues of each pupa have shown colonies of L 85.

The yeast assimilated by the larva has remained vital even after the
puparium formation and, therefore, it exists a source of inoculation also
inside puparium.

181

TABLE VI - Results concerning the experiment on the inner contamination
         of pupae.

	Plate 1	Plate 2	Control
Drosophila n. 9	+	+	-
Drosophila n. 10	+	+	-
Drosophila n. 11	+	+	-
Drosophila n. 12	+	+	-

(+) presence of yeast colonies
(-) absence of yeast colonies.

A similar test has been carried out on empty puparia and the results
confirm that even inside a puparium left by the adult, colonies of vital
yeasts exist.
Further tests are in progress to determine whether the adults coming
from larvae fed on a substrate containing yeasts could keep vital these microorganisms inside them.

7. Conclusions

Researches carried out in some vineyards of Lombardy allowed us to put
into evidence the constant presence of Diptera Drosophilidae on grapes affe
cted with grapevine sour-rot.
Drosophila fasciata Meig. is the most diffuse species, while less freque
nt are Drosophila funebris Fabr. and Drosophila fenestrarum Fall..
To understand the role of vinegar flies in the diffusion of grapevine
sour-rot, many olfactometric tests have been carried out, that have given
prominence to how, independently from the taking place of the phenomenon of
grapevine sour-rot, yeasts are able to attract these insects.
In particular, yeasts belonging to Hanseniaspora, Candida, Hansenula,
Zygosaccharomyces, Endomyces and Issatchenckia genera have resulted to be
particularly attractive towards D. fasciata, while those of Kluyveromyces,
Pichia and Torulaspora genera have a mean attractiveness.
As far as the transport of the yeasts themselves is concerned, laboratory tests were able to demonstrate that vinegar flies can easily transport
on their body vital propagules of the yeasts. Moreover, the researches we
carried out have allowed to establish that the larvae of these Diptera, fed
on an artificial substrate contaminated by yeasts, shelter in their feeding
channel these microorganisms essential to their metabolism. The larvae that
feed themselves with this substrate maintain a source of inoculation even
inside puparia. The adults, at the time of the flitting about can, in their
turn, contaminate themselves with yeasts present on the vegetal substrate
where the puparia are usually placed.

REFERENCES

1. ARK, P.A., THOMAS, H.E. (1936). Persistence of Erwinia amylovora in certain insects. Phytopatology, 26, 375-381.
2. BISIACH, M., MINERVINI, G., ZERBETTO, F. and VERCESI, A.(1982). Aspetti
   biologici ed epidemiologici di B. cinerea e criteri di protezione antibotritica in viticoltura. Vignevini, 12, 39-46.

3. CARTER, W. (1962). Insects in relation to plant disease. Interscience Publs., New York-London, 64-86.
4. CHAUVIN,R. (1956). Physiologie de l'insecte. I.N.R.A., Paris, 196-197.
5. DUDA, O. (1935). Familia Drosophilidae (in: Lindner, Die Fliegen der palär ktischen Region; Schweizerbartsche Verlagbuchmandlung, Stuttgart), Bd VII, vol. 58 G, 1-115.
6. GREENBERG , B. (1971). Flies and disease. Vol. 1. Princeton Univ. Press, 1-256 (cfr. 97-152; 216-218).

This work was partially supported by Assessorato Agricoltura of Regione Lombardia.

Mites of vineyards and control strategies

V.Girolami
Istituto di Entomologia Agraria dell' Università di Padova, Italy

Summary

The most important spider mite species on vine are the European red mite P.ulmi , the two-spotted mite T.urticae and the yellow spider mite E. carpini in Europe. T.pacificus and E.willimatei are present in North America.

Among Eriophydae in spring, shoots can be damaged by Eriophyes vitis and Calepitrymerus vitis . The latter species can also damage leaves, stopping shoot growth in summer.

Red varieties attacked by P.ulmi become uniformly bronzed and white ones yellow. This species does not produce webs.

T.urticae E.carpini and T.pacificus first colonise small areas near the veins which become necrotic and covered by webs.

A population of 10 P.ulmi per leaf, for one or two weeks, does not produce economic and "aesthetic" damage. Neither do similar population densities of the other species produce damage, but since they are smaller and difficult to count, the threshold level can be based on first visible damages.

Mite infestations are linked to the reduction of natural enemies produced by pesticides. In northern European vineyards, the side-effects of fungicides (Dithiocarbammates and Dinocap) on predacious mites (Phytoseiidae and Stigmaeidae) results the most important cause of imbalance.

The release of predacious mites on vineyards allows a satisfactory biological control of spider mites. Pruned twigs can be used to transport predators to new vineyards. The best time results the end of winter for Kampimodromus aberrans (Oud), commonly used in the commercial vineyards of North Italy.

1 PHYTOPHAGOUS MITES IN VINEYARDS

The species of economic importance vary in different parts of the world (Tab.1). Only P.ulmi can be considered a world wide problem, E.willamettei and T.pacificus are present only in N. America, according to their geographical distribution; T.urticae even though found all over the world is an important pest only in Western Europe.

1.1 Fam. Eriophydae

Eriophyes (= Colomerus) vitis (Pagenst.)
This species is known as the Grape Er.ineum Mite which produces "blisters" on the upper surface of leaves. On the other side of the blisters the hairs multiply abnormally and form a thick, whitish, furry spot called erinea. E.vitis has many generations per year but only in spring are the symptoms on leaves common. The damage is

Tab.1 Most common species of economic interest found in vineyards in different parts of the world.

SPECIES	ATTACKED ORGANS	GEOGRAPHICAL DISTRIBUTION
(FAM. ERIOPHYDAE)		
ERIOPHYES VITIS	SHOOTS AND LEAVES	WORLD WIDE
CALEPITRIMERUS VITIS	" "	WORLD WIDE
(FAM. TETRANYCHIDAE)		
PANONYCHUS ULMI	ALL THE LEF	WORLDWIDE (IN COOLER VINEYARDS)
TETRANYCHUS URTICAE	LEAVES NEAR VEINS	(WORLDWIDE) SPAIN FRANCE SWITZERLAND GERMANY
TETRANYCHUS PACIFICUS	" "	CALIFORNIA
EOTETRANYCHUS CARPINI	" "	ITALY, FRANCE, SWITZERLAND (IN WARMER VINEYARDS)
EOTETRANYCHUS WILLAMETTEI	ALL THE LEAF	CALIFORNIA

basically aesthetic. The mites overwinter inside the buds under the perules. A large number of E. vitis inside buds in spring can exceptionally cause shortened internodes, curled leaves and blossom destruction.

Calepitrimerus vitis (Nal).
This Eriophyid, different from E.vitis , overwinters outside the bud scales near the base. In spring, it can produce shortened internodes and blossom destruction. In summer it causes numerous whitish spots all over the leaf surface; in the middle of the whitish spot a small brown necrosis is visible. If the population is high the leaf may become blackish or dark purple (4). Following this, the leaves fall. This event only takes place in the absence of predators in oversprayed vineyards. Damage done by C.vitis in summer can be mistaken for the necrosis produced by the thrips Drepanothrips reuteri (Uz.). however, the necrosis of this thrips are wider and are more numerous along veins.

1.2. Fam. Tetranychidae

Panonychus ulmi (Koch)
This species attacks vine and fruit trees all over the world. Its red colour gave it the name of European red mite. The eggs overwinter on wood under the bark or even on one year old wood not protected by bark. The winter egg is red and onion-shaped, the summer eggs laid on leaves are pale red and roundish. Winter eggs hatch when

186

first leaves appear. The new mites reach maturity when the shoots have five or six leaves. During the year 6 - 9 generations occur.

Damages of P.ulmi are essentially linked to numerous small necrotic areas, produced by mite feeding, on all the leaf surface. On the first infested spring leaves, not yet fully open, reddish spots appear. In summer, heavily infested red varieties become uniformly "bronzed" and white ones "yellow". Prolonged infestation causes leaves to fall.

A spring and summer population of 10 active mites per leaf does not cause visible damage for at least one week. However, 15-20 mites per leaf can produce some bronzed leaves (the data reported in Fig.1 is in agreement with this evaluation). The threshold level to prevent "aesthetic" damages has been fixed at 10 mites/leaf (12). In August - September the threshold can be twice as high. The first spring leaves are occupied by numerous newly hatched mites from winter eggs; the shoots grow rapidly without an increase in the immmature mite population and therefore the young mites dilute on the increasing number of leaves.The young shoots can therefore support up to 30 mites per leaf without damage.

Vines with 'bronzed' but not shrivelled leaves give a normal yield in both quantity and sugar content. Miticides can be used when most of the vines have some 'bronzed' leaves if farmers are willing to accept some aesthetic damage.

Eotetranychus carpini (Oud.)

This oligophagous species, different to P.ulmi and T.urticae , is found on vines but not on other fruit trees. It causes damage in France, Switzerland, Germany and central and northern Italy. This species prefers warmer and less humid climates than P.ulmi . The mature females overwinter under the bark. Seven or eight generations per year can occur.

E.carpini show aggregative behaviour and first colonise small areas near the veins on lower leaf surface. The infested area is covered with a few silk threads (less abundant than for T.urticae), and becomes red in red varieties or yellow-brown in white ones. Serious infestations cause all the leaf to turn brown and fall down; this happens more often than in P.ulmi .

Small colonies on first leaves of shoots do not cause any damage to the yield but because the necrotic areas can be seen from a distance, sometimes farmers are not willing to accept any aesthetic damage. This can be prevented by treating when a mean of about 10 mites per leaf are present. However, E.carpini is very difficult to count even under a lens; therefore, contrary to P.ulmi , a threshold level based on the number of mites per leaf is not practical. A close relationship between leaves occupied by mites and necrosis was found. (3)

The distribution of E.carpini populations, different to P.ulmi , is very irregular even along the same row, so many samples should be taken to predict the population levels.

For this reason, the appearance of necrotic spots near veins, on the base leaf in spring and on the one opposite to the bunch in summer, of 30% of shoots can be considered an indicative "aesthetic" threshold (in absence of predators). Only visibly attacked vines can be sprayed.

It is most likely that, similar to T.urticae , only attacks which cause leaves to fall can be considered really damaging to grape quality. The economic threshold is very high.

Tetranychus   urticae   (Koch)
     The "two spotted spider mite" has a world-wide distribution and
attacks   nearly all cultivated plants. On vine it is very damaging in
Spain (2) and sometimes in France, Switzerland and West Germany. In
Italy it can be found on vine near infested maize. Mites leave the
maize when it dries and colonies can be found on vine leaves;
however, this infestation rapidly decreases and the mites disappear.
     T.urticae   unlike   E.carpini   can live on grass or weeds in
vineyards and produces large amounts of web. However, its life cycle,
behaviour and damage on vines are similar to E.carpini .
     Damage to harvest has been evaluated in Spain : each week of
defoliation of vines before harvest corresponds to a reduction in
grape sugar content of 1/4 degree Baume' (1), or 0.25 degree Baume
every 10% of defoliation at harvest (2). A relation was found between
the attacked leaves, their position on shoots, the season and the
probability of leaf shedding. The Spaniards advise an action
threshold when symptoms of attack are present on the 5% of vines in
May, 25% in June, 40% in July. (2)
     The damage evaluation is probably valid for vines in all
countries and other mite species, but the proposed action threshold
cannot be applied everywhere because it was established in vineyards
where both mite and insect predators are absent.

Tetranychus   pacificus McGregor
Eotetranychus   williamettei Ewing
     These species attack vines in California.   The life cycle,
behaviour and damages of T.pacificus are similar to that of
T.urticae and E.carpini (10). E.williamettei tends to disperse
over all the leaf surface more than the corresponding European
species and therefore infestations cause a yellowing of the entire
leaf. Biological control factors of this species are well-known
since the 1970's (10)

2. DISTRIBUTION OF MITES AND SAMPLING MATHODS

     P.ulmi is more frequent in spring on shoots near the trunk,
where most winter eggs are laid. In summer there are no longer any
significant differences between shoots. Inside shoots, the
population is higher on the base leaves in spring and early summmer
and on top leaves later in the season. The central leaves have a
population which normally does not present significant differences
between both base and top leaves.
     In general P.ulmi show no significant differences in vines of
the same row and of different rows in vineyards of the same
varieties, treated in the same way, of the same age and on the same
kind of soil (i.e. in homogeneous vineyards). The frequency
distribution of P.ulmi in homogeneous vineyards, relative to the
central leaves of shoots of different vines, fits the negative
binomial distribution regardless of the season and density (14).
     Based on the relatively high threshold level of 10 mites per
leaf and on a known frequency distribution which is not highly
contagious, a rapid sampling method was proposed for P.ulmi
populations to decide if treatment is necessary or not (14) (Tab.
2).
     If the cumulative number of mites on the observed leaves (left
column) is inferior to the number reported in the central column, the

Tab. 2. Sequential sampling method for deciding to treat or not, i.e. if the threshold of 10 mites leaf is reached or not in a homogeneous vineyard.

Observed Leaves	Cumulative	Number of	P. ulmi
(number)	NOT TREAT	(Observe other leaves)	TREAT
1	-		> 34
2	< 2		>52
3	< 6		>68
4	< 11		>83
5	< 17		>97
6	< 23		>111
7	< 30		>125
8	< 36		>138
9	< 43		>151
10	< 50		>189

threshold level is not reached; if the number is superior to the right column, the threshold is exceeded.

For example, if on 4 leaves less than 11 mites are found, there is less than a 1% probability that 10 mites per leaf or more are present.

(Confidence limits (1%) for increasing sample size for a mean = 10 and an aggregative coefficent K = 2 according to the negative binomial distribution.)

E.carpini and T.urticae show an irregular and more contagious distribution than P.ulmi . Significant differences in E.carpini populations are frequently found between vines of the same row. The small size and yellowish colouring of E.carpini and T.urticae makes counting difficult. The early appearance of small, visible necrosis on leaves allows a rapid indirect evaluation of population levels. An "aesthetic" threshold may be 30% of leaves with some spots. Localised treatment only on vines visibly attacked is possible. The economic threshold has already been reported for T.urticae (2).

3. PREDATORS OF MITES IN VINEYARDS

3.1 Predacious mites

The most important predators of phytophagous mites are predacious mites of the Phytoseiidae family: Kampimodromus aberrans (Oud), Amblyseius andersoni (Chant), and Typhlodromus pyri Scheuten in Europe (North Italy) and Metaseiulus occidentalis (Nesbitt) in America. Even in the Stigmaeidae family there are important predacious species not as well-known as the above-mentioned phytoseiids. A list of predatory mite species was given for America (10) and Europe (20) (22) (at this meeting).

3.2 Predacious insects

When mite populations increase winged predacious insects appear on attacked vines and can rapidly reduce mites. In north-east Italy

the more important insects are Anthocoridae , (pirate bugs) Orius
vicinus (Ribaut) and O.majusculus (Reuter) (6) which appear on
vines infested by a minimum of 3 - 5 P.ulmi per leaf and is an
important control factor in the absence of phytoseiids. Other
important species can be thrips (Haplothrips subtilissimus) (
Diptera Cecidomyiidae) and Coleoptera Coccinellidae (
Stethorus sp.), Staphylinidae (Oligota sp.) this last species
arrives when high densities (20 mites per leaf) are reached (12). In
America thrips Scolothrips sexmaculatus (Pergande) and the minute
pirate bug Orius tristicolor White (10) have been reported on
vine. The predators' arrival is not always synchronous with
population increase and some damage may be produced in vineyards
where Orius were able to reduce mite populations in previous
generations, indepent from the use of insecticides (8).

4. CAUSES OF MITE POPULATION IMBALANCE

Mites have become a problem of economic importance in the years
following World War II when the first synthetic insecticides and
fungicides were used.
As far as copper salts, sulphur and lead arseniate were employed
in vineyards spider mites were only of zoological interest.
Two theories developed to explain the mite imbalance. The first
one considered the elimination of predators due to pesticides the
cause of infestations. The second one (Physiological inducement or
trophobiotic theory) (5) considered mite imbalance was due to
fecundity increase linked to the modified plant metabolism, i.e. a
larger amount of sugars and amino acids in leaf juice; in Europe this
theory was generally accepted.
The importance of predators and particularly phytoseiids
Metaseiulus occidentalis (Nesbitt) in lowering population of
phytophagous mites, was clearly demonstrated in Californian vineyards
in the 1960's where the main cause of imbalance was the use of
insecticides which killed phytoseiid populations (10).
The importance of Phytoseids in phytophagous mite control was
considered in Europe by Mathys (17) Rambier (19) and Gambaro (15).
and many authors towards the end of the 1970's. However even though
the throphobiotic theory had been long discarded (11) it was
largely accepted in Southern Europe and the role of Phytoseiids
undervalued up to 1980 e.g. ditiocarbammates are considered
"freinent les pullulations: de P.ulmi E.carpini " while copper
salts are "Favorisent les pullulations" at low dosage (21) in OILB
reports.
As the predators were not taken into account, fungicides toxic
to mites were recommended, above all, dithiocarbammates and mancozeb
even if these products were also toxic to phytoseiids.
Traditional copper salts instead were not recommended as they
were considered a cause of mite imbalance, (21) but Gambaro (15)
considered copper to be a guarantee for preventing mite infestation.
Both the contrasting theories were based on experimental data and in
many vineyards the use of mancozeb was sufficient to prevent mite
infestation while copper salts had no effect. Therefore, the use of
copper salts and sulphur, the only fungicide used before the
appearance of mite infestation, was not, per se, a guarantee of
biological balance.
The different points of view were integrated in a general vision
of the problem by means of a simple experimentation (Fig.1).

In a vineyard where phytoseiids had been eliminated by previous use of pesticides, they were released on vine treated with the four following combinations: two fungicides against Grape Downy Mildew, Plasmopara viticola (Mancozeb or Copper) and two fungicides against the Grape Powdery Mildew Uncinula necator (Dinocap or Sulphur). Mancozeb was considered mite refraining and Dinocap is also a miticide. For each of the four fungicide combinations phytoseiids were released on some vines and not others.

Where copper and sulphur were used, and predacious mites were not introduced, the phytophagous mite population increased and leaves became bronzed.

The same fungicides on vines where predacious mites were released allowed a good biological control and phytophagous populations were low.

Where mancozeb and sulphur were used, phytoseiids were killed and no difference was found between vines with or without released phytoseiids, since phytophagous mites show some resistance to mancozeb, the population increases and vines become bronzed. Where dinocap was used either with mancozeb or copper both phytoseiids and phytophagous mites were killed and a good chemical control was reached (9) (13).

Therefore, the supposed stimulatory or inhibitory activity of copper or mancozeb has been explained by the presence or absence of phytoseiids and a selective toxicity of mancozeb on phytoseiids only.

These results are interesting because they demonstrate the importance of side effects of fungicides on phytophagous mites imbalance. It is noted that the "throphobiosis" theory was mostly supported by comparing data on the side effects of copper and dithiocarbammates on vines and ignoring the activity of phytoseiids (5) therefore this theory, already questionable for side effects of insecticides, (above all carbaryl), (10) now lacks experimental background.

5. TOXICITY OF FUNGICIDES ON PREDACEOUS MITES

5.1. Side effects of dithiocarbammates
The above-mentioned date refers to P.ulmi and T.aberrans but similar results have been obtained on E.carpini and other Phytoseiid species.

Ditiocarbammates are widely used on northern vineyards where Grape Downy Mildew is damaging. In the same vineyards P.ulmi is an economical problem. In southern vineyards P.ulmi is not a problem, at least in Italy. Only recently has an infestation been found in Sicilian plains vineyards; this is probably due to the use of pyrethroids for the control of grape moths (Vacante Personal communication).

All dithiocarbammates used in Italian vineyards result toxic to phytoseiid mites (Fig.2 and 3) and particularly to K.aberrans the populations of which do not seem resistant (in North Italy) to pesticides and this species disappears on vineyards (and orchards) where ditiocarbammates are used. A.andersoni can survive on vineyards sprayed with ditiocarbammates (7). (A.andersoni is also found on heavily sprayed apple trees). Ditiocarbammates also resulted toxic to T.pyri populations on vine.

The toxcity of fungicides on phytoseiids has been known since the 1940's (16) and the side effects of dithiocarbammates known since the 1950's (17).

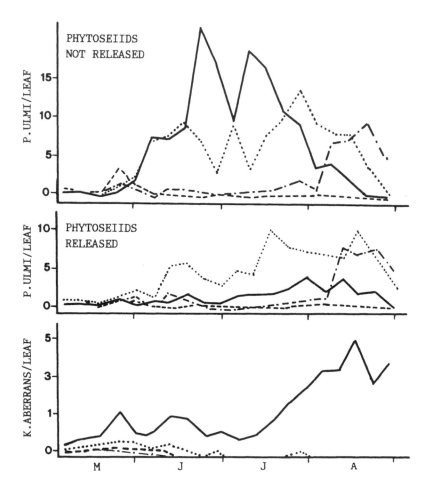

Fig. 1. Seasonal variations of European red mite populations on vines with or without released phytoseiids and treated with four different fungicide combinations: copper salts or mancozeb to control Grape Downy Mildew, and sulphur or dinocap to control Grape Powdery Mildew.

a) Where phytoseiids have not been released (top figure) P.ulmi populations increased mostly on copper and sulphur, a little less on mancozeb and sulphur. Where dinocap was used populations were chemically controlled (dinocap and sulphur show an increase at the end of the season when spraying is over).

b) Where phytoseiids were released (centre figure) dinocap and mancozeb gave the same results as above but copper and sulphur allowed a good control of mite populations.

c) Phytoseiids were destroyed by repeated sprays of mancozeb and dinocap. Sulphur and copper allow a good phytoseiid survival (lower figure) while guaranteeing a biological control of P.ulmi (centre figure).

Dinocap has destroyed both phytophagous and predacious mites, mancozeb, selectively toxic on phytoseiids only, caused an increase in P.ulmi populations even where phytoseiids were released (9).

192

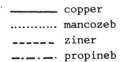

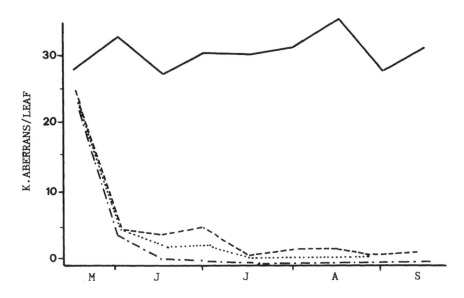

Fig. 2. The influence of different dithiocarbammates normally used in viticulture on the phytoseiid populations (K.aberrans), The vineyard was not trated with dithiocarbammates in previous years. The toxicity of the products was compared with copper salts. All dithiocarbammates resulted toxic to phytoseiids. Similar results have been obtained on populations of A.andersoni . The toxicity of dithiocarbammates on phytoseiids is evident.

5.2. Field evaluation of side effects on fungicides

The side-effects on phytoseiids of the most used fungicides in Italian vineyards have been field tested by comparing a well-known toxic product (mostly mancozeb) with a non toxic one (mostly copper salts); (sulphur was used in both cases to control Grape Powdery Mildew). Mite populations were counted every week in the different plots; as far as possible, to avoid effects linked to resistant populations, the products used had not been employed in the same vineyards in previous years. In this way, (fig. 1,2,3) it was possible to establish that in order to permit phytoseiid survival the use of dithiocarbammates (to control Grape Downy Mildew) and dinocap (to control Grape Powder Mildew) should be avoided and that old fungicides and new ones, even penetrating ones, do not kill predacious mites. Also the mixtures of dithiocarbammates, in this case used in low dosages, are not very dangerous for mite balance. Field research is in perfect agrement with recent laboratorial results. Therefore, the importance of fungicides in mite imbalance is well-known and there are no more contrasting points of view on side effects of different products.

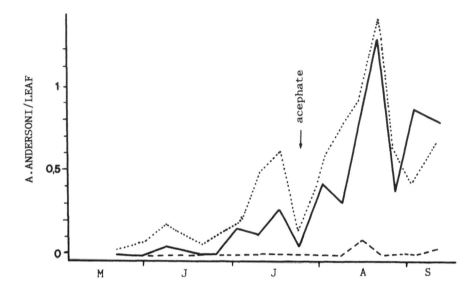

Fig .3. The influence of some penetrating fungicides on phytoseiids (A.andersoni) compared to copper salts and mancozeb. The modern fungicides usually have a low toxicity on phytoseiids. It is noted that a single insecticide spray (acephate) at mid July, having lowered phytoseiid populations, is not as destructive as repeated fungicide treatments with mancozeb.

Legend (top of figure):
———— copper-cymoxail + sulphur
·············· phosetil-al + triamedifon
— — — — mancozeb + sulphur

These results are not only restricted to vineyards and the side effects of widely used dithiocarbammates on mite imbalance has probably been underestimated in orchards where insecticides are widely used. In vineyards one spray per year is normally sufficient, (at least on wine grapes in northern Italy) (18) For this reason the side effects of fungicides are more evident in vineyards.

Furthermore, any field evaluation of side effects of a certain pesticide is questionable without taking into account fungicidal treatments carried out.

In vineyards it can be easily demonstrated (Fig.3) that some insecticidal sprays are less dangerous than repeated fungicide treatments, even if one spray only is not per se destructive.

6. BIOLOGICAL CONTROL

6.1. Phytoseiid release
In vineyards where Phytoseiids are absent, K.aberrans can be introduced on vines during the winter period.

Newly pruned vine branches from vineyards where Phytoseiids are abundant may be used. These branches must have an average of at least 5 overwintering females under the bark at the level of the ring of attachment of 1 year wood to 2 year wood. One half meter branch must be bound to each vine without mites and already pruned. Newly pruned wood must be used to prevent detachment of dried bark which is then abandoned by the phytoseiids. Only when it is already attached to the vine should the bark be abandoned by the phytoseiids.

The release of _Kampimodromus_ _aberrans_ is particularly simple; if not refrained by toxic pesticides, a complete and permanent biological control is guaranteed. Attention must be paid to the initial period, because a permanent balance can be reached when _K.Aberrans_ are more abundant than their victims. An initial ratio of one phytoseiid every 7-10 spider mites may be sufficient for establishing a balance. This ratio does not result sufficient in warmer summers, when temperatures reach over 30 degrees C.

Contrary to _K.aberrans_ , winter release of _A.andersoni_ has not always been successful, probably due to the fact that this species can migrate to weeds. However, where a balance is reached this last species keeps victims at low levels even if relatively less numerous than _K.aberrans_ (Gambaro personal communication). In some vineyards _A.andersoni_ can be found even if dithiocarbammates are used, some populations are probably resistant. It is possible to release this species in summer by collecting branches with leaves from peach or apple trees (unpublished data). This is probably the best method of release for this species and _T.pyri_ too.

Released phytoseiids (_K.aberrans_) control both _P.ulmi_ and _E.carpini_ . Also the Eriophyid mite _C.vitis_ does not damage leaves if phytoseiids are present and only rare necrotic spots can be found.

The balance will withstand one or two insecticide treatments used to control Grape moths, as long as toxic fungicides are not repeated too often. It is possible to state that, today, at least in Italy, mite infested vineyards are due to bad management.

REFERENCES

1 ARIAS A., 1981. Acarien jaune tisserand (Tetranychus urticae Koch). Rapport IV Réunion plénaire OILB, "Lutte integrée en viticulture". Gargnano, Mar.1981. Boll.Zool. agr.Bachic.(Ser.II), 16: 38-41.

2 ARIAS GIRALDA A., NIETO CALDERON J., 1983. Estimacion de las perdidas producidas por la "arana amarilla comun" (Tetranychus urticae Koch) en "Tierra de Barros" (Badajoz) y propuesta de un umbral de tolerancia economica. Bol. Serv. Plagas, 9: 227-252.

3 BAILLOD M., 1984 . Etude de distribution de population et symptomes de l' acarien jaune comun. Proc. Integrated Control in Viticulture, Vth Plenary Session. Bulletin SROP 1984/VII/2, p.19.

4 CIAMPOLINI M., ROTA P.A., CAPELLA A., LUGARESI C., 1984. Sensibile aumento della "acarosi" nella viticultura italiana. Informatore Agrario XL (32): 31-36.

5 CHABOUSSOU F., 1965. La multiplication par voie trophique
   des tétraniques a' la suite des traitements pesticides.
   Relation avec les phénomènes de résistence acquise.
   Boll. Zool. agr. Bachic., 7: 144-184.

6 DUSO C., GIROLAMI V., 1982. Ruolo degli Antocoridi nel
   controllo del Panonychus ulmi Koch nei vigneti.Boll.Ist.
   Entomol. Bologna, 37: 157-169.

7 DUSO C., LIGUORI M., 1984. Ricerche sugli Acari della vite
   nel Veneto:aspetti faunistici ed incidenza degli inteventi
   fitosanitari sulle popolazioni di acari fitofagi
   e predatori. Redia, 67: 337-353.

8 DUSO C., GIROLAMI V., 1985. Strategie di controllo biologico
   degli Acari Tetranichidi su vite. Atti XIV Congresso Naz.
   Entomol., Palermo :719-728.

9 DUSO C., GIROLAMI V., BORGO M., EGGER E., 1983. Influenza
   di anticrittogamici diversi sulla sopravvivenza di predatori
   Fitoseidi introdotti su vite. Redia, 66:469-483

10 FLAHERTY D. L., HUFFAKER C.B., 1970. Biological control of
   Pacific mites and Willamette mites in San Joaquin Valley
   vineyards. I. Role of Metaseiulus   occidentalis .
   II. Influence of dispersion patterns of Metaseiulus
   occidentalis . Hilgardia, 40 : 267-330.

11 HUFFAKER C.B., VAN DE VRIE M., MC MURTRY J.A.. 1970. Tetra-
   nichid populations and their possible control by predators:
   an evaluation. Hilgardia, 40: 331-389.

12 GIROLAMI V., 1981. Danni, soglie di intervento, controllo
   degli acari della vite. Proc. Incontro Difesa Integrata
   della Vite, 3-4 December/1981, Latina (Italy).
   Regione Lazio ed.(1983). pp. 111-143.

13 GIROLAMI V., DUSO C., 1984. Ruolo positivo del rame nelle
   strategie di controllo biologico degli acari della vite.
   Vignevini, 5: 90-94.

14 GIROLAMI V., MOZZI A., 1984. Distribution, economic thresholds
   and sampling methods of Panonychus ulmi (Koch) pp.89-101.
   In CAVALLORO R.,1984, Statistical and Mathematical Methods in
   Population Dynamics and Pest Control.
   A.A.BALKEMA, ROTTERDAM, BOSTON. pp. 243.

15 GAMBARO P., 1972. Il ruolo del Typhlodromus   aberrans Oud.
   (Acarina Phytoseiidae) nel controllo bilogico degli Acari
   fitofagi del Veronese. Boll.Zool.agr.Bachic.II, 11: 151-165.

16 LORD F.T., 1949 - The influence of spray programs on the
   fauna of apple orchards in Nova Scotia.III. Mites and their
   Predators. Can.Ent., 81: 217-230.

17 MATHYS G., 1958. The control of phytophagous mites in Swiss
   vineyards by Typhloromus species. Proc.10th Int. Entomol.
   Congr., 4: 607-610.

18 PAVAN F., 1985. Damage evolution,larval sampling and treatment
   period for grape moth (at this meeting).

19 RAMBIER A., 1958. Les Tétraniques nuisibles " la vigne en
   France continentale. Revue Zool.agr.Appl., 3: 1-20.

20 SCHRUFT G.A., 1985. Biological control of spider mites in
   viticulture. (at this meeting).

21 TOUZEAU J., 1981. Travaux du sous-groupe "effets secondaires".Rapport réunion plenaire OILB, Lutte integrée en
   viticulture, Gargnano, Boll.Zool.agr.Bachic., 16: 5.

22 VACCANTE S. 1985. Gli acari della vite in Sicilia:
   I contributo (at this meeting).

Mites of the grape-vine in Tuscany

M.Castagnoli & M.Liguori
Istituto Sperimentale per la Zoologia Agraria, Firenze, Italy

Summary

The mite fauna was studied in numerous samples from different viticultur
al areas of Tuscany. The known data on the biological characteristics of
the predatory mites are discussed in order to point out the research
still needed for their better use in integrated control programs.

1. Introduction

Over the past thirty years phytophagous mites have become an ever-increasing problem in vineyards pest management. Many of the new products used
to control pests and diseases have contributed to altering the existing relationship between phytophagous mites and their natural predators.

We have been examining mites of the grapevine more or less continually
for many years in the more important viticultural areas in Tuscany to ascer
tain the presence and the distribution of already identified phytophagous
species, and above all to understand which Phytoseiids, their predators, are
the most common.

2. Phytophagous Mites

As in other Italian regions *Eotetranychus carpini* (Oud.) and *Panonychus ulmi* (Koch) are the most important and widely distributed of the Tetranychid species in Tuscany. They may also be found together, although they
may differ in certain aspects of biological behaviour, as, for example, the
way they overwinter, the way they protect themselves under a web of silklike
threads, the way they distribute themselves on leaves and cause damage. *E.
carpini* is capable of causing evident symptoms on leaves at a lower density
than that needed by *P. ulmi*. In the case of heavy *E. carpini* infestation
necrotic areas more or less run into each other, whereas with large *P. ulmi*
populations the leaves appear more uniformly yellowed or reddened. The difference is also evident at a histological level: *E. carpini* attack causes
changes in lacunous tissue (LIGUORI, unpub.), that of *P. ulmi* in palisade
tissue (23). More recent sampling in Tuscany, and particularly in the Chian
ti area, has shown that *E. carpini* predominates over *P. ulmi* , while *Tetranychus urticae* Koch infestation is more sporadic.

Another group of phytophagous mites which may be of economic importance is that of the Eriophyids. *Calepitrimerus vitis* (Nal.) and *Eriophyes vitis* (Pgst.) have different biological behaviour. The first species is vagrant and deuterogynous, and the latter causes erineum, and at least morphologically, has only one kind of female. However, both of them overwinter between the scales of the buds, so that when the vegetative season restarts, if populations are already high, they cause the greatest damage by stunting the buds, shortening the internodes and deforming the young leaves. In Tuscany, as in other regions, there has been a remarkable increase in *Calepitrimerus vitis* populations, which at the height of summer reach a very high density (even more than 500 per leaf) causing the typical morphological and chromatic changes on the shoots and leaves. Other Eriophyids belonging to different genera have been collected, though more rarely, in this region. However, their precise identity and the damage they may cause have yet to be defined. On rare occasions another phytophagous mite has been found, a Tenuipalpid, *Hystripalpus lewisi* (McGregor). However, it does not appear to cause any damage.

3. Predatory Mites

In Tuscany the most common predatory mites on the vine are Phytoseiids, while Stigmaeids are only found occasionally.

The distribution and the frequency of the 12 species of Phytoseiids collected in Tuscan vineyards is shown in Table I. Many of these species (*Amblyseius andersoni* (Chant), *A. finlandicus* (Oud.), *Phytoseius horridus* Ribaga, *Thyphlodromus athenas* Swirski and Ragusa, *T. phialatus* Athias-Henriot, *T. talbii* Athias-Henriot, *T. triporus* Chant) have been found sporadically, and in very small numbers. *Typhlodromus kerkirae* (Swirski and Ragusa) is fairly well distributed, even if numbers are low, while *Phytoseius plumifer* (Can. and Fanz.), sensu Chant and Athias-Henriot, 1960, is sometimes found in abundance in coastal areas. *Kampimodromus aberrans* (Oud.), *Typhlodromus pyri* Scheuten, sensu Abbasova, 1970, and *Typhlodromus exhilaratus* Ragusa are the species which merit closer attention as they have been found constantly on the vine, and in reasonable numbers.

K. aberrans is common in vineyards in Switzerland (5), France (25) and various Italian regions (13,16,20,21,23). In the laboratory (11) it has been reared on *Eriophyes vitis* ; in the field (16) it feeds on eggs and immature stages of *E. carpini* and *P. ulmi*. A fair amount is known about its ecology and its biology, but less about the species' ability to develop resistance to the pesticides most commonly used in viticulture. In our regions it overwinters as a fertilised female adult, but in warmer climates males have also been found in winter (31). In vineyards in the Veneto region (16) *K. aberrans* has been shown to have a similar distribution to that of its prey, and to be able to contain this at low population levels. Moreover, it has appeared more suitable than other species (12) for reintroducing into vineyards with a low density of Phytoseiids by using pruned twigs from other more densely populated vineyards. It is important to note that in Tuscany this Phytoseiid was more widely distributed both in specialised and non-specialised vine-

Table I

Distribution and frequency of Phytoseiid species in Tuscan provinces

Species	Distribution	No.sample	Pop.density
Amblyseius			
andersoni	Siena	2	-
finlandicus	Firenze	1	-
Kampimodromus			
aberrans	Firenze,Siena,Pisa Pistoia,Grosseto	24	+ +
Phytoseius			
horridus	Siena	2	-
plumifer	Siena,Pistoia,Grosseto, Livorno	13	+
Typhlodromus			
athenas	Siena,Pistoia	2	-
exhilaratus	Firenze,Siena,Pisa, Pistoia,Livorno	29	+ +
kerkirae	Siena,Pisa,Pistoia, Livorno,Grosseto	12	+
phialatus	Firenze,Pistoia	3	-
pyri	Firenze,Siena	20	+ +
talbii	Siena,Pistoia	2	-
triporus	Siena	1	-

- low   + medium   + + high

yards at the beginning of the 70s. More recent sampling indicates that *K. aberrans* is less common, and perhaps it has been replaced by other species which are more resistant to the pesticides at present being used in viti-culture.

Unlike *K. aberrans*, *T. pyri* is also widely found on fruit trees. There are at least two different systematic interpretations (1,2,9) of this spe-

cies which complicate the evaluation of the numerous known biological and
ecological data. It has been the subject of research on the vine particular
ly in Switzerland (3,5,6,7) where it acts as a good predator towards *P. ulmi*, and like *K. aberrans*, it lends itself to being transferred from one
vineyard to another on pruned twigs. Observations in the field and in the
laboratory (6) have shown that it feeds on *T. urticae* and *P. ulmi* (not the
eggs), on Eriophyids and more rarely on Tydeids, as well as pollen and
plant juices. Its ability to spread has not been studied sufficiently, and
its distribution does not always coincide perfectly with that of its prey.
It may compensate for this, like many other Phytoseiids, on account of its
good active research ability. The speed at wich it develops is not high in
comparison with other Phytoseiids that are considered good Tetranychid predators (15,29). However, it is sufficiently higher than that of *P. ulmi* ,
and is compensated for by its appetite. Moreover, *T. pyri* appears particular
ly suitable for use in integrated control programmes on account of its abil
ity to develop strains which are resistant to some organophosphorus compounds (8,10,24). In Tuscan vineyards, however, it has only recently been
establishing itself in more internal areas of the region where it is found
with reasonablysized populations.

   T. exhilaratus has been described in Sicily on various spontaneous and
cultivated plants, including the vine, but up to now it had appeared as established only in citrus orchards (26,27). In the past it has been found in
various areas of Tuscany, but generally not with a high number of specimens
(20). Over recent years the species has become more frequent and often populations have been associated with *P. ulmi* , *E. carpini*, *C.vitis* in vineyards in full production. The dynamics of its populations have been followed in the 'Chianti Classico' area. The graph refers to samplings made in
1982 and 1983 at Gaiole (Siena) in a specialised vineyard where fungicide
treatments had been carried out using Mancozeb and Copper oxychloride in
the first year, and in the second year Zineb, Sulphur, Cymoxanil and Triadimefon together with Mancozeb. In this vineyard there was a contained pres
ence of Tetranychids (*P. ulmi* and *E. carpini*), while populations of *C.
vitis* were more significant, more so in 1982 (late summer about 180 specimens per leaf) than in 1983 (about 40 specimens per leaf at the same period). *T. exhilaratus* is already present when the vine starts growing again;
it diminishes with the increase in vegetation, only to rise again in late
summer when phytophagous populations are at their height. In the vineyard
under consideration this development is particularly evident in 1983, when
in September/October it reached a density of 5.7 specimens per leaf, in
spite of the greater variety of pesticides used against downy mildew and
powdery mildew. This may indicate, together with the wider distribution
recorded in Tuscan vineyards in recent years, that *T. exhilaratus* is capable
of developing strands that are resistant to certain pesticides and/or of
reacting quickly to environmental changes. The possibility of using it ef-
fectively in Tuscany in integrated control programmes in viticulture, however, is dependent on a deeper knowledge of its biological characteristics.
Postembrionic survival, the rate of oviposition, the length of the immature
stages of *T. exhilaratus* have been studied in the laboratory (27,28) with

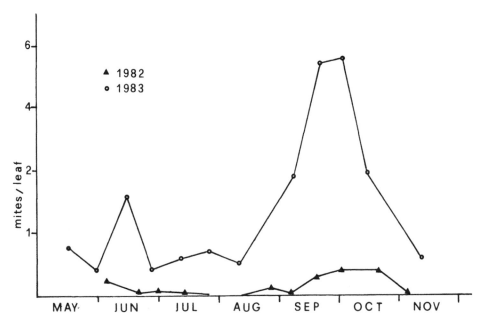

Fig. I - Population trends of *Typhlodromus exhilaratus* in specialised vineyard (Gaiole - Siena).

various kinds of food (*Panonychus citri* (McGregor), *T. urticae*,some pol-
lens). As yet no data are avaiable either on its behaviour in the vineyards or on what are its preferred prey in that environment. Larger populations of this Phytoseiid have been found together with large Eriophyid
populations.

4. Other Mites

As with other cultivations a fair number of Tydeid populations are also found on the vine.The greatest number of specimens and the greatest variety of species are to be found in non -specialised vineyards.The species so
far identified are *Orthotydeus caudatus* (Dugés), *O. kochi* (Oud.),*Pronematus
ubiquitus* (McGregor) and *Homeopronematus anconai* (Baker). The role of these
species in the biocenosis of the vine is still not clear:next to references
to damage (22,30), there are others in which they are indicated as active
predators of Eriophyids (30). In Californian vineyards (18,19), when selective pesticides have been used, *H. anconai* , not only predates Tetranychid
eggs, but also plays an important role as an alternative prey for *Metaseiulus occidentalis* (Nesbitt) when phytophagous species are lacking. In
fact, *H. anconai* shows a high sensibility to sulphur and other pesticides
used in viticulture. In Italy information on Tydeids of the grapevine are
rather fragmentary; the most common species and their biological characteristic are still to be defined. Precisely because they are mites with no

specialised feeding habits, Tydeids play an important role in maintaining a balance between phytophagous mites and predators; and this importance is not to be underestimated in integrated control programmes.

REFERENCES

1. ABBASOVA, E.D. (1970). Little known and new species of predatory mites from the Phytoseiidae of the fauna of Azerbaijan. Zool. Zh.,XLIX:45-55 (in Russian).
2. ABBASOVA, E.D. (1980). Genus *Typhlodromus* (Parassitiformes,Phytoseiidae) in Azerbaijan. Zool. Zh.,LIX: 830-837(in Russian).
3. BAILLOD, M. (1984). Lutte biologique contre les acariens phytophages. Rev. Suisse Vitic. Arboric. Hortic., 16(3): 137-142.
4. BAILLOD, M., BASSINO, J.P. and PIGANEAU, P. (1979). L'estimation du risque provoqué par l'acarien rouge (*Panonychus ulmi*) et l'acarien de charmilles (*Eotetranychus carpini* Oud.) en viticulture. Rev. Suisse Vitic. Arboric. Hortic., 11(3): 123-130.
5. BAILLOD, M. and VENTURI, I. (1980). Lutte biologique contre l'acarien rouge en viticulture. I. Répartition, distribution et méthode de contrôle des populations de prédateurs typhlodromes. Rev. Suisse Vitic. Arboric. Hortic., 12(5): 231-238.
6. BAILLOD, M. et al. (1982). Lutte biologique contre l'acarien rouge en viticulture. II. Equilibres naturels, dynamique des populations et expe riences de lâcher de typhlodromes. Rev. Suisse Vitic. Arboric. Hortic., 14(6): 345-352.
7. BAILLOD, M. and GUIGNARD, E. (1984). Résistance de *Typhlodromus pyri* Scheuten à l'azimphos et lutte biologique contre les acariens phytophages en arboriculture. Rev. Suisse Vitic. Arboric. Hortic., 16(3): 155-160.
8. BAILLOD, M. et al. (1985). Essais de lutte biologique en 1984 contre les acariens phytophages en vergers de pommiers, sensibilité et résistance aux insecticides de *Typhlodromus pyri* Scheuten. Rev. Suisse Vitic. Arboric. Hortic.,17(2): 129-135.
9. CHANT, D.A., HANSELL, R.I.C. and YOSHIDA, E. (1974). The genus *Typhlodromus* Scheuten (Acarina: Phytoseiidae) in Canada and Alaska. Can. J. Zool., 52:1265-1291.
10. CROFT, B.A. (1982). Arthropod resistance to insecticides: a key to pest control faillures and successes in North American apple orchards. Ent. Exp. Appl., 31: 88-110.
11. DAFTARI, A. (1979). Studies on feeding, reproduction, and development of *Amblyseius aberrans* (Acarina: Phytoseiidae) on various food substances. Z. ang. Ent., 88: 449-453.
12. DUSO, C. and GIROLAMI, V. (1985). Strategie di controllo biologico degli acari Tetranichidi su vite. Atti XIV Congr. Naz. Ital. Ent.: 719-728.
13. DUSO, C. and LIGUORI, M. (1984). Ricerche sugli Acari della vite nel Veneto: aspetti faunistici ed incidenza degli interventi fitosanitari

sulle popolazioni degli Acari fitofagi e predatori. Redia LXVII:
337-353.

14. GIROLAMI, V. and MOZZI, A. (1983). Distribution, economic threshold
and sampling methods of *Panonychus ulmi* (Koch). Proceedings EC
Experts' Meeting, Parma, 26-28 October 1983: 90-101.

15. HERBERT, H.J. (1961). Influence of various numbers of prey on rate of
development, oviposition and longevity of *Typhlodromus pyri* Scheuten
(Acarina: Phytoseiidae) in the laboratory. Can. Ent., 93: 380-384.

16. IVANCICH-GAMBARO, P. (1973). Il ruolo del *Typhlodromus aberrans* Oudm.
(Acarina: Phytoseiidae) nel controllo biologico degli Acari fitofagi
dei vigneti del veronese. Boll. Zool. agr. bachic. Ser.II,11:151-165.

17. IVANCICH-GAMBARO, P.(1982). Le infestazioni di acari sulla vite vent'
anni dopo. Inf. agrario, XXXVIII(35): 22377-22380.

18. KNOP, N.F. and HOY, M.A. (1983). Factors limiting the utility of
Homeopronematus anconai (Acari: Tydeidae) in integrated pest management in San Joaquin Valley vineyards. J. Econ. Entomol.,76:1181-1186.

19. KNOP, N.F. and HOY, M.A. (1983). Biology of a Tydeid mite,*Homeopronematus anconai* (n.comb.)(Acari: Tydeidae), important in San Joaquin
Valley vineyards. Hilgardia, 51(5): 1-30.

20. LIGUORI, M. (1980). Contributo alla conoscenza degli acari della vite
in Toscana. Redia, LXIII: 407-415.

21. LOZZIA, G.C., NEPOMUCENO, R. and RANCATI, M.A. (1984). Presenza e distribuzione di Acari Fitoseidi in alcuni vigneti lombardi. Vignevini,
11: 31-35.

22. MAL'CHENKOVA, N.I. (1967). A mites of the genus *Tydeus* (Acariformes,
Tydeidae) a pest of the grape vine in Moldavia. Ent. Oborz., 46:
117-121 (in Russian).

23. NUCIFORA, A. and INSERRA, R. (1967). Il *Panonychus ulmi* (Koch) nei vigneti dell'Etna. Entomologica, 3: 177-236.

24. PENMAN, D.R., FERRO, D.M. and WEARING, C.H. (1976). Integrated control
of apple pest in New Zealand. VII. Azimphosmetyl resistance in
strains of *Typhlodromus pyri*from Nelson. N.Z.J.Exp.Agric.,4: 377-380.

25. RAMBIER, A. (1958). Les Tétranyques nuisibles à la vigne en France
continentale. Rev. Zool. Agric. et Appliq., 57: 1-20.

26. RAGUSA, S. (1977). Notes on Phytoseiid mites in Sicily with a description of a new species of *Typhlodromus* (Acarina: Mesostigmata).Acaro-
logia, XVIII: 379-392.

27. RAGUSA, S. (1979). Laboratory studies on the food habits of the pre-
daceous mite *Typhlodromus exhilaratus* in Recent Advances in Acarology,
J.G. Rodriguez ed., Academic Press, N.Y.I: 485-490.

28. RAGUSA, S. (1981). Influence of different kinds of food substance on
developmental time in young stage of the predacious mite *Typhlodromus
exhilaratus* Ragusa (Acarina: Phytoseiidae). Redia, LXIV: 237-243.

29. SABELIS, M.W. (1981). Biological control of two spotted spider mites
using phytoseiid predators. Part. I., Agricultural Research Repts.910,
Wegeningen, 242 pp.

30. SCHRUFT, G. (1972). Les tydéides (Acari) sur vigne. OEPP/EPPO Bull.,
3: 51-55.

31. WYSOKI, M. and SWIRSKI, E. (1970). Studies on overwintering of predacious mites of the genera *Amblyseius* Berlese, *Typhlodromus* Scheuten and *Iphiseius* Berlese (Acarina: Phytoseiidae) in Israel. Contribution Volcani Instit. of Agricultural Research, 1726-E: 265-292.

Grape mites in Sicily – Contribution I*

V.Vacante & G.Tropea Grazia
Istituto di Entomologia Agraria dell' Università di Catania, Italy

Summary

The authors present the results of a survey carried out on grape mites in eastern Sicily. Thirty three species were observed, belonging to ten families. Of these species, two are newcomers to the Italian fauna. The phytophagous mites observed are less numerous (a total of nine species) than the predators (a total of 13 species) and than those with a varied diet (11 species in all).

1. INTRODUCTION

Over the past ten years the growing and irrational use of plan health products against pathogens, mites and insects which attack vines in Sicily, has had serious harmful effects on the biological balances which would exist in the agro-ecosystem as it adapts to soil and climate if these all-purpose pesticides were not used. One of the unfortunate effects, which is certainly a result of this state of affairs, is increasingly frequent infestation by phytophagous mites, which were never observed previously. Many of our vine-growers use chemicals to tackle this problem. It is well know, that this method has serious economic, toxicological and ecological effects. It has therefore become necessary in vine-growing also to rationalize pest control generally. It is vital to establish a mandatory programme for the rational control of pathogens and arthropods which attack crops.

The present survey is concerned with this situation and constitutes an initial contribution with the preliminary aim of making a census of the grape mite fauna on the island in order to obtain a fuller knowledge of which species (both useful and harmful) are present, i.e. in order to evaluate, at a subsequent stage, the contribution which the predators observed can make to a programme for the integrated control of harmful arthropods.

The research was carried out in different localities at various altitudes. Random sampling of leaves was carried out in vineyards located in hilly regions at high and medium altitudes and in the coastal plains in the east of the island.

---
* This study is part of an inter-university programme financed by the MPI to study the mite fauna and related symbionts of the main agricultural crops in the Mediterranean basin.

The species observed were studied morphologically. The observations were recorded in each case and provide information on the frequency of occurrence and distribution of the species. The data discussed here refer to the first year of the survey.

2. LIST OF SPECIES OBSERVED

So far the existence of 33 species belonging to 10 families has been observed. There are 9 phytophagous species, among which the Tetranichidae Tetranychus urticae (Koch) and Panonychus ulmi (Koch) are of particular importance. The first species is very widespread and is harmful only on crops which have been treated for animal and vegetable pests with all-purpose chemicals. In cases where the crop is treated exclusively with sulphur or copper-based products the presence of the mite causes no serious problems and normally is not observed by the vine growers. P. ulmi, which so far has been observed only in the vineyards on Etna, was observed in various areas of Messina, Ragusa and Syracuse. Populations of this latter tetranichida, which is commonly known to be influenced by climatic factors resulting from the altitude, temperature, relative humidity and rainfall, were also found in the lowland areas which have a typically Mediterranean climate. Here too serious infestation was observed only in crops on which irrational pest control methods are practised. The other phytophagous species (P. citri (Mc Gregor), Eotetranychus pruni (Oudemans), E. carpini vitis (Oudemans) Colomerus vitis (Pagenstecher), Brevipalpus phoenicis (Geijskes), B. pulcher (Canestrini and Fanzago) and Tenuipalpus granati (Sayed)) are not a serious problem in the vineyards of east Sicily.

However, some species are found sporadically.

Among the predatory species the following are of particular interest because of their frequency and spread : the phytoseidae Amblyseius stipulatus Athias-Henriot, Typhlodromus exhilaratus Ragusa, Phytoseius finitimus Ribaga ans Phytoseiulus persimilis Athias-Henriot.

A. stipulatus and T. exhilaratus were observed in all the areas under survey. The first species is common in the cold months of the year and the second in summer. P. persimilis was observed in frequent association with populations of T. urticae. The Stigmaeidae Zetzellia graeciana Gonzales and Z. mali (Ewing) are common throughout the eastern part of Sicily. Z. mali was observed serveral times in the act of preying on mobile forms of C. vitis.

Of the species whose feeding habits are varied or not well known the tydeida Orthotydeus californicus (Banks) is certainly of interest. The bio-ethology of this species is not sufficiently known but its widespread occurrence and its constant presence on crops supports the assumption that it plays an important role in the adaptation of the agro-ecosystem to soil and climate.

THE FAMILY TETRANYCHIDAE DONNADIEU

Tetranychus urticae Koch

1836. Tetranychus urticae Koch, mentioned by Koch in "Deutschlands Crustaceen, Myriapoden und Arachniden, Ein Beitrag zur Deutschen Fauna", Regensburg, Vol. 1 N° 10.

A phytophagous species which occurs in all the vine-growing areas in eastern Sicily. The attacks of the mite are particularly serious only in areas where irrational methods of controlling pathogens and other phytophages have been used.

Panonychus citri (McGregor)

1916. Tetranychus citri McGregor, mentioned by McGregor in Ann. ent. Soc. Am., 9.28.

A phytophagous species found on one occasion only on the leaves of a vine in the area of Vittoria (Ragusa). In eastern Sicily no damage by this mite has ever been reported.

Panonychus ulmi (Koch)

1836. Panonychus ulmi Koch described by Koch in "Deutschlands Crustaceen, Myriapoden und Arachniden, Ein Beitrag zur Deutschen Fauna", Regensburg, Vol. 1 n° 11.

A phytophagous species found in different vine-growing areas of east Sicily. The mite was observed up to a few years ago only on the slopes of Mount Etna. Recently it has also been observed in vineyards in lowland areas which have always been free of the tetranichida.

Eotetranychus pruni (Oudemans)

1931. Tetranychus pruni Oudemans, mentioned by Oudemans in "Ent. Ber., Amst.", 8 (177): 195.

A phytophagous species found sporadically in vineyards on the slopes of Mount Etna associated with fruit crops. This mite has not shown itself particularly harmful to vines.

Eotetranychus carpini vitis (Oudemans)

1905. Tetranychus carpini Oudemans, mentioned by Oudemans in "Tijdschr. voor Ent.", 48 : LXXIX.

A phytophagous species found on the slopes of Mount Etna and in some hill areas in the province of Enna. So far no seriously harmful effects have been observed.

THE FAMILY TENUIPALPIDAE BERLESE

Brevipalpus phoenicis (Geijskes)

1939. Tenuipalpus phoenicis (Geijskes), mentioned by Geijskes in "Beiträge zur Kenntnis der Europäischen Spinnmilben (Acari,

Tetranychidae), mit besonderer Berücksichtigung der Nieder-
ländischen Arten", Veenman and Zonen, Wageningen, 42 (4), 23.

This is a phytophagous species which is found sporadically in various
areas of south-east Sicily. So far this mite has not caused serious
damage to crops.

Brevipalpus pulcher (Canestrini and Fanzago)

1876. Caligonus pulcher Canestrini and Fanzago, mentioned by Canestrini
and Fanzago in Atti Acad. scient. veneto trent.-istriana, 5, 134.

This tenuipalpida was observed on one occasion only on the leaves of
vines under glass in Pachino (Siracusa). No serious damage to the crops
was observed.

Tenuipalpus granati Sayed

1946. Tenuipalpus granati Sayed, mentioned by Sayed in Bull. Soc. Fonad
1st Ent., 30, 100.

This mite is found on the leaves of vines in the province of Catania.
The presence of this phytophage has not produced any serious damage to
crops.

FAMILY ERIOPHY DAE NALEPA

Colomerus vitis (Pagenstecher)

1857. Phytoptus vitis Pagenstecher, mentioned by Pagenstecher in
"Naturhist. Med. Ver. der Heidelberg Verhandl.", 1 (2), 46.

A phytophagous species widespread in Sicily. The vine-growers are not
concerned by the damage done by this mite because it is considered to be
of secondary importance.

FAMILY TYDEIDAE KRAMER (*)

Pronematus ubiquitus (McGregor)

1932. Tydeus ubiquitus (McGregor), mentioned by McGregor in "Proc. ent.
Soc. Wash.", 34, 62.

This species is widespread on vines in Sicily. Its feeding habits are
little known, it may be microphagous, saprophagous or coprophagous.

---

* The Tydeidae referred to in this paper are classified in accordance
with the systematic revision proposed by André (1980).

Orthotydeus mississippiensis (Baker) Comb. n.

1970. Tydeus (Tydeus) mississippiensis Baker, mentioned by Baker in
"Ann. ent. Soc. Am. ", 63 (1), 171

This species was sampled on one single occasion on the leaves of vines
in Catania. The mite is a newcomer to Italian fauna. Its eating habits
are not known.

Orthotydeus californicus (Banks)

1904. Tetranychoides californicus Banks, mentioned by Banks in "J.N.Y.
Ent. Soc.", 12, 54.

This species occurs in all the vine-growing areas in Sicily. Its feeding
habits are little known. It is probably of the heterogeneous type –
predator, microphage, coprophage and saprophage.

Orthotydeus foliorum (Schrank)

1776. Acarus foliorum Schrank, mentioned by Schrank in "Beytrage zur
Naturgeschichte", Augsburg, 33.

This species is found sporadically on leaves in various areas in eastern
Sicily. Its feeding habits are little known.

Orthotydeus caudatus (Dugés)

1834. Tetranychus caudatus Dugés, mentioned by Dugés in "Annls. Sci.
nat.", Ser. 2, 2 (Zool.), 29.

This species is found on the leaves of vines under glass in Pachino
(Syracuse). Its feeding habits are unknown.

Tydeus teresae (Carmona)

1970. Lorryia teresae Carmona, mentioned by Carmona in "Acarologia", 12
(2), 310.

This species is found sporadically on leaves of vines on the slopes of
Mount Etna. Its feeding habits are not known.

Tydeus formosa (Cooreman)

1958. Lorryia formosa Cooreman, mentioned by Cooreman in "Bull. Inst. r.
Sci. nat. Belg.", 34 (8), 7.

This microphagous species occurs in various vineyards.

FAMILY <u>PHYTOSEIIDAE</u> BERLESE

Amblyseius aberrans (Oudemans)

1930. <u>Typhlodromus aberrans</u> Oudemans mentioned by Oudemans in "Ent. Ber.", Amst. <u>8</u>, 48.

This is a predatory species found on the leaves of vines in various areas. The mite is more common in vineyards on Mount Etna and in hilly areas.

Amblyseius stipulatus Athias-Henriot

1960. <u>Amblyseius</u> Athias-Henriot, mentioned by Athias-Henriot in "Acarologia", <u>2</u>, 294.

This predatory species is commonly found in all the areas surveyed. It is probably the most widespread phytoseida in Sicily. It occurs most frequently in the cold months.

Amblyseius andersoni (Chant)

1957. <u>Typhlodromus (Amblyseius) andersoni</u> Chant, mentioned by Chant in "Can. Ent.", <u>90</u>, 296.

This predatory species was found on a single occasion on the leaves of vines in a vineyard on the slopes of Mount Etna in Fornazzo (Catania).

Typhlodromus exhilaratus Ragusa

1977. <u>Typhlodromus exhilaratus</u> Ragusa, mentioned by Ragusa in "Acarologia", <u>18</u> (3), 380.

This predatory species is widespread in Sicilian vineyards. The mite is observed more frequently in the summer months. This phytoseida together with <u>A. stipulatus</u> is probably a major factor in containing the populations of phytophagous mites on the island.

Phytoseius finitimus Ribaga

1902. <u>Phytoseius finitimus</u> Ribaga, mentioned by Ribaga in "Riv. Pat. Veg.", Padua, <u>10</u>, 178.

A predatory species common in vineyards in eastern Sicily.

Iphiseius degenerans (Berlese)

1889. <u>Seius degenerans</u> Berlese, mentioned by Berlese in "Acari, Myriapoda et Scorpiones hucusque in Italia reperta", Padua, fasc. 54, N° 9.

A predatory species which occurs commonly in all the vine-growing areas in eastern Sicily. The mite occurs more frequently in the summer months.

Phytoseiulus persimilis Athia-Henriot

1957. Phytroseiulus persimilis Athias-Henriot, mentioned by Athias-Henriot in "Bull. Soc. Hist. nat. Afr. N.", 48, 347.

This predatory species of phytophagous mite is of the type Tetranychus Koch. The mite is commonly found on crops which are infested by T. urticae and which are not treated with all-purpose products.

Seiulus amaliae Ragusa and Swirski

1976. Seiulus amaliae Ragusa and Swirski, mentioned by Ragusa and Swirski in "Redia", 59, 1983.

A predatory species observed infrequently in various localities in eastern Sicily.

FAMILY ASCIDAE VOIGTS AND OUDEMANS

Proctolaelaps pygmeus (Muller)

1895. Gamasus pygmaeus Muller, mentioned by Muller in "Lotas", 9, 26.

A predatory species commonly occurring in vineyards in eastern Sicily.

FAMILY CHELETIDAE LEACH

Cheletogenes ornatus (Canestrini and Fanzago)

1876. Cheletogenes ornatus Canestrini and Fanzago, mentioned by Canestrini and Fanzago in "Atti Accad. scient. veneto-trent.-istriana", 5, 106.

A predatory species of mites and insects occurring sporadically in vineyards in the south-east of Sicily.

Eutogenes citri Gerson

1967. Eutogenes citri Gerson, mentioned by Gerson in "Acarologia", 9 (2), 363.

A predatory species found sporadically in vineyards in eastern Sicily.

FAMILY STIGMAEIDAE OUDEMANS

Zetzellia graeciana Gonzalez

1965. Zetzellia graeciana Gonzalez, mentioned by Gonzalez in "Univ. Calif. Publ. Ent.", 41, 22.

A predatory species of insects and mites common in the vineyards in eastern Sicily.

Zetzellia mali (Ewing)

1917. Caligonus mali Ewing, mentioned by Ewing in "J. econ. Ent.", 10, 499.

A predatory species common in the vineyards in eastern Sicily. On various occasions it was observed that the mite was preying on mobile forms of the eryophida C. vitis.

FAMILY ACARIDAE EWING AND NESBITT

Tyrophagus putrescentiae (Schrank)

1781. Acarus putrescentiae Schrank, mentioned by Schrank in "Enumeratio Insectorum Austriae indigenorum", Augsburg, 521.

A microphagous or saprophagous species occurring sporadically on the leaves of vines in a number of aereas in eastern Sicily.

Tyrophagus longior (Gervais)

1844. Tyrophagus longior (Gervais), mentioned by Walckenaer in "Hist. Nat. Ins. Apt.", 3, 262, t. 35.

This species has a heterogenous diet, since it may possibly be microphagous, saprophagous and phytophagous. This mite was observed on one single occasion on the leaves of vines in the area of Vittoria (Ragusa).

FAMILY TARSONEMIDAE KRAMER

Tarsonemus aurantii Oudemans

1927. Tarsonemus aurantii Oudemans, mentioned by Oudemans in "Tijdschr. voor Ent.", 70, 34.

A microphagous species found infrequently on the leaves of vines in the area of Catania.

Tarsonemus waitei Banks

1912. Tarsonemus waitei Banks, mentioned by Banks in "Proc. ent. Soc. Wash.", 14, 96.

A microphagous species found on the leaves of vines in various parts of the island.

3.   CONCLUSIONS

The survey showed the existence of a small number of species, some of which have a particular ecological place in the agro-ecosystem under survey. Some of these have a role which is easy to determine and is well defined - phytophagous or predatory, but others are not well known from the bio-ecological point of view, although this does not make them less important. Of these latter mites we consider that the Tydeidae are of particular interest and should be studied at a later date.

The work discussed above forms a starting point for further research to extend the list of species observed, which is certainly not complete, and investigate their bio-ethology in order to provide further knowledge on which to base future programmes for the biological and integrated control of crop pests.

BIBLIOGRAPHY

1. ANDRE H.M. (1980). A generic revision of the family Tydeidae (Acari, Actinedida). IV Generic descriptions, keys and conclusions. "Bull. Ann. Soc. r. belge Ent.", 116, pp. 103-168.

Comparison of two control strategies of *Panonychus ulmi* (Koch) on vineyards

C.Duso

Istituto di Entomologia Agraria dell' Università di Padova, Italy

Summary

Two control strategies of Panonychus ulmi (Koch) on vine were compared for two years. A threshold of 10 mites per leaf is adopted following the first method (Girolami, 1981). Vineyards are sprayed when the percentage of leaves infested by one or more mites is higher than 30-45 % in summer (60-70 % in spring) following the second method (Baillod et al., 1979). Research was carried out in vineyards in the Veneto region heavily infested by P.ulmi . Five miticide treatments were applied in 1983 and 7 in 1984, in the plots where the percentage of infested leaves method was used. When the threshold level of 10 mites per leaf was used, 1 spray was applied in 1983 and 2 in 1984. Following the first strategy, vineyards are sprayed at low density levels (maximum 1-4 mites per leaf), therefore prematurely; these treatments prevent the possible increase of predacious mites and insects; a delayed increase of phytophagous mite population is not prevented in favourable climatic conditions. The threshold level of 10 mites per leaf allows, in many cases, the arrival of winged predators, mostly Anthocoridae , (Orius vicinus Ribaut) and an increase in population of predacious mites (Phytoseiidae and Stigmaeidae); Tetranychid mites are controlled without damages and a low number of treatments are required.

1. Introduction

The establishment of an economic threshold of Panonychus ulmi (Koch) on vine is an important aspect in the integrated control of the tetranychid mite. Research still in progress on Panonychus ulmi (Koch) in the vines of north-east Italy has shown that populations of over 20 mites per leaf, for 2 or 3 weeks, have not caused serious damage to production, even though leaves were clearly bronzed (8, 6). The amount of sugar in the grapes decreases only if the vines become defoliated some weeks before harvest (1); data refers to Tetranychus urticae Koch. Many experiments have shown that populations of 10-12 mites per leaf for 1 or 2 weeks do not cause bronzing on vine (8, 6). Therefore a threshold level of 10 mites per leaf, as proposed by Girolami (1981), prevent any economic and esthetic damage.

In the U.S.A. economic thresholds of 6 to 30 mites per leaf are adopted on apple trees (11). In research carried out in Italy on apple trees (13) no obvious damage to yield resulted even with high infestations of P.ulmi (42 mites per leaf).

The percentage of leaves infested by mites can be used as an indirect parameter to evaluate the population density (2); it is advisable to spray when more than 60-70 % of leaves are infested in

spring and more than 30-45 % in summer. These thresholds levels
correspond to a population mean of 3-4 mites per leaf in spring and
0.5-1 mites per leaf in summer (10). A sequential method was later
proposed to reduce the sampling time (3). This method is adopted in
Switzerland, France and Germany and accepted by some Italian authors
(7).

The method based on presence or absence of mites on leaves
(2) tries to simplify sampling work, avoiding the actual counting of
active forms. However, it does not allow a precise estimate of
populations higher than 3 mites per leaf, and advises treatments when
mite density is far from causing concrete damages (10). The method
proposed by Girolami (8) even though it is based on a count of active
forms, allows a decision by examining a limited number of leaves (3 -
10). The aim of this work is a comparison between the two methods of
control mentioned above.

2. Materials and methods

Trials were carried out in 1983 and 1984 in four vineyards in
north-east Italy, Treviso area. Some characteristics of vineyards are
reported in Table 1.

Table No.1:

No.	Place	Soil	Cv.	Fungicides (No. sprays)
1	Bibano	medium-textured coating	Merlot	metiram, sulphur (6-8)
2	Maserada	medium-textured coating	Merlot	mancozeb, sulphur (7-9)
3	Lancenigo	clay coating	Merlot	mancozeb-copper, sulphur (7-9)
4	Villorba	medium-textured tilled	Merlot	copper, sulphur (6-8)

In each vineyard a surface of about 5000 sq m was subdivided
in two homogeneous parts where the different control strategies were
applied:

- Method A (Plot A): threshold level of 60-70% of infested
leaves in spring-early summer and 30-45% of infested leaves in summer
(Baillod et al.,1979). The highest percentages reported were used
(70% and 45%).
- Method B (Plot B): threshold level of 10 active mites per
leaf (Girolami, 1981).

Every 15 days, samples of 50 leaves were chosen for each plot
(threshold levels not sampling methods were compared). The leaves
were examined by a stereoscope microscope, counting phytophagous
mites (Panonychus ulmi (Koch); Eotetranychus carpini (Oud.),
predacious mites (Phytoseiidae and Stigmaeidae) and predacious
insects (only Anthocoridae have been found).

218

Miticide treatments were carried out when threshold level was excedeed with a mixture of dicofol+tetradifon (0.25%). No insecticide treatments were applied to avoid interference.

3. Results and discussion

P.ulmi constitutes the most common species in examined vineyards; rare specimens of E.carpini were found but not reported.

Vineyard No.1 (Fig.1): In 1983 P.ulmi populations did not reach high levels during spring and early summer. About mid August the threshold level was exceeded in plot A (57% of infested leaves) requiring a miticide application (17 Aug.); the threshold level was exceeded at the end of August in plot B (11.2 mites per leaf). Rare specimens of predacious Anthocorids (Orius vicinus Rib.) were found only in plot B.
In 1984 the situation was similar to the previous year; the threshold level was exceeded on 6 August in plot A (53% of infested leaves) and on 21 August in plot B (13.4 mites per leaf). Anthocorids appeared in August both in plot A and B but at low levels (1-2 individuals on 10 leaves). No predacious mites were observed.

Vineyard No.2 (Fig.2): P.ulmi density reached moderate levels only at mid August; the threshold level in plot A was exceeded on 17 August (56% of infested leaves). Tetranychid populations peaked (6 mites per leaf) at the end of August in plot B but they were controlled by Anthocorids (4-5 individuals on 10 leaves).
In the following year P.ulmi populations exceeded the threshold level in plot A at the end of June (77% of infested leaves) requiring treatment (26 June); then a new Tetranychid infestation occurred in early September with an excess of the threshold level (50% of infested leaves). In plot B mite populations increased at the end of summer; in early August some specimens of O.vicinus appeared but they were unable to control tetranychid mites under threshold level. A mean density of 10.5 mites per leaf was reached and miticide was applied. No predacious mites were observed.

Vineyard No.3 (Fig.3): In 1983 mite populations exceeded the threshold level in early July in plot A (75% of infested leaves); a new increase occured in early September requiring miticide treatment. In plot B P.ulmi populations peaked at mid July (7 mites per leaf) but the action of Anthocorids resulted useful in the control of Tetranychids at low levels.
In 1984 the situation in plot A was similar to the previous year. The threshold level was exceeded twice (4 July and 3 September). In plot B tetranychid population increased moderately due to the progressive presence of O.vicinus . This predator which also appeared in plot A was killed by miticides. The appearance of Phytoseiids (Amblyseius andersoni Chant)* was observed in plot B at the end of August.

Vineyard No.4 (Fig.4): In 1983 the P.ulmi population remained at low levels in spring and early summer; then the threshold level was exceeded in plot A (58% of infested leaves). In plot B Phytoseiids and Stigmaeids were able to control mites during all the season. Stigmaeids observed in plot A were limited by miticide. In the following year P.ulmi density exceeded threshold level in

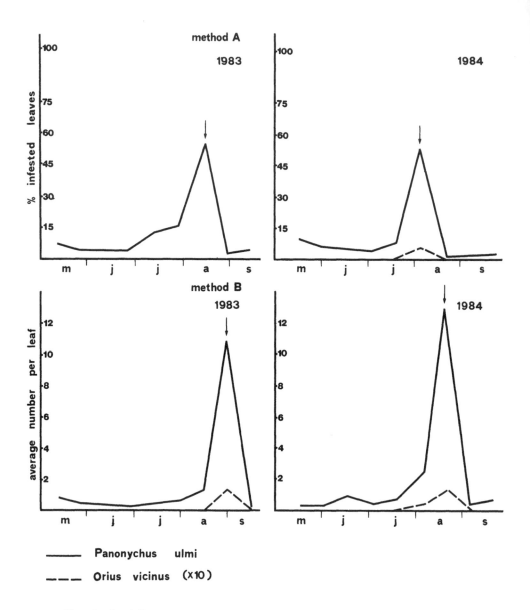

Fig. 1: Populations of <u>P.ulmi</u> and <u>O.vicinus</u> in a vineyard treated with metiram, copper after blooming and sulphur as fungicides. In 1983 <u>P.ulmi</u> populations did not reach high levels during spring and early summer. About mid August the threshold level was exceeded in plot A (57% of infested leaves) requiring a miticide application (17 Aug.); the threshold level was exceeded at the end of August in plot B (11.2 mites per leaf). Rare specimens of <u>Orius</u> <u>vicinus</u> Rib. were found only in plot B. In 1984 the situation was similar to the previous year; the threshold level was exceeded on 6 August in plot A (53% of infested leaves) and on 21 August in plot B (13.4 mites per leaf). Anthocorids appeared in August both in plot A and B but at low levels (1-2 individuals on 10 leaves). No predacious mites were observed.

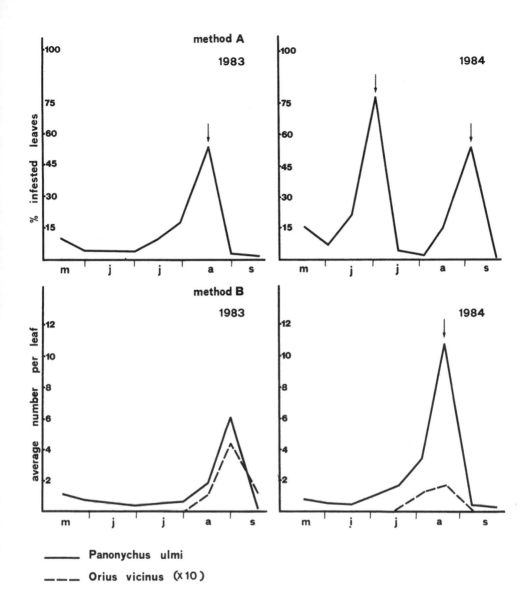

Fig. 2: Populations of P.ulmi and O.vicinus in a vineyard treated with mancozeb, copper after blooming and sulphur as fungicides. P.ulmi density reached moderate levels only at mid August; the threshold level in plot A was exceeded on 17 August (56% of infested leaves). Tetranychids peaked (6 mites per leaf) at the end of August in plot B but they were limited by Anthocorids (4-5 individuals on 10 leaves). In the following year P.ulmi populations exceeded the threshold level in plot A at the end of June (77% of infested leaves) requiring treatment (26 June); then a new P.ulmi infestation occurred in early September with an excess of the threshold level (50% of infested leaves). In plot B mite populations increased at the end of summer; in early August some specimens of O.vicinus appeared but they were unable to control phytophagous mites under threshold level. A mean density of 10.5 mites per leaf was reached and miticide was applied. No predacious mites were observed.

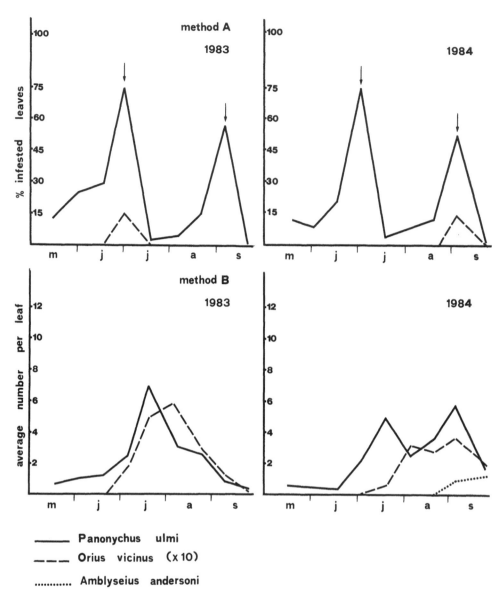

Panonychus ulmi
Orius vicinus (x 10)
Amblyseius andersoni

Fig. 3 : Populations of <u>P.ulmi</u> , <u>O.vicinus</u> and <u>A.andersoni</u> in a vineyard treated with mancozeb-copper, copper after blooming and sulphur as fungicides. In 1983 mite populations exceeded the threshold level in early July in plot A (75% of infested leaves); a new increase occurs in early September requiring miticide treatment. In plot B <u>P.ulmi</u> populations peaked at mid July (7 mites per leaf) but the action of Anthocorids resulted useful in the control of tetranychids at low levels. In 1984 the situation in plot A was similar to the previous year. The threshold level was exceeded twice (4 July and 3 September). In plot B <u>P.ulmi</u> population increased moderately due to the progressive presence of <u>O.vicinus</u> . This predator which also appeared in plot A was killed by miticides. The appearance of Phytoseiids (<u>Amblyseius   andersoni</u> Chant) was observed in plot B at the end of August.

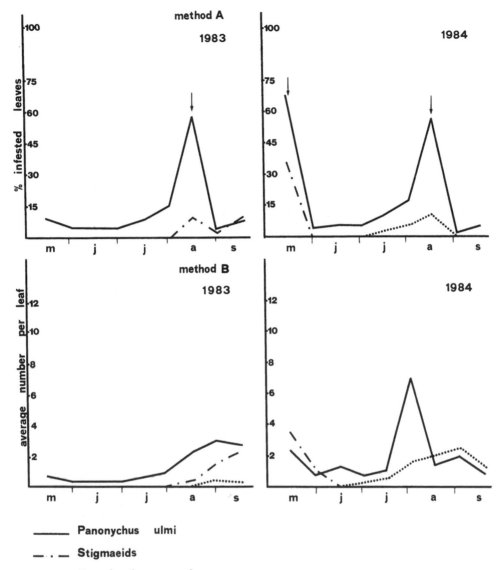

Fig. 4 : Populations of P.ulmi , K.aberrans and Stigmaeids in a vineyard treated with copper and sulphur as fungicides. In 1983 the P.ulmi population remained at low levels in spring and early summer; then the threshold level was exceeded in plot A (58% of infested leaves). In plot B Phytoseiids and Stigmaeids were able to control mites during the season. Stigmaeids observed in plot A were limited by miticide. In the following year P.ulmi density exceeded threshold level in spring (72% of infested leaves) and mid August (56% of infested leaves); Phytoseiids and Stigmaeids were present at low levels. In plot B in spring Tetranychids were efficiently controlled by Stigmaeids; in early August a rapid growth of mite populations (7 mites per leaf) was limited by Phytoseiids (Kampimodromus aberrans Oud.) which reached moderate levels (2-3 active forms per leaf). Phytoseiids and Stigmaeids were present for all the season in the following year (data not reported).

spring (72% of infested leaves) and mid August (56% of infested leaves) in plot A; Phytoseiids and Stigmaeids were present at low levels. In plot B, in spring, Tetranychids were efficiently controlled by Stigmaeids; in early August a rapid growth of mite populations (7 active mites per leaf) was limited by Phytoseiids (Kampimodromus aberrans Oud.)* which reached moderate levels (2-3 active mites per leaf). Phytoseiids and Stigmaeids were present for all the season in the following year (data not reported).

4. Conclusions

        Comparison between the two methods of P.ulmi control gave interesting results. Numerous sprays (12 miticides in the 4 vineyards for 2 years) were necessary for keeping Tetranychid populations at the very low levels proposed by Baillod et al.(1979). Following the method proposed by Girolami (1979) a limited number of miticides (3) was applied. In plots treated when the threshold level of 10 mites per leaf was exceeded no bronzing was observed; this is a further demonstration that this method prevents leaf damage.
        The number of miticide treatments used in the two different plots is linked to the high or low presence of predacious mites or insects. The population density of Tetranychids, in the plot treated at over 10 active mites per leaf, has allowed in many cases the active reproduction of Anthocorids; adults are attracted in vineyards with populations of at least 3-5 mites per leaf. If these densities must not be reached, as proposed by Baillod et al.(1979), Anthocorids will be rarely seen in vineyards. From reported data (fig. 2,3,4) O.vicinus can often control P.ulmi populations under a mean of 10 mites per leaf. Nevertheless sometimes Anthocorids are not able to control Tetranychid populations under the threshold level proposed by Girolami (1981) in unsprayed vineyards; both agronomic (single-crop system, chemical treatments) and ecological (alternative prey, migrations) factors can influence these predators in controlling phytophagous mites (4).
        In two of the examined vineyards (No.3 and No.4) Phytoseiids and Stigmaeids were observed in the two plots, but only in plot B were predacious mites able to limit Tetranychid populations under the threshold level. The appearance of predacious mites in these vineyards is probably due to fungicides non-toxic to predators (5, 9, 12); however, the reproduction of predators and the successful control of P.ulmi was obtained only by application of a threshold of 10 mites per leaf.
        Phytoseiids in vineyards are very important in keeping Tetranychid mites at low non-damaging levels for long periods of time. (12, 6, 5). Threshold levels proposed by Baillod et al. 1979 has brought about the need for a high number of miticides and the destruction of predacious mites. This causes a consequent repeat of infestation in the following year.
        The application of pesticides which are non toxic for predators (5, 9) and the use of threshold levels of 10 or more mites per leaf, allow a re-establishment of biological balance and a progressive reduction of infestations.

_____

        *   The species were identified by Dr. Marialivia Liguori (Istituto Sperimentale per la Zoologia agraria, Firenze).

Acknowledgements

        The author wishes to thank Mr. A. Procida and Mr. A. Venturin
(Centro Quadrifoglio, Treviso) for assistence in field trials.

        REFERENCES

        1. ARIAS, A. (1981). Acarien jaune tisserand (Tetranychus
urticae   Koch). Rapport IV Réunion plénière OILB, "Lutte integrée en
viticulture".   Gargnano,   Mar.1981.   Boll.Zool.agr.Bachic.(Ser.II),
16:38-41.
        2. BAILLOD,   M.,   BASSINO,   J.P.,   PIGANEAU, P. (1979).
L'estimation du risque provoque' par l'acarien rouge (Panonychus
ulmi Koch) et l'acarien des charmilles (Eotetranychus carpini Oud.)
en viticulture. Revue suisse Vitic.Arboric.Hortic., 11(3): 123-130.
        3. BAILLOD, M. and SCHLAEPFER, R. (1982). Technique simplifiée
de controle pour l'acarien rouge (P.ulmi Koch) et les vers de la
grappe (1re génération). Revue suisse Vitic.Arboric.Hortic., 14(4):
211-215.
        4. DUSO, C. and GIROLAMI, V. (1982). Ruolo degli Antocoridi nel
controllo   del   Panonychus   ulmi   Koch   nei   vigneti.
Boll.Entom.Bologna, 37:157-169.
        5. DUSO, C. and GIROLAMI, V., (1985). Strategie di controllo
biologico   degli   Acari   Tetranichidi   su   vite.   Atti   XIV
Cong.Naz.Ital.Ent.,Palermo, pp.719-728.
        6. DUSO,   C.,   GIROLAMI,   V.,   BORGO,   M.,   EGGER, E. (1983).
Influenza di anticrittogamici diversi sulla sopravvivenza di
predatori Fitoseidi introdotti su vite. Redia , 66: 469-483.
        7. EGGER, E. (1981). Lotta antiparassitaria in viticoltura.
L'informatore agrario, 36 (16): 15201-15209.
        8. GIROLAMI, V. (1981). Danni, soglie di intervento, controllo
degli acari della vite. Atti III Incontro La difesa integrata della
vite.Latina, 3-4 Dicembre 1981.
        9. GIROLAMI, V. and DUSO, C. (1985). Controllo biologico degli
acari nei vigneti. Inform.Agrario, LXI (18):83-89.
        10. GIROLAMI, V. and MOZZI, A. (1983). Distribution, economic
threshold and sampling methods of Panonychus ulmi (Koch)., In
CAVALLORO R., 1983. Statistical and Mathematical Methods in
Population Dynamics and Pest Control.90-101.
        11. HOYT, S.C., LEEPER, J.R., BROWN, G.C., CROFT, B.A. (1983).
Basic biology and management components for insect IPM. In CROFT,
B.A. and HOYT, S.C. (1983). Integrated management of insect pests of
pome and stone fruits.
        12. IVANCICH GAMBARO, P. (1973). Il ruolo del Typhlodromus
aberrans Oud. (Acarina : Phytoseiidae) nel controllo biologico
degli acari fitofagi nei vigneti del veronese. Boll.Zool.agr.Bachic.
(Ser.II) 11: 151-66.
        13. PASQUALINI, E., BRIOLINI, G., MEMMI, M. (1981). Indagini
preliminari sul danno da Panonychus ulmi Koch (Acarina :
Tetranychidae) su Melo in Emilia Romagna. Boll.Entom.Bologna, 36:
173-190.

Biological control of spider mites in viticulture

G.A.Schruft
Staatliches Weinbauinstitut, Freiburg, FR Germany

Summary
Spider mites are in all grapevine regions of Europe very
important pests. In the northern parts there are mostly
the species Panonychus ulmi and Tetranychus urticae; in
the mediterranean vineyards the species T.urticae and
Eotetranychus carpini vitis are in majority. The chemical control with miticides is more or less effective in
dependence of the seasonal appearence of a species and
on the premises that resistance do not exist. With regard to integrated pest management, the strategy differs
with the control program of the diseases and other pests.
Biological methods for spider mite management are in de-
velopping in some viticulture regions of Europe. The
most important predators are mites, esp.of the phytoseiid family, chrysopids, bugs, and ladybird beetles. The
bionomics and the capacity of predation of some species
are given. For an effective biological control of spider
mites it is necessary to find methods for preserving an
original or released predator population, to study the
side-effects of pesticides against it, and to give wat-
ching the pests and the predators in the vineyards.

1. Introduction

Spider mites of the family Tetranychidae are already
described on grapes in the begin of our century, but not before 1950 they have been a serious pest in all viticultural
areas of Europe. We believe that particularly some reasons
have been responsible for this event. One is the common use
of DDT for insect control, which increases the amount of
spider mite eggs laid by DDT-treated females (Löcher 1958,
Seifert 1961). A similar effect is happend by the periodical
sprays with copper containing fingicides against the powdery
mildew (Plasmopara viticola) and some other diseases (Chaboussou 1966). In addition, the modification of some cultural
practices has favored the living conditions of the spider
mites. In the meantime the broadspectrum insecticides have
destroied the natural antagonists of this pest, and morever
resistance against a lot of pesticides exists.
The future of an economical and ecological control of
spider mites or other pests and diseases will be the integrated management by reducing the use of chemical pesticide
products and favoring the natural control agents, which

227

means the parasites and predators of the pests, respectively the natural prevention mechanisms and the antagonists of the diseases.

For the spider mites in viticulture we like to point out the current state of the biological control.

2.The spider mite species of grapes

In Europe, some spider mite species of different importance occur on grapes. In the moderate climate areas the red spider mite, Panonychus ulmi (Koch), and the two-spotted spider mite, Tetranychus urticae Koch, are predominant, whereas in the mediterranean vineyards and in warmer areas the species T.urticae and the yellow vine mite, Eotetranychus carpini vitis (Oudemans) exist on grapevines. Occasionally the species Tetranychus turkestani Ugarov & Nikolski (= T. atlanticus McGregor) is damaging the grapes in vineyards of the mediterranian coast (Rambier 1982).Some tetranychid species are only of local or inferior significance, so the recently found species Tetranychus mcdanieli McG. in vineyards of the Champagne (Rambier 1982) or Eotetranychus pruni (Oudemans) in certain grapevine areas of Bulgaria (Balevski 1980).

The injury of all spider mite species is more or less comparable, caused by sucking the cells of the leaves and sometimes of the grape berries,too.Generally, a discoloration of the foliage, in totality of in spots, occurs,followed by its fall,sooner or later.Depending on the time and on the degree of the attack, the number of berries, their maturity and the quality are affected.

With regard to biological control possibilities,it is important to know some bionomic statements of the spider mite species.So it is necessary to distinguish to groups concerning the form of hibernation.The species Tetranychus urticae and Eotetranychus carpini vitis go in the winter as mated female unter the bark of the trunk and/or in the declined foliage.In spring and in summer,in dependence on the species and the climatic situation, the mites invade the grape leaves attacking it.Females of T. urticae lay 80 - 120 eggs during its life of about 30 days, and 6 - 8 generations occur in the season.For Eotetranychus carpini vitis 6 - 8 generations was observed too, but the female of this species produces 30 - 40 eggs.In contrast, Panonychus ulmi overwinters as winter egg on the canes nearby the buds and on the head of the trunk.In spring,the young leaves are attacked, and a second dangerous situation occur in late summer.The reproduction rate is 20 - 80 eggs per female during an average duration of life from about 19 days.

3.The biological control agents

The biological control agents of spider mites on grapes until now consist of predators only.These are predator mites and predatory insects,which are found feeding on spider mites in vineyards, but they are of different importance.

Generally we can distinguish between the so-called protective predators and the cleaning predators (Rambier 1976). Protective predators live on a plant organ independent of the presence or absence of the prey.If the prey invades, the predator pick it up and the plant is protected from the pest at-

tack.But this mode of living brings discredit when prey is
missing, so the protective predator needs an amend food of
animal or vegetable origine.On the other hand, the cleaning
predators usually have a small food-spectrum, why they look
for the specific prey.This is the reason that most of the
cleaning predators are able to fly and settle a plant organ
not before numerous amounts of prey individuals are present.
Under this conditions, the predator is active and able to
clean up the plant organ.Most of the predator mites are of
the protective typ,whereas the majority of the cleaning predators belongs to the predator insects.

3.1 The predator mites

A lot of predator mites is found on grapes.The different
species belong in majority to the family Phytoseiidae, but
Cunaxids, Stigmaeids and Anystids are found in immediate connection with spider mite populations,too.

The phytoseiid mites found on grapes in France,Germany,
Italy and Swizzerland are listed by Günthart (1956), Schruft
(1967), Ivancich Gambaro (1973) and Baillod & Venturi (1980)
(Table I).Most of it are of low significance, since they are
in small numbers.Only the species Typhlodromus pyri and
Amblyseius aberrans are predominant, completing more than 90
percent of the first in the northern, of the second in the
southern viticultural areas of Europe.

The best informations concerning the life table we have
from the species Typhlodromus pyri.The developpment from egg
to female takes 10 - 12 days, and the female lives about 30
(15 - 54) days in dependence on the food.One female lays an
average number of 20 (7 - 40) eggs during its life.There are
2 - 4 generations per year (Overmeer 1981,1982).

Specific life table statements for Amblyseius aberrans
we did not find in exception those of Ivancich Gambaro (1973).

The most important Stigmaeid predator mite is the
species Zetzellia mali.We could find it on leaves of grapes
colonized by the grape rust mite Calepitrimerus vitis (Nal.)
(Schruft 1969), butthis species is known to be a predator of
tetranychid mites in orchards.The predation capacity is not
so good as in phytoseiid mites, and this species has a high
sensitivity against pesticides, so it is to find especially
in vineyards with reduced spray programme.

Species of Cunaxidae and Anystidae are well known predators of spider mites, but its number of individuals and
generations are to small for an effective control without
any other predators.

3.2 Predatory insects

Some predatory insects assoviated with spider mites are
found on grapes.Günthart (1956) enumerates unidentified species of Scymnus and Oligota,the Scolothrips longicornus Prie-
sner, Anthocoris nemorum (L.) and Orius minutus (L.), but he
do not give any information on its predatory capacity.Two
other anthocorids, Orius vicinus Ribaut and Orius majusculus
Reuter, are found in Italy on grapeleaves attacked by the
red spider mite and the yellow vine mite (Duso & Girolami
1982). A list of 21 species of neuroptera from 4 families
captured in various vineyards of southern Germany is publish-

Table I : List of phytoseiid mites on grapes
(CH - Swizzerland, F - France, G - Germany, I - Italy,
after references 3, 11, 13, 19, 20)

Amblyseius aberra s (Oudemans)	CH	F		I
Amblyseius agrestis (Karg)	CH			
Amblyseius andersoni Chant	CH		G	I
Amblyseius cucumeris (Oudemans)			G	
Amblyseius finlandicus (Oudemans)	CH		G	I
Amblyseius marinus (Willmann)		F		
syn. A. californicus McGregor				
Amblyseius zwoelferi Dosse	CH			
Paraseiulus soleiger (Ribaga)	CH	F	G	I
Paraseiulus subsoleiger Wainstein	CH		G	I
Phytoseius macropilis (Banks)	CH	F	G	I
Typhlodromus longipilus (Nesbitt)	CH			
Typhlodromus pyri Scheuten	CH	F	G	I
Typhlodromus tiliarum (Oudemans)	CH			

Table II : Side-effects of some fungicides against
the predator mite Typhlodromus pyri (after several
references)

harmless - slightly harmful	harmful - moderately harmful
Al-phosethyl + Folpet	Dinocap
Captafol	Mancozeb
Captan	Propineb
Copper	Zineb
Dichlofluanid	
Folpet	
Iprodione	
Metalaxyl	
Procymidone	
Sulfur	
Triadimefon	
Vinchlozolin	

ed by Schruft et al.(1983), and Haub et al.(1983) give some
informations on the preying of natural found and released
lacewings of the species Chrysopa carnea against the red spider mite species.

4. The strategy of the biological spider mite control

From all biological control agents at present only two
are interesting for a promotion: the phytoseiids Typhlodromus
pyri and Amblyseius aberrans.The principal occurrence of both
is different; the first is spreading in the northern grape
areas of Europe, the latter is dominant in the southern vine-
yards.Concerning the prey, T.pyri is subsisting on Panonychus
ulmi, whereas Eotetranychus carpini vitis is the prefered
prey of A.aberrans.

For a successful biological control of spider mites in
vineyards the follow strategy is necessary,
- predator mite release,if they are absent, and
- predator mite preservation, if they are present.
For Typhlodromus pyri all both proceedings are in practical
use,so we will explain it.

4.1 Predator mite release methods

For releasing predator mites you collect canes pruned
off from an occupied grape plant and attach two or three of
it to the trunk in the new vineyard (Boller 1978,Baillod et
al. 1982).An other method uses stripes of wool blankets,which
are wraped up the trunk nearby the head of the occupied grape
in the late summer befor the foliage drop.For overwintering,
the predator mites hide in the stripes.Early in spring, the
occupied stripes are taken and attached in the same way
(Baillod 1984).In summer,you can take occupied leaves and fix
it on the leaves and canes of the colonizing vineyard.

4.2 Estimation of the predator mite population

The biological control success depends upon the population density of the predator mite species.For Typhlodromus
pyri at least 1 - 2 individuals per leaf are necessary to
control the spider mite Panonychus ulmi.Baillod & Venturi
(1980) have studied the correlation between the average number of predator mite per leaf and the percentage of occupied
leaves in the vineyard, and they could find a simple method
for the population estimation during the season,which is
identical with the spider mite estimation method.In winter, a
check for presence or absence of hibernating predator mite
females is possible by washing a number of pruned canes in
surfactant water, then filter it and inspect the predator
mites.The same method is to use for counting the number of
predator mites on leaves (Boller 1984).

4.3 Predator mite preservation

If predator mites are present in a vineyard, it is necessary to preserve it.They are especially the pesticides,
which can be dangerous for these natural enemies, often more
sensitive against it as the pest,the spider mites.A lot of
studies were done looking for the effects and side-effects
of pesticides against the predator and its prey.Not only the
insecticides/miticides con be harmful for the predator mite,

and each species may be of different sensitivity,the fungicides and weed killers sometimes have an effect,too (Boller
et al.1984).So it is necessary to find a spray programme for
managing the pest and diseases without damage of the preda-
tors.A list of the side-effects of fungicides against the
predator Typhlodromus pyri,compiled from diverses authors,
is given in table 11.The effects of insecticides and miticides is mostly harmful for the predator mites, in exception
of any products and in the case of local resistance, so it
is not suitable to give a list here.

5.Conclusions
     The biological control of spider mites is in different
viticultural areas of Europe in development.Only the system
Typhlodromus pyri - Panonychus ulmi and Amblyseius aberrans -
Eotetranychus carpini vitis seem to be successful.For its
advance, it is necessary to find new strategies in the control of the insect pests of grapes, or the rearing and releasing of these predator mites,resistant to some insecti-
cides.Altogether, all pest and disease mechanisms need to integrate together and with the cultural practices.On the other
hand, the vine growers have to learn and undertake the new
methods of integrated pest management,which are more economical and ecological than the previous used.

REFERENCES

1. BAILLOD, M. (1984).Lutte biologique contre les acariens
    phytophages.Rev.suisse Vitic.Arboric.Hortic.16,137 - 142
2. BAILLOD, M.,SCHMID, A., GUIGNARD, E.,ANTONIN, Ph., CAC-
    CIA, R. (1982). Lutte biologique contre l'acarien rouge
    en viticulture.II.Equilibres naturels,dynamique des popu-
    lations et expériences de lâchers de typhlodromes. Rev.
    suisse Vitic.Arboric.Hortic.14,345 - 352
3. BAILLOD, M.and VENTURI, I. (1980).Lutte biologique contre
    l'acarien rouge en viticulture.1.Répartition,distribution
    et méthode de contrôle des populations de prédateurs ty-
    phlodromes.Rev.suisse Vitic.Arboric.Hortic.12,231 - 238
4. BALEVSKI, A. (1980).How we control the second generation
    of the codling moth and the yellow apple mite on vines
    (bulg.).Rast.Zashch.28,42 - 43
5. BOLLER, E. (1978).Zunehmende Aktualität von Raubmilben
    im Weinbau.Schweiz.Zeitschr.Obst Weinbau 114,87 - 91
6. BOLLER, E. (1984).Eine einfache Ausschwemm-Methode zur
    schnellen Erfassung von Raubmilben,Thrips und anderen
    Kleinarthropoden im Weinbau.Schweiz.Zeitschr.Obst Wein-
    bau 120,16 - 17
7. BOLLER, E.F., JANSER, E. and POTTER, C. (1984).Prüfung
    der Nebenwirkungen von Weinbauherbiziden auf die Gemeine
    Spinnmilbe Tetranychus urticae und die Raubmilbe Typhlo-
    dromus pyri unter Labor- und Semifreilandbedingungen.
    Zeitschr.Pflanzenkrh.Pflanzenschutz 91,561 - 568
8. CHABOUSSOU, F. (1966).Nouveaux aspects de la phytiatrie
    et de la phytopharmacie.Le phénomène de la trophobiose.
    FAO Symp.Integrated Control,Vol.1,33 - 61
9. DUSO, C. and GIROLAMI, V. (1982).Ruolo degli Antocoridi
    rel controllo del Panonycnus ulmi Koch nei vigneti.Boll.

Istituto Entomol.Univ.Bologna 37,157 - 169
10. GÜNTHART, E. (1956).Das Rote-Spinne-Problem im Weinbau.
    Schweiz.Zeitschr.Obst Weinbau 65,14 - 20
11. GÜNTHART, E. (1957).Neues über Auftreten und Bekämpfung
    der Spinnmilben an Reben.Schweiz.Zeitschr.Obst Weinbau
    66,231 - 236
12. HAUB, G., STELLWAAG-KITTLER, F. and HASSAN, S.A. (1983).
    Zum Auftreten der Florfliege Chrysopa carnea Steph. als
    Spinnmilbenräuber in Rebanlagen.Die Wein-Wissenschaft
    38,195 - 201
13. IVANCICH GAMBARO, P. (1973).Il ruolo del Typhlodromus ab-
    errans Oud.(Acarina Phytoseidae) nel controllo biologico
    degli Acari fitofagi dei vigneti del Veronese.Boll.Zool.
    agr.Bachic.Ser.II.11,151 - 165
14. LOCHER, F.J. (1958).Der Einfluß von Dichlordiphenyltri-
    chlormethylmethan (DDT) auf einige Tetranychiden (Acari,
    Tetranychidae).Zeitschr.angew.Zool.45,201 - 248
15. OVERMEER, W.P.J. (1981).Notes on breeding phytoseiid
    mites from orchards (Acarina:Phytoseiidae) in the labo-
    ratory.Med.Fac.Landbouw.Rijksuniv.Gent 46,503 - 509
16. OVERMEER, W.P.J., KOODEMAN, M. and VAN ZON, A.Q. (1982).
    Copulation and egg production in Amblyseius potentillae
    and Typhlodromus pyri (Acari,Phytoseiidae).Zeitschr.ang.
    Entomol.93,1 - 11
17. RAMBIER, A. (1976).Beziehungen zwischen den schädlichen
    Milben und ihren Feinden.IOBC/WPRS (Ed.).Nützlinge in
    Apfelanlagen,Wageningen,107 - 109
18. RAMBIER, A. (1982).Un acarien,sur vigne en Champagne,nou-
    veau en France:Tetranychus mcdanieli McGregor 1931 du
    groupe Pacificus.Progr.Agric.Vitic.99,261 - 266
19. RAMBIER, A. (1982).Protection intégrée contre les aca-
    riens de la vigne.Progr.Agric.Vitic.99,267 - 271
20. SCHRUFT, G. (1967).Das Vorkommen räuberischer Milben aus
    der Familie Phytoseiidae (Acari,Mesostigmata) an Reben.
    Die Wein-Wissenschaft 22,184 - 201
21. SCHRUFT, G. (1969).Das Vorkommen räuberischer Milben aus
    den Familien Cunaxidae und Stigmaeidae (Acari) an Reben.
    Die Wein-Wissenschaft 24,320 - 326
22. SCHRUFT, G., WEGNER, G., MÜLLER, K.-D.and SAMPELS, J.
    (1983).Das Auftreten von Florfliegen (Chrysopidae) und
    anderen Netzflüglern (Neuroptera) in Rebanlagen.Die Wein-
    Wissenschaft 38,186 - 194
23. SEIFERT, G. (1961).Der Einfluß von DDT auf die Eiproduk-
    tion von Metatetranychus ulmi Koch (Acari,Tetranychidae).
    Zeitschr.angew.Zool.48,441 - 452

The biological control of phytophagous mites on vines: Role of *Kampimodromus aberrans* (Oud.) (Acarina: Phytoseiidae)

P.Ivanicich Gambaro

Istituto di Entomologia Agraria dell' Università di Padova, Italy

SUMMARY

Longitudinal research on the dynmics of Phytophagous mites and Phytoseiidae populations have shown that fluctuations of vine Tetranychidae are closely linked to anti-mildew programmes.

Studies on the biology and ecology of Kampimodromus aberrans (Oud.) has revealed that this species can both reduce high infestation levels and regulate phytophagous mites to a permanent low density owing to:

    a. the rapid growth rate,

    b. the high fecundity,

    c. the good capacity of diffusion,

    d. the high density levels during all the period of vegetation, due to its capacity to survive and reproduce even in the absence of prey,

    e. the high overwintering population which occupies the leaves at the beginning of vegetation.

The strategy of the introduction of Phytoseiidae in the vineyards is reported.

1. INTRODUCTION

Mites infestation in vineyards, which appeared in the 50's and were very widespread in vineyard areas, represents one of the most obvious phenomenoms due to the use of organic synthetic pesticides. In fact, its connection with the spread of grapevine mildew in viticulture was soon evident. All experts in this field were convinced of this, but what was not understood was the active mechanism of such preparations. There were two lines of research at that time. One line researched that the cause of infestation was an increase of life and fecundity of mites, due to the use of such insecticides. This was the theory of "trofobiosis" (2) followed by many authors (5, 14, 15). The other line attributed it to the destruction of enemies. Among the supporters of this theory are (12) and later (4).

When my research started in 1963, mite infestation was already widespread in vines all over Verona; however, my first observations referred to two vineyards when first signs of infestation appeared. The initial conditions were, in fact particularly interesting in understanding the activity and population dynamics of both phytophagous mites and predators.

Figure 1 shows the development of these populations in a vineyard where copper salts were reintroduced at the first appearance of infestation: the development of the Phytoseiidae population is

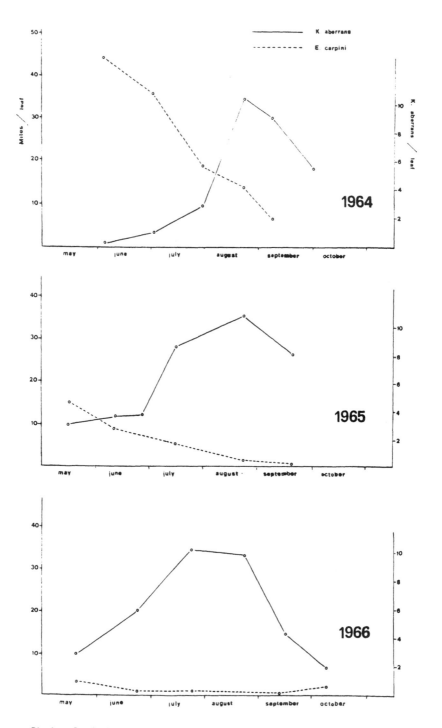

Fig.1. Population dynamics of E.carpini and K.aberrans during the years 1964, 1965, and 1966 after returning to the use of copper salts in a vineyard infested by E.carpini .

evident here already in the first season (1964) and a
re-establishment of balance in the two following years.

Afterwards, repeated longitudinal research also in other
vineyards confirmed the results of my first observations, i.e. the
value of Phytoseiidae in reducing and refraining the increase of
mites and, on the other hand, their sensibility to anti-mildew
fungicides, especially dithiocarbamates (7, 9).

2. PHYTOPHAGOUS MITES AND PREDATORS

The two species of phytophagous mites usually widespread in the
Verona area are: Panonychus ulmi Koch and Eotetranychus
carpini Oud. distributed variously at the beginning of infestation
and often found together on the same plant. Following this, a gradual
reduction of E.carpini with a greater diffusion of P.ulmi
resulted.

Consistent populations of Stethorus punctillum Weise or
Anthocoridae rarely appear in vineyards among the predators. The
only mite enemies, therefore, are some species of Phytoseiidae .

Research in 1970 and 1983 has shown that the species generally
widespread in the Verona area is Kampimodromus aberrans Oud.:
other species existed in small quantities, the order of frequency
being Typhlodromus pyri Scheuten, Amblyseius finlandicus
Oudemans, and Seiulus subsoleiger Wainstein. More recent research
has shown a reduction of K.aberrans Oudemans and an increase in
other species.

The frequency and spread of K.aberrrans is found in other
Italian viticulture environments: in the Venetian region (16), in
Sicily (13), in Tuscany (11), in the Marche region (PEGAZZANO,
personal communication). Furthermore, it is noted in the vineyards of
Southern France, Canton Ticino and other countries (Austria,
Bulgaria, Russia, Georgia and Iran).

My research refers to K.aberrans and is exclusively field
research, because of the obvious difficulty of rearing this species
in a laboratory. The methods used in various investigations are
described in my previous pubblications on this topic (8, 9, 10).

3. BIOLOGICAL AND ECOLOGICAL CHARACTERISTICS OF KAMPIMODROMUS
ABERRANS (OUD.).

3.1. The Biological Cycle and Characteristics

K.aberrans , like other Phytoseiidae , live on the undersurface
of leaves, along the midrib and in the corner between two nervations
where they form a group of 8-10. Also the eggs are placed along the
midrib, or more frequently, in the angle formed by their branching
off, even in groups of 3 or 4; also the small, newly hatched larvae
lie there.

The time required for development in a laboratory at 25 C. is
10 days (3) but in a natural environment this varies, due to various
factors but above all to the temperature and nature of feeding: in
the prescence of prey the average numbers of population increase
quickly from a low number to 25-30 per leaf (Fig.1). It is not
possible, therefore, to give an exact number of generations per
season, which also varies according to environmental conditions,

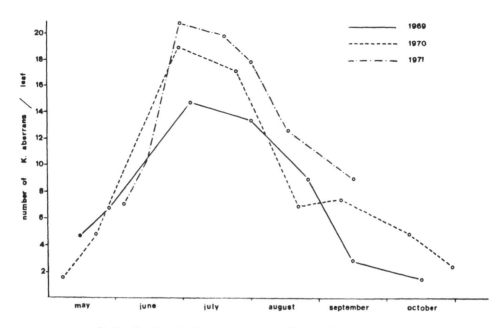

Fig.2. Development of K.aberrans populations in the absence of prey.

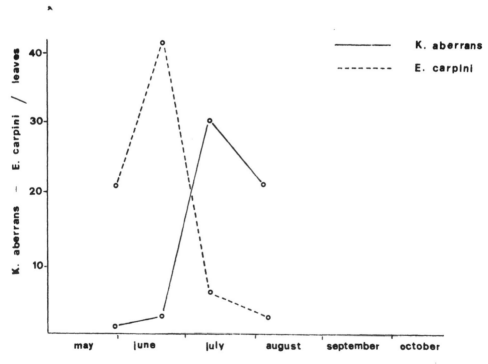

Fig.3. Rapid increase of K.aberrans populations and consequent reduction of E.carpini infestations.

because of the rapid multiplying of the individual from the 2nd and 3rd generations, and the continuous laying of eggs during the month of June, even in the absence of prey.

In our climate, K.aberrans Oud. is present on the plant from early in spring in the 2nd decade of April when the overwintering females reach the first shoots. Due to its remarkable fecundity, even in the absence of prey, the number per leaf increases continously in May and June, reaching its peak in July when 8-10 eggs can be found per leaf and medium population density of 30-35 per leaf. Reproduction decreases in the 3rd decade of July even though it continues until September (Fig.2) and in the presence of prey, until the 1st decade of October. The mated adult females are found on rare plant leaves, which have remained green, until the 2nd decade of November.

3.2 Dispersion Behaviour

K.aberrans is able to move very rapidly, so continually moves around the leaf surface, going from one leaf to another, from one part of the plant to the other. On infested vines, this dispersion corresponds to that of phytophagous mites. In fact, in April-May it is most likely found on the leaves at the base of the vine branches; in June-July on the leaves in the central areas, and in autumn on the top leaves.

Even on vines where prey is nearly absent, the type of dispersion during the season shows similar characteristics. It has been observed that the laying of eggs spreads upwards from the leaves at the base to the distals, and, in autumn, the highest number of K.aberrans per leaf is found on the apicals.

Research still in progress has also revealed consistent daily movement of phytoseiidae from one part of the plant to the other, this movement being determined by the sunlight. Such movement takes place only on plants in absence of prey, because in other cases, as has been noted, the phytophagous determine the dispersion of predators.

3.3. Prey Capacity

Studies of the population dynamics of K.aberrans in the presence of P.ulmi and E.carpini infestations, have shown how this predator is able to reduce serious infestation levels in a relatively short time. This is, above all, due to its high fecundity which, from a low number, can reach a medium density of 25-30 and maximum of 70 mites per leaf (Fig.3).

Both the young and adult mites are active predators: they prefer to feed on young phytophagous, but also feed on both P.ulmi and E.carpini adults. After having fed on the prey their abdomen takes on colour from the red or yellow juices of the Tetranychids.

A very important characteristic of this species is that it is possible for them to live and survive in large numbers on the host plant, even in the absence of prey, therefore, it allows them to refrain any eventual phytophagous regeneration, thus keeping them at an endemic density.

In the absence of prey K.aberrans can use other sources of nourishment: (3) in laboratory tests was able to make them grow on Colomerus vitis (Pgst) (Acarina : Eriophyidae), on pollen of D.stramonium Vl. and on Podosphaera leucotricha (Ell.and

239

Ev.)Salm. Some of my studies on the number of mites on cultures in the absence of prey has demonstrated that this species can use the juices of the host plant.

3.4. Overwintering

This takes place on the host plant as mated adult females. When reproduction ceases many females still remain on the leaves for about two months. Research carried out in autumn 1984 has revealed that in the 2nd decade of October these females still occupy 100% of the leaves, however in this season they prefer the top leaves of the vine branch. In the 3rd decade of October the lower leaves were occupied by 9 females per leaf while the top leaves had 35 per leaf. In the 1st and 2nd decade of November they are found only on the few remaining green leaves.

The females overwinter under the bark of vine branches (preferably two year old ones) in small groups or colonies of 30-50, often attached to Tydeidae masses. Groups of 6-8 can be found under the perules of the flower buds.

In such overwintering sites, these females reach a medium density of 15-30 per metre. The density varies on different parts of the vine branch: with the training system "pergola" on two year vine branches, 50% of the population can be found on the part of the branch near the vinestock, 19% on the central part and 22% on the last remaining part. BAILLOD has revealed a different overwintering population localisation in different training systems (1).

This abundance of well-protected shelters facilitates overwintering of vine Phytoseiidae hosts which, in fact, always reappear in large numbers on the new spring vegetation.

4. RELEASE OF PHYTOSEIIDAE

K.aberrans , like all other Phytoseiidae , is very sensitive to pesticide treatments. On vines it is, above all, fungicides (especially dythiocarbamates) which brought about large decreases of predators and are the basis of mite infestation. The insecticides used against grape moths do not result in large reductions. The reason for this is that the treatments are limited (1 or 2 per season) and carried out at a time of year when these predators reproduce again, being able to rebuild populations from eggs (not damaged by insecticides) or from the few females which escaped treatment.

Therefore, in many cases, it is sufficient to establish a control programme to obtain the natural reappearance of Phytoseiidae on the vine. However, in other cases where the reduction is nearly complete, it is necessary to recolonise the vineyard by transporting some Phytoseiidae from cultures which have an abundant supply.

On the contrary, as verified in orchards, the transport of vine leaves during summer is not successful. Transfer trials on young 3 year old vines with 20 leaves occupied by 30-50 mites per leaf, have not obtained a sufficient population reduction of E.carpini to avoid the symptoms of leaf damage, even if predator transmission onto the new host results. In fact, at the end of August, 68% of the leaves were occupied by K.aberrans with a medium density of 2 mites per leaf.

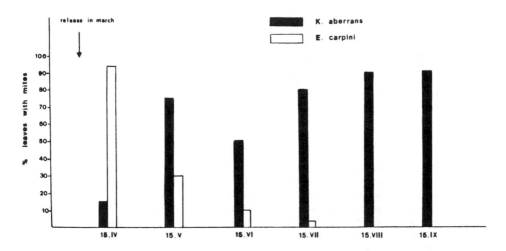

Fig.4. Development and diffusion of <u>K.aberrans</u> in a vineyard infested by <u>E.carpini</u> . Predacious mites were released, by transporting them with pruned wood, at the end of winter. Verona, 1984.

Transfer, at the beginning of the season, of shoots occupied by overwintering females and eggs, has given better results.

The best success, however, has been obtained with wood at the end of winter; with this method, which I have now used for 5 years, the results can be already seen at the end of May.

Figure 4 describes the results of a trial carried out in a 3 year old vineyard seriously infested by <u>E.carpini</u> and controlled for all the season. The release of <u>Phytoseiidae</u> was carried out at the end of winter with 2 year old wood, by collecting vine branches 60-80 cm long from near the vinestock, which have the highest number of overwintering predators. Such vine branches are tied 2-3 per vine on alternative rows: this results in a medium release of about 100 mites per vine.

Females were found in large numbers on the small leaves (15-20 per shoot) at the beginning of vegetation. Following this, the number of occupied leaves decreased because of the rapid growth of leaves. Later, in the months of July, August and September it increased again, reaching the highest population density with a medium of 15 predators per leaf in the 2nd decade of September.

The infestation was quickly reduced after the first months, so that during all the season there were no signs of leaf damage.

This, together with other trials, shows that the release of <u>Phytoseiidae</u> on vines is simple and practical.

The tendency of <u>K.aberrans</u> to spread is clearly evident here, in fact after some weeks following the release of the overwintering females, the predator is also present on intermediate rows where no release was carried out. Periodical controls over many years have shown that once the colonisation has taken place, the population increases and becomes dispersed in the culture. This keeps the phytophagous at a permanent low density, even in vineyards infested by <u>E.carpini</u> which is more difficult to control.

As seen from research on the biology and population dynamics of <u>K.aberrans</u> on infested vines and vines with a mite per predator

241

balance, this species can either respond to an elevated density of prey or regulate the phytophagous mites to an endemic density, owing to some of its important characteristics: the rapid growth rate and high fecundity which permit a rapid population increase; the ability to actively disperse; the long soggiorn on the plant, from the beginning of vegetation to late autumn at a continuous high density; the ability to overwinter in large numbers near shoots which become occupied early in the spring.

5. THE PRESENT SITUATION

In recent years with the use of copper salts in viticulture the situation has greatly improved. This demonstrates how mites have been a direct consequence of the destruction of their enemies, following the pluriennal use of dithiocarbamates fungicides.

In 1982 a research was carried out on some vines in Valpolicella (9). It showed a remarkable reduction in the number of mites following the reappearance of Phytoseiidae , and in many vineyards it showed a completely re-established biological control.

This unexpected solution of the problem of mites of vines comes from the new line of anti-mildew control that pathologists have been indicating for many years.

In fact, for various reasons (lack of colateral activity against Grey mould and Grape powdery mildew, late maturing of vine branches, etc) the pathologists advise to limit the use of diothiocarbamates fungicides to the first vegetative period only and use copper salts in treatments following.

The introduction of 'mixed' anti-mildew treatments (copper+zineb, copper+mancozeb etc) has consequently weakened the noxious action of the products on Phytoseiidae mites which remain undisturbed from July on and can build up a certain population density towards the end of the season. This is the reason for the surprising reapperance of Phytoseiidae mites in vineyards whereas, about 10 years ago there was no sign of any mites.

Therefore, the fluctuations of phytophagous mites which we have observed during 25 years are shown to be directly connected with anti-mildew treatment programmes which can either stop or protect the balance of mites per predators. Therefore the biological control, as seen in the vineyards, and different from other cultures, can be done without difficulty.

REFERENCES

1. BAILLOD, M. and VENTURI, I. - (1980) - Lutte biologique contre l'acarien rouge en viticolture. Rev. Suis. Vitic. Arboric. Hortic. 12, 2321 - 238.

2. CHABOUSSOU, F. (1969) - Recherches sur les facteurs de pullulation des acariens fitofages de la vigne a' la suite des traitements pesticides du feuillage. These Fac. Sci. Paris, 1 - 238.

3. DAFTARI, A. (1979) - Studies on the feeding, reproduction and development of Amblyseius aberrans (Acarina : Phytoseiidae) on various food substances. Zet. Ang. Ent.88, 85-89.

4. FLAHERTY D.L. and HUFFAKER, C.B. (1970) - Biological control of Pacific Mites and Willamitte Mites in San Joaquin Valley vineyard. I. Role of Metaseiulus occidentalis II. Influence of dispersion patterns of Metaseiulus occidentalis Hilgardia 40, 267 - 330.

5.   GUNTHART, E. and VOGEL, W. (1965) - L'influence des produits antiparasitaires sur les araignees rouges. Boll.Zool. Agr. Bach 2, 131-141.

6.   IVANCICH GAMBARO, P. (1965) - Considerazioni sulle inestazioni di acari della vite. L'inform Agr. 11, 979.

7.   IVANCICH GAMBARO, P. (1972) a - I trattamenti fungicidi e gli acari della vite. L'inform. Agr. 8141-43.

8.   IVANCICH GAMBARO, P. (1972) b - Il ruolo del Typhlodromus aberrans (Acarina : Phytoseiidae) nel controllo biologico degli acari fitofagi dei vigneti del Bernese. Boll. Zool. Agr. Bach, 11, 151 - 165.

9.   IVANCICH GAMBARO, P. (1982) - Le infestazioni degli acari sulla vite: venti anni dopo. L'Inform. Agr.38, 22277-81.

10. IVANCICH GAMBARO, P. (1984) - Rame e acari della vite - Richerche nei vigneti del Veronese. Vignevini, 2, 85-89.

11. LIGUORI, M. (1980) - Contributo alla conoscenze degli acari della vite in Toscana. Redia, 63,407-15.

12. MATHYS, G. (1958) - The control of Phytophagous mites in Swiss vineyard by Typhlodromus species. 10 Int. Congr. Ent.4, 607-610.

13. NUCIFORA, A. and INSERRA, R. (1967) - Il Panonychus ulmi Koch nei vigneti dell'Etna. Entomologica, 3 177-236.

14. RUI, D. and MORI, P. (1968) - Interferenze fra le applicazioni terapeutiche e le nuove infestazioni di acari fitofagi sulla vite. Ati Accad. ital. Vite vino Siena, 20, 3-19.

15. SCHRUFT, G. (1972) - Effects secondaires de fongicides agissant sur les acariens (Tetranychidae : Acarina) sur Vigne. OEPP/EPPO Bull.3, 57-63.

16. ZANGHERI, S. and MASUTTI, L. (1962) Osservazioni e considerazioni sul problema degli acari della vite nelle Venezie. Riv. Enol. Vitic. 15,75-89.

Session 2
Diseases and weeds
Chairman: M.Clerjeau

The evolution of major vine diseases and methods of control against them

M.Clerjeau

INRA Bordeaux Research Center, Plant Pathology Station, Pont de la Maye, France

Summary

European vines are sensitive to some fifteen diseases (excluding viruses), the extent varying according to region for primarily climatic reasons. Man influences the relative importance of these diseases directly or indirectly through international exchanges, varietal or cultural choices, spraying, etc... Substantial progress in fighting against them has been made in the last hundred years. However it must be admitted that phytosanitary protection of vines, especially by chemical means, imposes a heavy financial burden whilst risking to induce either pest resistance of undesirable secondary effects. In addition, the epidemology of certain fungi remains unclear and efficient protection against root diseases is still unobtained. In order to guarantee vineyard protection and quality production as well as to accompany progress in viticulture with a fair estimation of epidemologic consequences, research effort should be intensified in several directions : impact studies of prophylactic measures, harm thresholds, prediction methods and models, biological treatments and chemical treatment strategies (minimizing risks of resistance and secondary effects). These actions should be taken in several fields and deserve international cooperation.

1. Introduction

The listing and describing of vine diseases in 1985 risks being superfluous due to the numerous works on the subject published since the turn of the century. However, phytopathologists responsible for establishing research priorities necessary for the future of viticulture may be assisted by an analysis of why, where and how these diseases appeared and developped as well as by an evaluation of the advantages and disadvantages of the available means of fighting against them. Such is the aim of this exposé.

2. Major Vine Diseases

2.1. Cryptogamic parasites

In the treatises of Arnaud (1931) or Galet (1977) more than thirty distinct diseases are noted. Many of them are rare or of secondary importance at least in our temperate climate. Actually, a dozen diseases preoccupy to various degrees the European vinegrowers :

2.1.1. Leaf and Berry diseases

Geographic distribution and the extent of these diseases is determined essentially by their climatic needs. Powdery mildew **(uncinula necator)**, Downy Mildew **(Plasmopara viticola)** and Grey mold **(Botrytis cinerea)** are the most widespread.

Black Rot **(Guignardia bedwellii)** appears especially in rainy regions (preferably warm). Thus it is not found in Mediterranean or Northern vineyards.

Brenner Rot **(Pseudopeziza tracheiphila)**, on the contrary, needs cold and dry winter conditions and appears mainly in northern or balkanic regions or occasionally in southern regions.

Anthracnose **(Sphaceloma ampelinum)** occurs mainly in humid areas and only appears sporadically in non-sulphated vineyards, principally on hybrids.

White Rot **(Coniothyrium diplodiella)**, ancient European disease, occurs only occasionally in certain regions after hail damage.

2.1.2. Cane or trunk diseases caused by ligneous fungi.

All of these are widespread in European vineyards, where vine variety, age and preventative measures are in relation to damage : budding defects, slow growing, vine weakness and death. Implicated are dead arm disease **(Phomopsis viticola)**, Eutypa die back **(Eutypa armeniacae)** and Black measles **(Stereum hirsutum** and **Phellinus igniarius** probably) with aerial dissemination in addition to Root rot, whose agents **(Armillaria mellea, Rosellinia necatrix)** are soil inhabiting.

2.2 Phytopathogenic bacteria and mycoplasma like organisms (MLO).

There are no widespread bacterial or MLO-caused disease in the entire European vineyard. Bacterial necrosis or Oleron disease, caused by **Xanthomonas ampelina**, is at present the most dreaded of these. It is limited to a few Atlantic (Cognac) or Mediterranean vineyards.

Crown-gall, responsible for surface tumors, is caused by **Agrobacterium tumefaciens**. It is considered to be an occasional disease connected with winter frosts.

Flavescence dorée (caused by a MLO) has always remained localized in the south-west region of France.

Pierce disease, serious in North America, has not yet been reported in Europe. It presents however, a potential risk.

2.3 Virus diseases

Some thirty identified virus diseases attack vines, in addition to several others, poorly defined. They are transmitted by aphids, soil fungi or nematodes (Fanleaf or Panachure virus). The transmission means of some is not yet known. In all cases contamination can occur by grafting, rendering indispensable the indexing of mother plants used for stock production.

Thanks to sanitary selection these diseases are becoming rarer in vineyards. Because of the general inefficacy of treatments for this type of disease, the topic will not be treated in this exposé.

3. Evolution factors of relative disease importance.

If we were to describe the sanitary state of the European vineyard of 150 years ago our work would be rather simple. Viticulture lived its golden age until the mid-1800s, troubled only by a few relatively harmless indegenous diseases such as Grey mold, White Rot, Dead arm disease and the most dangerous Anthracnose (fig. 1)

Fig. 1
Anthracnose (Black spot):
Main disease of Grapevine
in Europe until 19th cen-
tury, it has now prati-
cally disapeared.

Lignivorous parasite causing Black measles and Root rot were also part of
the tribute paid to an otherwise generous Nature.

The evolution of plant parasite problems since that time dominated by
the appearance and development of new diseases in certain regions, was
unwittingly determined by Man. Several factors played an essential role.

3.1 Transport of contaminated plant matter.

The growth of trade amongst regions and nations, particularly between
the Old World and the New resulted in the introduction to Europe during the
Second half of the XIXth Century of four vine enemies previously present
only in North America. First was oïdium imported with cuttings in 1845 and
the first to make necessary a vine plant protection programm. Fortunately,
the preventive action of sulphur soon became evident. However, the heavy
importation of American vines from the 1860's on resulted in the successive
introduction of Phylloxera in 1858, Downy mildew in 1878 and Black Rot in
1885, leading to unprecedented socio-economic crises.

These examples are striking in that they involve the worst enemies of
vines and their transcontinental development over a limited period of
time. However, other more recent diseases must be noted : the appearance of
Brenner Rot in the United States, identified in 1985, and the spread of
Bacterial necrosis in the South of France through the use of vine plant
cuttings of sensitive varieties (Grenache-Alicante Bouschet).

3.2 Varietal changes.

Different factors lead vinegrowers to reorient their choice of vine
variety (market demand, quality improvement, adaptation to new cultivation
techniques, etc...). The most noted change has been the reconstruction of
vineyards after the **Phylloxera crisis** and the use of hybrid varieties.
Taking into account the particular sensitivity of each variety to different
diseases, it may be said that the varietal profile of a given vineyard
associated with the region climate type strongly influence its
phytopathological risk profile. Many examples illustrate this point.

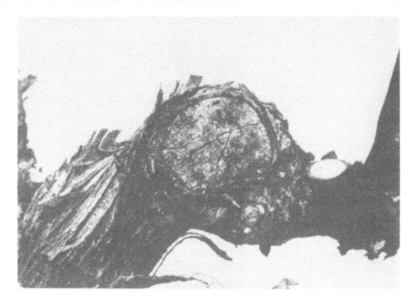

Fig. 2 : Prune woundings : a way for penetration of lignivorous fungi, such as **Eutypa, Stereum** or **Phellinus.**

3.3 Change of cultivation methods.
      Cultivation methods can influence plant receptivity or different phases of parasite biology, and hence favour or not vine diseases.
      Among the more significant effects are :

3.3.1. Those which act on conservation stage of pathogenic agents.
            Ex. the presence of **G. bidwellii** in spring is determined by contaminated leaf abundance on soil surface, hence by soil-working practice. Likewise, mechanical harvesting increases the level of inoculum by leaving contaminated bunch stalks on the vine.
            Ex. the generalization of grafting has led to the production of plants from buds situated sometimes at the base of vine-shoots, frequently contaminated by **Ph. viticola,** agent of Dead arm disease. Before the **Phylloxera crisis** rooted plants were produced from healthy buds, being located in the middle or ends of vine-shoots. Thus grafting has indirectly contributed to Dead arm spread.

3.3.2. Those acting upon parasite penetration and transmission.
      These involve mainly parasites inoculated by vine wounds. Such is the case for **E. armeniacae,** Eutypiose agent whose recent development is attributed to the severe pruning practiced on vines whose training is changed to accommodate mechanical harvesting. Likewise, harvesting machines and pruning tools are notorious agents of Bacterial necrosis extension (Fig. 2).
3.3.3. Those which modify the vegetal microclimate and hence parasite
                  ecological condition.
      Training systems (tied to pruning methods) operations affecting leaf volume (suckering, leaf removal,...) or plant vigor (fertilizer, rootstock) have well-known effects on the development of humidity-sensitive parasites. Such is the case for Downy Mildew or Grey mold. Also, leaf mass humidity is increased by irrigation or cultivation in lowland areas.

250

4. Evolution of fungicides used on vines.

Fungicides act not only on the parasites they are intented to control. As biocidal or biostatic substances they can have a considerable influence on the ecosystem. Thus they may have a use against parasites considered as secondary as well as perturb the florae in a sense favorable to the development of others parasites. The complexity of these phenomena and their slow evolution render necessary long-term experiment programs.

Since the discovery of Bordeaux mixture a century ago, vineyards have benefited from constant discoveries in phyto-pharmaceutics. They have also been subjected to the effects of new fungicide families. These progressive changes have been accompanied by a concommitant evolution of parasitic problems. For example, cupric treatments against Downy Mildew which became generalized at the beginning of the century had led to the quasi-disappearance of Anthracnose. Their progressive replacement towards the 1950s by synthesized organic fungicides has led to increase attack by Powdery mildew and a regression of Mites attack. Likewise, the development of Bacterial necrosis in the Charentes region of France is in part attributed to the abandon of cupric treatments. There is also a parallel between Black measles attack and the substitution in post-budding treatments by synthesized organic compounds of sodium arsenite, used in Winter against Dead arm disease.

The examples presented above, far from being exhaustive, were chosen to demonstrate that viticulture is an evolving system in rapport with equally evolving parasite factors. If in the past the phytopathologist was mostly a knowledgeable spectator, occupied with solving the newly posed questions, it is today that he has become a co-actor of viticulture's technological progress. For this he must evaluate and war against phytopathological consequences of new viticultural practices, together with phytotechnicians and breeders. What consequences are there for research in the future ?

5. Future research to develop on vine phytopathology.

It cannot be denied that in the last hundred years, ever since vine diseases first became economically harmful, constant progress has been made to fight against them. Thanks to varietal selection, grafting and chemicotherapy, self rooting hybrids have been eliminated, thus permitting both vine protection and product quality. Familiarity with parasite biology (Downy Mildew, Black-rot, Dead arm) has led to a better organisation of agriculture forecastings.

In addition, new weapons have come to light concerning antagonist characteristics of certain microorganisms used in biological control (ex. **Trichoderma harzianum**). Rapid immunology virus detection methods have improved sanitary selection traditionally based on the indexing of mother plants. Thus a greater guarantee against vine decline disease is possible. In spite of the progress made, it must be recognized that phytosanitary protection of vines constitutes a sizeable part of production costs (5 or 7 per cent in France). Treatments are rarely programmed in relation to objective risk criteria, thus raising the cost burden. Root rot diseases have still not been mastered, nor has the epidemiology of the most dangerous parasites (notably Oïdium). In addition, chemical control can induce parasite resistance or other undesirable effects, as a consequence, several research guidelines should be favored in the future.

Fig. 3
Trichoderma harzianum :
an antagonistic fungus
for **Botrytis cinerea.**

5.1 Improve risk evaluation methods.

Although it is necessary to continue studying the epidemiology of certain parasites whose transmission and dissemination conditions are not well known (eg. lignicole agents as well as **U. necator** and **G. bedwellii**), it would be wise to further research regarding risk modeling. Results already obtained in attack simulation by **B. cinerea** and **P. viticola** with ACTA models perfected by STRIZYK indicate a new approach to agricultural warnings and treatment decision-making. These models should be improved and adapted to particular variety or vineyard conditions in order to cover other parasites, especially Oïdium.

5.2 More exact appreciation of prophylactic measures and danger thresholds.

The prophylactic role of certain cultural practices (suckering and clipping against Downy mildew, leaf removal against Grey mold) is generally well krown. So are the consequences of certain agro-technical factor on diseases development (fertilizer, rootstock, pruning, etc.). But this is after empirical knowledge. It is mostly unknown whether or not cultural pratices can influence the number of treatments. It is important to engage in experimentation in this area, defining a code of "savoir-vivre" with the diseases. This implies as well a greater knowledge of the danger threshold of the different vine parasites.

5.3 Better use of fungicides.

Chemical control is the backbone of vineyard protection. The perfection of on the spot experimental methods using artificial contamination of plants with fogging or spraying to accentuate diseases development today permits knowledge of vine fungicides action (systemicity, curative power, persistance, vapor effect etc.). Instructions for use of products are the result of knowledge acquired from these experiments.However,two aspects were regulated up until now :

Fig. 4
Monitoring of vineyards to estimate the development of resistant strains to fungicides needs suitable laboratory tests : here inoculation of **P.viticola** sporocysts on leaves discs-treated with metalaxyl.

5.3.1. The adaptation of treatment strategies to the risk of resistant strains development.

For the last fifteen years viticulture has been confronted with problems of **B. cinerea** resistance to fungicides (benzimidazoles since the 1970s, cyclic imides since 1978). More recently the development of **P. viticola** strains resistant to phenylamides has compromised the future use of these new systemic anti-mildew fungicides. To solve these problems concerning very active molecules having a unisite mode of action, vineyard monitoring systems must be established, especially for detecting any eventual resistance to new fungicides (Fig. 4). For instance no monitoring exists for the anti-oïdium of ergosterol biosynthesis inhibitors. Likewise the incidence of alternation or active matters association on resistant strains frequency and on protection efficiency in resistance situations must be examined.

5.3.2. Secondary effects.

An important goal of integrated control against vine diseases is awareness of the major repercussions of treatments applied against a given disease, notably their direct or indirect (microflora modification) action on other diseases. Work has already begun in this field. The length of time needed for tests, the need for randomisation of treatments and the diversity of active matterials used in viticulture form sizeable obstacles. Realism obliges us to admit that every thing desired cannot be undertaken in this area and that specialists must decide together on priorities.

5.4 Taking advantage of biological control perspectives.

Among the microorganisms having antagonistic or hyperparasitic properties, **Trichoderma harzianum** appears to be a potentially interesting agent of biological control against **B. cinerea,** this according to experimental results obtained over the last ten years. Efficiency results obtained in France are in general inferior to those observed with dicarboximides (average of to 70 per cent efficiency). However, fungicide resistance phenomena should lead to clearer reasoning in treatment strategies. Test by INRA of Bordeaux have shown that mixed control based

253

Fig. 5 : Cultivars tested for their diseases susceptibility: artificial
contamination and fogging.

on **T. harzianum** in stages A and B with dicarboximides in stages C and D
insure satisfactory protection. The practical development of biological
control must base itself on intensifying research in several directions :
strains selection, inoculum production technology and the biology of **T.
harzianum** in aerial ecosystems.

5.5 Making use of genetic varietal resistance.
    Sensitivity differences between varieties could only with difficulty
permit agricultural forecasting stations to modify their treatment warnings
for the different grape varieties. Nevertheless, vinegrowers know that
chemical protection is easier to insure for certain varieties than for
others. It would be desirable in the future to improve the inventory of
varietal sensibility to the greatest number of diseases especially for new
varieties. This knowledge is essential for the realization of zoning
operations as well as for genetic improvement programs for **V.
vinifera** vines.

    If it is desired in the future to reduce the number of treatments
through varietal selection, the creation of new varieties must be
accelerated. Selection programs underway in several European countries aim
to associate the organoleptic quality of grapes with vine resistance to the
major diseases. For example the goal of research at INRA of Bordeaux is to
create varieties of high agronomic value with high resistance to Downy
mildew, less sensitivity to Oïdium, Black rot, Anthracnose and Grey mold,
capable of being protected by three treatments annually. To obtain this,
selection programs should be established in close collaboration with
epidemiology and chemical studies.

Conclusion

It is inevitable that in the future decisions taken for Vine protection must result from global reflection concerning possible operations. Positive effects (protection level) and negative consequences (cost, pesticide residues, etc.) will more and more be weighed one against the other for economic or ecological reasons. Until recently, the work of researchers presented an analytic character poorly adapted to the needs of integrated protection (eg. action of a fungicide on a given parasite, influence of pruning method on harvest quality and quantity, agronomical value of a rootstock). Now the guidelines should be more synthetic andinvolve several fields of research. The diversity and the complexity of the problems at hand,due to the great variety of ecological systems and cultural practices of Europes vineyards, necessitate close collaboration between research and advances technology organisations. This knowledge exchange must benefit from extensive international cooperation.

Biological and integrated control of gray mould of grape

M.L.Gullino, M.Mezzalama & A.Garibaldi
Istituto di Patologia Vegetale dell' Università di Torino, Italy

Summary

Several isolates of Trichoderma spp., used as biocontrol agents, alone or in combination with benzimidazole and dicarboximide fungicides, showed a good activity against gray mould of grape incited by Botrytis cinerea Pers.
During field experiments, carried out throughout the years 1978--1984, the alternation of one chemical spray with 2-3 treatments with the antagonists gave satisfactory results.
The importance of the selection of strains of Trichoderma resistant to the most common fungicides used against the most important grape pathogens is discussed.

1.1 Introduction

Gray mould of grape, caused by Botrytis cinerea Pers., causes considerable losses when environmental conditions are favourable to its development. At present fungicides still play a major role in the control of this disease.
The general need of reducing the dependance on chemicals and, particularly, the development and spread of resistance to benzimidazoles and dicarboximides, the two groups of fungicides originally the most active against gray mould, strongly support the search of alternative control measures.
The antagonistic capacities of the fungus Trichoderma are well known: several Authors described the activity of different species of Trichoderma against gray mould of apple, strawberry, grape and lettuce (Wood, 1951; Tronsmo and Dennis, 1979; Tronsmo and Ystaas, 1980; Dubos et al., 1982; Lantero et al. 1982).
The work carried out under laboratory and field conditions during

Work carried out with a grant from Ministero Pubblica Istruzione (40%) "Lotta biologica contro i patogeni delle piante".

the past years in order to develop and test alternative control strategies against gray mould of grape is here summarised.

1.2 Laboratory experiments

Hundreds of strains of Trichoderma spp. isolated from soil, grape leaves and bunches were tested on grape berries under laboratory conditions (Lantero et al., 1982). About 20 isolates, showing the highest antagonistic activity, were selected and used for further experiments.

In order to develop integrated control strategies, based on the use of both fungicides and antagonists, resistance to dicarboximides and benzimidazoles was induced in the most active Trichoderma strains. Generally Trichoderma growth is reduced by 300 μg/ml of vinclozolin: at this concentration sporulation is poor. Benomyl, at 1 μg/ml, completely inhibits Trichoderma growth.

Resistant strains of Trichoderma were obtained from the wild type by mass selection of conidia treated with UV radiation (30 minutes, 240 nm) on malt agar amended with 1,000 μg/ml of vinclozolin and 100 μg/ml of benomyl. Resistant isolates developed at variable frequency in the case of both fungicides (table 1).

Table 1 - Frequency of appearance of mutants of Trichoderma spp. resistant to dicarboximides and benzimidazoles after a 30' exposure to UV radiations (from Gullino et al., 1985)

Wild type	Frequency of appearance ($\times 10^{-8}$) of mutants resistant to	
	dicarboximides	benzimidazoles
T 5	4	11
T 4	4	6
T 15	3	3
T 14	12	7
T 13	5	3
T 2	3	2

Dicarboximide and benzimidazole resistant mutants, coded respectively as RD and RB, were able to grow in the presence of high concentrations of the fungicides to which they are resistant (M.I.C. respectively > 1,000 μg/ml for vinclozolin and > 300 μg/ml for benomyl). Both dicarboximide and benzimidazole resistant strains of Trichoderma spp. profusely sporulated in the presence of the fungicides to which they showed resistance and maintained high antagonistic activity, as evaluated under laboratory conditions on grape berries (table 2).

Particularly interesting appears the high activity shown by the iso-

258

Table 2 - Antagonistic activity of isolates of _Trichoderma_ spp. resistant to dicarboximides (RD) and to benzimidazoles (RB) against gray mould evaluated on grape berries under laboratory conditions.

Antagonistic activity	Number of strains	
	RD	RB
Low	4	9
Medium	13	18
High	21	64

lates of _Trichoderma_ spp. resistant to benzimidazoles and dicarboximides when used against strains of _Botrytis cinerea_ resistant to these fungicides on detached grape bunches under laboratory conditions.   Actually the antagonist actively controlled a dicarboximide resistant isolate of the pathogen while vinclozolin alone, at normal dosage, was only partially active (table 3).

Further induction of resistance by using the dicarboximide and benzimidazole resistant strains previously described led to the development of mutants showing double resistance to both groups of fungicides. When resistance to benzimidazoles was induced in isolates already resistant to dicarboximides, mutants were coded as RDB; when resistance to dicarboximides was obtained in mutants already resistant to benzimidazoles, strains were coded as RBD (table 4).   Also these strains maintained a good antagonistic activity (table 5).

More recently, new mutants of _Trichoderma_ spp., resistant to fungicides active against grape powdery and downy mildew have been selected (Gullino et al., unpublished).

1.3 Field experiments

Under field conditions _Trichoderma_ was applied as conidial suspensions (10^7 conidia/ml) prepared in 1°/°° malt solutions by spraying the bunch region.   Malt should offer a nutritional support to the antagonist helping to start colonising flowers and berries.   Isolations carried out at regular intervals from sprayed bunches by means of a spore trapping device directly in the field, confirmed the presence of high amount of _Trichoderma_.

The antagonist was used as single isolates during the first years, then as mixtures of several (4 to 8) isolates: the use of mixtures should avoid insuccesses due to failure of single antagonists and exploit eventual synergism among different isolates.

Trichoderma alone and better in mixture (table 6) or alternation with a dicarboximide (table 7) significantly reduced the incidence of gray mould in experiments carried out during 1981.

Table 3 - Effectiveness of an isolate of Trichoderma sp. resistant to dicarboximides, against an isolate of B. cinerea resistant to dicarboximides, evaluated on grape berries under laboratory conditions

Grape inoculated with		Vinclozolin mg of a.i./l	% infected berries
B. cinerea RD20	--	--	21.8 c
--	Trichoderma sp. MIRD	--	0.0 a
B. cinerea RD20	Trichoderma sp. MIRD	--	9.5 a
B. cinerea RD20	--	750.0	13.2 b
B. cinerea RD20	--	375.0	15.1 bc
B. cinerea RD20	--	187.5	18.7 bc
B. cinerea RD20	Trichoderma sp. MIRD	375.0	6.1 a
B. cinerea RD20	Trichoderma sp. MIRD	187.5	5.8 a

Table 4 - Frequency of appearance of mutants of Trichoderma spp. showing double resistance to dicarboximides and benzimidazoles after a 30' exposure to UV radiations.

Isolate	Frequency of appearance $(x\ 10^{-8})$ of mutants resistant to	
	dicarboximides (RBD)	benzimidazoles (RDB)
5/7 RB	3.2	--
5/9 RB	6.7	--
5/12 RB	3.1	--
13/4 RB	5.2	--
4/14 RB	6.2	--
4/18 RB	4.8	--
4/11 RD	--	4.0
13/5 RD	--	0.0
13/3 RD	--	6.7
15/2 RD	--	12.5
2/4 RD	--	10.9
2/5 RD	--	5.9
5/7 RD	--	0.0

The development of field resistant isolates of B. cinerea showing a low-moderate level of resistance to dicarboximides (Gullino and Garibaldi, 1982) led us to insist, during the following years, on control strategies based on the alternation of the antagonist and the fungicide, because the use of reduced dosages of dicarboximides, as employed in mixtures, should further select for resistance.

A very interesting activity was exhibited during 1984 by a mixture

Table 5 - Antagonistic activity of isolates of Trichoderma sp. showing double resistance to dicarboximides and benzimidazoles (RBD and RDB) against gray mould evaluated on grape berries under laboratory conditions.

Antagonistic activity	Number of strains	
	RBD	RDB
Low	5	2
Medium	4	6
High	10	15

of isolates of Trichoderma resistant to benzimidazoles when applied in a vineyard where the low activity of these fungicides alone suggested a consistent incidence of benzimidazole resistance in B. cinerea, as confirmed by monitoring (table 8).

1.4 Discussion

Trichoderma spp. when used as biocontrol agent against gray mould of grape offers a partial control of the pathogen, similar to that offered by "conventional" fungicides such as dichlofluanid and phthalimides. Unfortunately, expecially in the case of varieties very susceptible to the pathogen, a partial control is not sufficient from an economic point of view.

The integration of chemical and biological control measures seems much more promising: it allows a reduction of the total amount of chemicals used offering, at the mean time, a very good control of the pathogen. A reduction of the amount of fungicides used, besides any ecological implication, is also fundamental in order to delay and reduce the risk of development of fungicide resistant strains of the pathogen in the field.

In our country up to now resistance to dicarboximides in the case of grape is not a practical problem and resistance to benzimidazoles, though present, is less dramatic than in other countries (Gullino and Garibaldi, 1985). In this situation a rational and limited use of chemicals so called "at risk" would delay the failure of disease control induced by the use of fungicides alone.

From a practical point of view, the alternation of the use of Trichoderma and fungicides offers several advantages:
1) the antagonist, sprayed at the earlier treatments, grows on flower residues and berries, avoiding colonisation by B. cinerea;
2) the fungicides, employed at the most critical moment for gray mould infection (changing of colour of the berries, in our cultural conditions) perfectly integrate the antagonistic activity of Trichoderma. The present situation of benzimidazole-resistance in italian vineyards still allows one spray/year with such group of compounds;

Table 6 - Effectiveness of different treatments against gray mould (cv. 'Grignolino', 1981) (from Lantero et al., 1982).

Treatment	Moment of application	Dosage of vinclozolin g of a.i./ha	% infected berries
Trichoderma sp. MIRD	ABC	--	15.9 ab
Trichoderma sp. MIRD	ABCD	--	19.8 b
Trichoderma sp. 3RD	ABC	--	21.7 b
Trichoderma sp. 3RD	ABCD	--	16.8 b
Vinclozolin	ABC	750	12.5 a
Vinclozolin	ABCD	750	8.3 a
Vinclozolin	ABC	375	13.7 ab
Vinclozolin	ABCD	375	13.9 ab
Vinclozolin	ABC	187	16.4 b
Vinclozolin	ABCD	187	21.3 b
Trichoderma sp. MIRD+Vinclozolin	ABC	375	16.3 ab
Trichoderma sp. MIRD+Vinclozolin	ABCD	375	10.2 a
Trichoderma sp. MIRD+Vinclozolin	ABC	187	11.4 a
Trichoderma sp. MIRD+Vinclozolin	ABCD	187	14.4 ab
Trichoderma sp. 3RD+Vinclozolin	ABC	375	7.9 a
Trichoderma sp. 3RD+Vinclozolin	ABCD	375	11.8 a
Trichoderma sp. 3RD+Vinclozolin	ABC	187	18.6 b
Trichoderma sp. 3RD+Vinclozolin	ABCD	187	12.1 a
--	--	--	34.5 c

Table 7 - Effectiveness of different treatments against gray mould (cv. 'Grignolino', 1981) (from Lantero et al., 1982)

Antagonist	Moment of application	Moment of treatment with Vinclozolin (750 g of a.i./ha)	% infected berries
Trichoderma sp. MIRD	ABC	--	14.1 ab
Trichoderma sp. MIRD	ABCD	--	15.4 ab
Trichoderma sp. 3RD	ABC	--	14.8 ab
Trichoderma sp. 3RD	ABCD	--	14.0 ab
Trichoderma sp. MIRD	AB D	C	17.0 b
Trichoderma sp. MIRD	AB	C	14.4 ab
Trichoderma sp. 3RD	AB D	C	13.8 ab
Trichoderma sp. 3RD	AB	C	10.2 a
--	--	C	14.2 ab
--	--	ABC	12.5 a
--	--	ABCD	8.3 a
--	--	--	34.5 c

Table 8 - Effectiveness of different treatments against gray mould (cv. 'Barbera', 1984) (from Mezzalama et al., 1985).

Antagonist	Moment of application	Fungicide applied at C	% infected berries
Trichoderma RD (*)	ABCD	--	13.2 b
Trichoderma RD (*)	AB D	Vinclozolin (750 g/ha)	3.2 a
Trichoderma RB (*)	ABCD	--	17.3 c
Trichoderma RB (*)	AB D	Benomyl (500 g/ha)	6.3 a
--	--	Vinclozolin (750 g/ha)	6.3 a
--	--	Benomyl (500 g/ha)	21.2 cd
--	--	--	25.3 d

(*) Mixtures of isolates resistant to dicarboximides (RD) and benzimidazoles (RB).

3) a reduced use of chemicals sprayed at full dosages, compared with repeated treatments or with mixtures of fungicides (at reduced rates) + antagonists, restricts the selection pressure exerted, thus reducing the risk of a further selection for resistance.

On the contrary resistance to fungicides, in the case of Trichoderma, can, once in a while, represent a helpful tool: the induction of mutants of the antagonists resistant to fungicides commonly used against powdery and downy mildew permits early application which can improve colonisation of the host.

Additional studies in order to find new strains of Trichoderma and, eventually, other microorganisms well adapted to colonise and survive on grape berries, are necessary.

Moreover nutritional substrates favourable to the antagonist and not to the pathogen should be developed in order to improve berry colonisation by Trichoderma.

Besides scientific problems, practical problems too still need to be solved. The preparation of large amounts of bioformulations, stable and active for satisfactory lengths of time, represents a practical problems. Moreover in Italy, where the legislation concerning biological compounds is very strict, the registration of such preparations will not be easy to obtain.

REFERENCES

1. DUBOS, B., JAILLOUX, F. and BULIT, J. (1982). L'antagonisme microbien dans la lutte contre le pourriture grise de la vigne. Bulletin OEPP. Vol. 12 171-175.
2. GULLINO, M. L. and GARIBALDI, A. (1982). Risultati di una indagine effettuata per verificare la presenza di ceppi di Botrytis cinerea Pers. resistenti ai benzimidazolici e ai dicarbossimidici nei vigneti italiani. Notiziario sulle Malattie delle Piante. Vol. 103 145-150.
3. GULLINO, M.L. and GARIBALDI, A. (1985). Present situation of resistance to fungicides in italian vineyards. Fungicides for Crop Protection, BCPC. Monograph nr. 31. Vol. 2 319-322.
4. GULLINO, M. L., MEZZALAMA, M. and GARIBALDI, A. (1983). Selection of mutants of Trichoderma viride antagonist to Botrytis cinerea of grapes and resistant to several fungicides. Proceedings of the International Conference of Integrated Plant Protection, Budapest.
5. GULLINO, M. L., MEZZALAMA, M. and GARIBALDI, A. (1985). Biological and integrated control of Botrytis cinerea in Italy: experimental results and problems. Quaderni di Viticoltura ed Enologia, Università di Torino. Vol. 9 299-308.
6. LANTERO, E., BAZZANO, V. and GULLINO, M. L. (1982). Tentativi di impiego della lotta integrata nei confronti della Botrytis cinerea della vite. La Difesa delle Piante. Nr. 5 11-20.
7. MEZZALAMA, M., ALOI, C. and GULLINO, M. L. (1985). Attività antagonistica nei confronti di Botrytis cinerea di mutanti di Trichoderma spp. resistenti ai benzimidazolici. La Difesa delle Piante. Nr. 8 169-174.
8. TRONSMO, A. and DENNIS, C. (1979). The use of Trichoderma species to control strawberry fruit rots. Netherlands Journal of Plant Pathology. Vol. 83 449-455.
9. TRONSMO, A. and YSTAAS, J. (1980). Biological control of Botrytis cinerea on apple. Plant Diseases. Vol. 64 1009.
10. WOOD, R. K. S. (1951). The control of diseases of lettuce by the use of antagonistic organisms. I. The control of Botrytis cinerea Pers. Annals of Applied Biology. Vol. 38 203-216.

In all the tables, values followed by the same letter do not differ significantly (P = 0.05) following Duncan's test.

Possible integrated control of grape-vine sour-rot

M.Bisiach, G.Minervini & F.Zerbetto
Istituto di Patologia Vegetale dell' Università di Milano, Italy

Summary

Sour rot of grapes is a serious disease mainly affecting varieties
with dense bunches, near the harvest. Disintegration of bunch, acetic
acid smell and presence of fruit flies are the typical symptoms. In
North and Central Italy yeasts and acetic bacteria, which enter the
berries through wounds of any kind, are the pathogens. The most frequent yeasts include Kloeckera apiculata, Saccharomycopsis vini, Hanseniaspora uvarum and so on. Because of the biological nature of the
sour rot pathogens, no direct chemical control is possible particularly
on wine grape near the harvest time. Integrated chemical control of
berry wound agents, such as powdery mildew, grey mould and grape cater
pillars produced a significantly reduction in the disease. Use of insecticides against fruit flies such as Drosophila spp., which are important vectors of the pathogenic agents of the sour rot, is not always
satisfactory. Use of plastic nets or cheesecloth to prevent the infestation of fruit flies was one of the best methods to control the sour
rot in fruit-store during the raisining for the production of strong
sweet wines. A suitable cropping system to prevent skin wounds and the
use of clones with less dense bunches and more resistant berry skins
are, in any case, recommended.

1.1 Introduction

Within the range of grape bunch diseases, grey mould and sour rot must
be mentioned. These two diseases are quite frequent and severe in the north
and central regions of Italy, and, in particular years, can cause large los
ses. Their occurrence varies depending on the climatic conditions, the susceptibility of the vines and other predisposing factors. These two rots ra
rely cause damage before veraison; from this stage onwards the risk of serious infection increases step by step with ripening.

Because of present day knowledge of the biology and epidemiology of
grey mould caused by Botrytis cinerea Pers. and the introduction of specific
chemicals which control this pathogen, a plant protection programme could be
worked out which is able to keep damage to the grape bunch within acceptable
limits (BISIACH and ZERBETTO, 1983).

However in regard to sour rot, the studies have begun more recently
(BISIACH et al., 1982a)and must still be completed.

Up to a few years ago, the two diseases were confused indeed so much so
that sour rot was thought to be the final and most distructive stage of grey
mould; recent research has shown that the two rots are caused by different
aetiological agents and that the oenological consequences of grapes infected
with sour rot are much more serious and difficult to remedy than those with
grey mould (BISIACH et al., 1982a; ZIRONI et al., 1982).

TABLE 1 – DIRECT ACTIVITY AGAINST GREY MOULD AND INDIRECT ACTIVITY AGAINST
SOUR ROT ON BARBERA

S. MARIA DELLA VERSA (PV) – 1981

ACTIVE INGREDIENT	g/ha a.i.	GREY MOULD			SOUR ROT		
		%I.I.	P>95%	%P.I.	%I.I.	P>95%	%P.I.
VINCLOZOLIN (Ronilan)	750	0.49	c	98.33	10.65	b	55.99
DELTAMETHRIN (Decis)	42	24.38	ab	16.71	11.39	b	52.93
CAPTAN	350	17.96	b	38.64	6.53	c	73.02
UNTREATED		29.27	a	–	24.20	a	–

% I.I. : % Infection index

% P.I. : % Protection index

TABLE 2 – DIRECT ACTIVITY AGAINST GREY MOULD AND INDIRECT ACTIVITY
AGAINST SOUR ROT ON MERLOT

GRUMELLO (BG) – 1981

ACTIVE INGREDIENT	g/ha a.i.	GREY MOULD			SOUR ROT		
		%I.I.	P>95%	%P.I.	%I.I.	P>95%	%P.I.
VINCLOZOLIN (Ronilan)	750	4.52	b	91.73	6.90	b	61.45
MYCLOZOLIN	375	3.05	b	94.42	8.81	ab	50.78
MYCLOZOLIN	500	1.57	b	97.13	9.43	ab	47.32
UNTREATED		54.66	a	–	17.90	a	–

1.2 Symptomatology

The initial stages of the disease are difficult to distinguish and are not altogether different from grey mould as they both show a similar colour change usually beginning at the point of pedicel attachment to the berry. Later, however, and especially nearing harvest time, the two diseases show distinct and clearly recognisable symptoms. The grape berries, whether white or black, take on a brown colour of varying intensity but their turgor remains unaltered. Successively the skin become thin and extremely fragile, the berries empty and the juice drips over the lower berries. Finally the whole bunch more or less disintegrates and releases a pungent odor of acetic acid at times mixed with ethyl acetate. In severly damaged vineyards this smell can be distinguished at quite a distance. All around and on these bunches can be found various species of Drosophila, commonly called fruit flies, in every stage of development: eggs, larvae, pupae and adults. In the final stages, the berries are completely empty and the skins dried out; this pheno menon occurs as a results of the complete fermentative breakdown of the pulp, while, in berries infected by B. cinerea, the typical grey mould can be seen and the alteration process is limited to the tissue immediately under the skin.

1.3 Aetiology

Using isolation and experimental inoculation techniques, the aetiological agents were identified (BISIACH et al., 1982a)

As regard the vineyards of Northern and Central Italy, sour rot is principally induced by yeasts, sometimes associated with acetic acid bacteria. Other fungi, such as Penicillium spp., Aspergillus spp., Alternaria spp., Cladosporium spp., Rhizopus spp., Mucor spp., Coniothyrium (Coniella) diplodiella (Spef.) Sacc., Phomopsis viticola (Redd.) Goid., Monilia fructigena (Aderh. et Ruhl) Honey, etc.,though isolated from grapes with sour rot, cannot induce the disease. In the vineyards of Southern Italy some of these microorganisms are responsible for serious rotting phenomena but in these cases it is a disease quite different from sour rot (LACCONE et al., 1982).

Among the yeasts causing sour rot, the strongly pathogenic species are: Kloeckera apiculata (Rees emend. Klöker) Janke, Saccharomycopsis vini (Kreger - Van Rij) van der Walt et Scott, Hanseniaspora uvarum (Niehaus) Shehata et al., Kluyveromyces lactis (Dombrowski) van der Walt, Candida pseudotropicalis (Cast.) Basgal, Candida valida (Leberle) van Uden et Buckley, Candida steatolytica Yarrow, Torulaspora delbrueckii Lindner, Issatchenkia terricola (van der Walt) Kurtzman et al., Hansenula jadinii (A. et R. Sartory, Weill et Meyer) Wickerham and Zygosaccharomyces bailii (Lindner) Guill.; among the bacteria: Acetobacter spp., Gluconobacter spp. and Bacillus spp.. Though the yeasts and acetic bacteria mentioned above can individually cause sour rot, in nature the disease is probably due to simultaneous or successive infections by the two types of microorganisms.

From experimental inoculation tests (BISIACH et al., 1982a)it has found that yeasts and bacteria cannot attack the substances making up the berry skin; in other words these pathogens, before beginning their alteration processes of the pulp, must first come in direct contact with it through wounds of the skin.

Even though the infective yeast propagules are not very mobile often in vineyards the epidemic spreads rapidly especially as harvest time approaches. The constant presence of fruit flies, even in the adult stage, on infected bunches and the possibility that yeast cells are carried both internally and externally on their bodies (CANTONI, 1984) suggests that these insects be considered the most important vectors of sour rot agents.

267

TABLE 3 - DIRECT ACTIVITY AGAINST GREY MOULD AND INDIRECT ACTIVITY
AGAINST SOUR ROT ON BARBERA

VINCLOZOLIN (Ronilan) g/ha a.i.	GREY MOULD			SOUR ROT		
	%I.I.	P>95%	%P.I.	%I.I.	P>95%	%P.I.
1000	1.25	d	97.23	4.61	d	85.71
750	4.18	c	90.73	8.82	c	72.65
400	16.86	b	62.59	19.36	b	39.97
UNTREATED	45.07	a	-	32.25	a	-

CORRELATION COEFFICIENT (R) BETWEEN GREY MOULD AND SOUR ROT**(0.995)

TABLE 4 - DIRECT ACTIVITY AGAINST GREY MOULD AND INDIRECT ACTIVITY SOUR
ROT OF SINGLE OR COMBINED CHEMICAL TREATMENT ON BARBERA.
IMOLA (BO) - 1983

TREATMENTS	GREY MOULD		SOUR ROT	
	%I.I.	%P.I.	%I.I.	%P.I.
B	1.37	85.03	5.08	38.87
T	8.90	2.73	6.60	20.58
D	9.18	- 0.32	6.51	21.66
O	9.13	0.22	7.47	10.11
B + T	1.42	84.48	4.40	47.05
B + D	1.45	84.15	4.07	51.02
B + O	1.46	84.04	4.65	44.04
T + D	8.76	4.26	6.13	26.23
T + O	9.13	0.22	6.12	26.35
D + O	8.81	3.72	6.10	26.59
B + T + D	1.36	85.14	3.63	56.32
B + T + D + O	1.38	84.92	3.32	60.05
B + T + O	1.39	84.81	3.97	52.23
B + D + O	1.32	85.57	4.07	51.02
T + D + O	8.34	8.85	6.04	27.32
UNTREATED	9.15	-	8.31	-

B = Treatments against grey mould
O =   "   "   powdery mildew
D =   "   "   fruit flies
T =   "   "   caterpillars

1.4 Control

Preliminary studies have shown that, for the moment, to control this disease by means of chemicals, no treatments can be proposed due to the biological nature of the aetiological agents. In fact only antibiotics are effective against the bacteria but their use in agriculture is forbidden in Italy; against the yeasts,for example, thiophthalimides are effective, but these, an in particular captan, are highly persistent. As these chemicals are not specific for pathogenic yeasts, they can seriously interfer with alcoholic fermentation, slowing or even stopping it completely. For these reasons, the disease can only be controlled using indirect methods so as to eliminate or reduce all causes of wound to the berry and avoid, as far as possible, the spread of disease by repressing the infestation of insect vector. Various factors can cause wounds to grape berries: in the first place some morphological and anatomic characteristics of the bunch such as its compactness and the thickness of the berry skin. An excessive plant vigour due to the rootstock or excessive nitrogen fertilization causes an increase in the volume of the berries and thus greater compactness of the bunch. Training system, light but persistent rain after dry weather, hailstones, etc., can induce wounds indirectly or directly. Fungal disease such as powdery mildew and grey mould or infestation of the vineyard with grape caterpillars and wasps can also cause them.These observations, made during preliminary studies (BISIACH, 1981; BISIACH, 1982; BISIACH et al., 1982a;BISIACH et al., 1982 b), have indicated an experimental design to find possible means of reducing the disease to acceptable levels.

1.5 Recent experimental research

Eight trials were made on grapevine cultivars in different environments. The chemicals were used at the usual doses and times. The experimental schemes are reported in Tables 1÷8.

In 1981 (Table 1) in a Barbera vineyard, treatments with Vinclozolin, dicarboxymide with a specific antibotrytic activity, reduced sour rot by 55.99%; the fungicide, ineffective against yeasts and bacteria, acted indirectly by eliminating B.cinerea which cause wounds. Two treatments, at 30 and 15 days before harvest, with Deltamethrin, an insecticide active against fruit flies, reduced the sour rot by almost half. Greater protection (73.02%) of grapes was obtained with Captan at 350 g/ha. This chemical, a well know toxic of yeasts, has shown, even at low concentration, a strong direct activity against sour rot and weaker against grey mould. Its practical use depends on acceptable residue level in the must.

In the same year, on Merlot, it was found that the almost complete elimination of B.cinerea obtained using antibotrytic specific treatments resulted in a 50-60% reduction in sour rot (Table 2).

In 1982, in a Barbera vineyard, the role of B.cinerea as a predisposing agent, was further studied using decreasing doses of Vinclozolin and completely eliminating powdery mildew and grape caterpillars from the vineyard, both of which can cause wounds in the berries (Table 3). At harvest time a high indirect activity against sour rot was noted whose infection index was found to be in close statistic correlation with that of grey mould.

In 1983, a complex trial was carried out on Barbera, in which grey mould, powdery mildew and grape caterpillars were considered as wounding agents and fruit flies as vectors. A design of single or combined treatments against the above mentioned predisposing factors gave the results shown in Table 4; in spite of the vast amount of uncontrollable variability due to the spontaneous formation of wounds following the hyperhydration of the berries, it is possible to state that by control the single wounding agents and the vector using suitable treatments, a 10-39% reduction in sour rot was obtained,while

TABLE 5 – DIRECT ACTIVITY AGAINST GREY MOULD AND INDIRECT ACTIVITY AGAINST
SOUR ROT OF SINGLE OR COMBINED TREATMENTS ON ALBANA
SASSO MORELLI (BO) - 1984

TREATMENTS	GREY MOULD			SOUR ROT		
	%I.I.	P>95%	%P.I.	%I.I.	P>95%	%P.I.
B	7.95	gh	84.77	15.60	bcd	74.86
T	28.65	e	45.11	41.45	b	33.20
D	32.95	d	36.88	61.40	a	1.05
O	37.00	c	29.12	61.85	a	0.32
B + T	4.90	il	90.61	15.75	bcd	74.62
B + D	5.90	i	88.70	17.30	bcd	72.12
B + O	3.85	l	92.62	14.65	d	76.39
T + D	39.60	b	24.14	61.30	a	1.21
T + O	27.50	ef	47.32	42.20	b	31.99
D + O	25.95	f	50.29	61.15	a	1.45
B + T + D	3.90	l	92.53	15.40	cd	75.18
B + T + D + O	7.95	gh	84.77	18.60	bc	70.02
B + T + O	8.70	g	83.33	14.40	d	76.79
B + D + O	4.85	il	90.71	19.25	b	68.98
T + D + O	36.75	c	29.60	60.35	a	2.74
UNTREATED	52.20	a	-	62.05	a	-

TABLE 6 – INDIRECT ACTIVITY AGAINST SOUR ROT ON PINOT GRIGIO
VILLAGA (VI) - 1984

TREATMENTS	FIRST CHECK			SECOND CHECK		
	%I.I.	P>95%	%P.I.	%I.I.	P>95%	%P.I.
MANCOZEB	15.32	a	-	53.26	a	-
COPPER OXYCHLORIDE	1.47	b	90.67	38.96	b	26.85
TRIADIMEFON	1.19	b	92.23	26.55	c	50.15
MANCOZEB + TRADIMEFON	1.38	b	90.99	35.46	b	33.42
COPPER OXYCLORIDE + TRIADIMEFON	1.00	b	93.47	17.65	d	66.86

the combined effect of treatments against grey mould, powdery mildew, grape caterpillars and fruit flies produced a protection index of 60%.

The same experimental design with 16 different treatment shemes was tried on Albana in 1984. From Table 5 it can be seen that the sour rot infection index was very high (62%) greatly favoured by wounds caused by grey mould. This can be confirmed by the fact that with all combinations, in which the antibotrytic treatment was used, sour rot was reduced to 69-76%, while treatments against other factors gave a protection index varying from 0.32-33%. In particular, it was noted that while the elimination of the grape caterpillar resulted in a 33% protection of sour rot, treatments against powdery mildew and fruit flies appeared, in this trial, to have no effect; this result is only in part justifies by the low incidence of powdery mildew under the experimental conditions. Statistical analysis, carried out according to Duncan test, of the sour rot infection index showed a consistent variability in the plots of the same treatment perhaps due to uncontrollable wounds on thin skinned berries produced by hyperhydration.

Examining Tables 1÷5 from time to time differing activity of the antibotrytic treatments can be noticed. In fact these differences are caused by the impossibility of preventing wounds connected with the characteristics of the clones and the cultivars or with other causes such as powdery mildew or grape caterpillars.

In 1983, during a trial on Pinot grigio, a very high susceptibility to sour rot was observed, even in absence of wounds caused by grey mould. Taking into account the fragility of the berry skin and the susceptibility of this grapevine to powdery mildew, the following year, in the same vineyard, a complex experimental design was carried out to compare the effect, among others, of copper compounds against downy mildew and dithiocarbamates both with or without Triadimefon, very active against powdery mildew. The results can be seen in Table 6. On vines treated only with Mancozeb the incidence of sour rot found at the first examination (27.8.84) rapidly increased up to an infection index of 53.26% at the second (10.9.84). The use of copper oxychloride instead of Mancozeb at first gave 90.67% protection which later fell to 26.85% as the disease got worse. Triadimefon initially gave similar amount of protection but later its effect was found statistically superior (50.15%). The importance of protection against powdery mildew is confirmed once more by the combination Mancozeb+Triadimefon in spite of the fact that Mancozeb predisposes the vine to the disease, Triadimefon reduced sour rot in a statistically significant way. The most important effect on the disease control is achieved by integrating copper oxyxhloride with Triadimefon which gives a protection index of 93.47% and 66.86% respectively at the initial and final stages of sour rot.

Sour rot is also very serious disease for raisining grapes used in the production of strong sweet wines as the refinement of the final product does not permit the slightest anomaly introduced by the disease.

In a trial on Garganega cultivar, apparently healthy bunches from a vineyard treated against powdery mildew, grey mould and grape caterpillars were placed in fruit-store in aerated plastic fruit boxes as such or covered with fine plastic net to prevent fruit flies infestations. Shortly before winemaking, the sour rot infection index was determined (Table 7). It was found that accurate plant protection in the vineyard against wounds reduces the disease by 77.57% while the protective nets gave 94-95% protection indipendent of whether the grapes had been protected or not with chemicals. These results demonstrate the importance of wounds caused by parasites but show also that, under experimental conditions, the role of the vector Drosophila is predominant.

In a further trial on Garganega, apparently healthy bunches from untrea

271

TABLE 7 - SOUR ROT ON RAISINING GARGANEGA IN FRUIT-STORE

GAMBELLARA (VI) - 1984

TREATMENTS (*)		SOUR ROT			
		%I.I.	P>95%	P>99%	%P.I.
B + O + T	+ NET	0.64	c	C	95.43
UNTREATED	+ NET	0.79	c	C	94.36
B + O + T		3.14	b	B	77.57
UNTREATED		14.00	a	A	-

(*) All plots where treated against downy mildew
    B + O + T: Treatments against grey mould (B), powdery mildew (O)
        and caterpillars (T).

TABLE 8 - SOUR ROT ON RAISINING GARGANEGA IN FRUIT-STORE

GAMBELLARA (VI) - 1984

POST-HARVEST TREATMENTS	SOUR ROT			
	%I.I.	P>95%	P>99%	%P.I.
DELTAMETHRIN (Decis)	2.57	c	C	78.85
VINCLOZOLIN (Ronilan)	9.79	b	B	19.42
DELTAMETHRIN + VINCLOZOLIN	3.86	c	C	68.23
CHEESECLOTH + DELTAMETHRIN	0.71	d	D	94.16
UNTREATED	12.15	a	A	-

ted vines were placed in aerated plastic fruit boxes and sprayed with Delta-methrin, Vinclozolin and a mixture of the two. Other boxes were covered with fine cheesecloth impregnated with Deltamethrin. From the results shown in Table 8 it can be concluded that in this trial, as well as in the previous one, the role of the vector is of the utmost importance. A very high protection index (94%) was obtained by using cheesecloth to exclude the insects; sprayed insecticide gave less satisfactory results as the fruit flies, before felling the lethal effects of the chemical, could, even, if to a limited extend, contaminated the berries with pathogenic yeasts. The effectiveness of the antibotrytic compound is slight because the predisposing role of the grey mould is, in this case, of little importance as its growth is very reduced under the experimental conditions.

1.6 Conclusion

With the aim of extending previous investigations on sour rot (BISIACH, 1981; BISIACH, 1982; BISIACH et al., 1982 a; BISIACH et al., 1982 b), trials were carried out, over a four year period, 1981-84, in different vineyards and on different cultivars in order to determine an appropriate method for controlling the disease. Taking into account its aetiological agents, principally certain yeasts and in the second place acetic bacteria and the need for the grape berry skin to be wounded before they can penetrate, we studied the possibility of preventing or limiting the formation of wounds of a parasitic nature produced by fungi (Botrytis cinerea and Oidium tuckeri) and insects (Lobesia botrana and Eupoecilia ambiguella) using suitable fungicides and insecticides. At the same time, a study was made of the possibility of reducing the number of wounds, of non-parasitic origin, caused by the hydration of thin skined berries in dense bunches, using copper compounds against downy mildew in place of dithiocarbamates. In addition an examination was made of the possibility of preventing infestations of Drosophila spp., considered the principal vectors of the agents of sour rot, using insecticides or protective net or cheesecloth.

In some cases it was possible to study simultaneously the single or combined effect of the treatments on different wounding factors and on the vector and consequently the indirect effects on sour rot.

The amount of infection observed on untreated vines confirms that sour rot is a serious disease which, in certain occasions, can reach very high levels (Barbera 1981 and 1982, Albana 1984, Pinot grigio 1984). In particular, analysing the indirect effects on sour rot of treatments carried out against grey mould it can be seen that the complete elimination of this disease reduces significantly the sour rot. This fact is more obvious when one considers the importance of grey mould as a cause of wounds. It is obvious that, when a different cause of wounds (for example powdery mildew or grape caterpillars) prevales over grey mould or otherwise the wounds are due to clonal characteristics, the antibotrytic treatments have a limited effect. In our trials the effects of treatments against grey mould were widely studied and, except in the cases where the disease was absent or scarse, they caused a reduction in sour rot varying between 47.32 and 85.71%.

The predisposing role of wounds caused by powdery mildew was studied in three trials; in the first two (Barbera 1983 and Albana 1984) no powdery mildew infection occurred and consequently the treatments against powery mildew had no effect while in the third (Pinot grigio 1984) the complete control of the disease using Triadimefon reduced sour rot by 50.15%.

In regard to the role of treatments against grape caterpillars we were able to note a 20.58-33.20% protective effect against sour rot (Tables 4 and 5). Fungicide and insecticide treatments have been used with positive results against sour rot also by other authors (MORANDO et al., 1983; MORANDO et al., 1984; HAAS, 1985).

The effect of the copper compounds against downy mildew as compared to the dithiocarbamates demonstrated that the pysiological activity of copper on the cuticular barrier by reducing the number of wounds, resulted in a lower susceptibility to sour rot.

Treatments against fruit flies, in particular effective in 1981 on Mer lot were on the other hand of little or no use in 1983 on Albana and in 1984 on Barbera, while the exclusion of the insects in 1984 from Garganega during raisining was of fundamental importance in preventing sour rot.

Fruit flies are always associated with sour rot so it is possible that on Albana in 1984, where the treatments were found to have no effect, the disease developed without the aid of the vector.This could be explained by the presence of a very large amount of pathogenic yeast inoculum on grape berries which was spread by rain or dew. However the problem requires further studies.

The integration of each single treatment shows its usefulness in the various combinations. This aspect is particularly evident from the comparison of the single and combined traitments on Barbera (1983) and on Pinot grigio (1984). On Barbera while the single treatments against grey mould, powdery mildew, grape caterpillars and fruit flies gives protection varying between 10.11 and 38.87%, the integration of all four treatments gives as much as 60.05% protection. For Pinot grigio the integration of copper oxychloride,to strengthen the berry skins, and Triadimefon, to prevent wounds caused by powdery mildew,gave a protection index of 66.86%, which is signifi cantly different from that obtained using each single chemical.

From trials carried out during the four year period 1981-84 it has been found that sour rot can be reduced to acceptable levels by protecting the vines from every parasitic cause of wounds (grey mould, powdery mildew, gra pe caterpillar) combining the treatments with other copper compounds after setting to make the berries less fragile and thus prevent wounds occurring due to rainfall after veraison. Also avoiding excessive nitrogen fertilization and adopting all the agricultural methods which moderate plant vigour in vines will aid in the prevention of sour rot by reducing the incidence of spontaneous wounds. However in certain vine varieties and clones it will be difficult to prevent the disease. In such cases the only proposable solution remains in genetic improvement wich even at the risk of a losing productivity gives priority to sour rot resistence.

We hold that the investigation has not yet been completed particularly in regard to the _Drosophila_ spp. vectors, some trials gave different results; their tentative interpretation must be confirmed.

Research on sour rot, especially in regard to the ecology of the patho genic yeasts is still incomplete. Also the entomological aspects of its epi demic progress must be cleared. The solution of these problems will improve, without doubt, the possibility of sour rot control.

REFERENCES

1. BISIACH, M. (1981). Sauerfäule an den Trauben. Obstbau-Weinbau, 18, (12), 422-423.
2. BISIACH, M. (1982). Il marciume acido del grappolo. Terra e Vita, 3, 47-49.
3. BISIACH, M. and ZERBETTO, F. (1983). Criteri di impiego dei nuovi fungicidi antibotritici in viticoltura. Not.Mal.Piante, 104, 85-102.
4. BISIACH, M., MINERVINI, G. and SALOMONE, M.C. (1982 a). Recherches expérimentales sur la pourriture acide de la grappe et sur ses rapports avec la pourriture grise. Bull. OEPP, 12, (2), 15-27.
5. BISIACH, M., MINERVINI, G., ZERBETTO, F. and VERCESI, A. (1982 b). Aspetti biologici ed epidemiologici di _Botrytis_ _cinerea_ e criteri di protezio ne antibotritica in viticoltura. Vignevini, 9, 39-46.

6. CANTONI, A. (1984). Osservazioni sulla distribuzione di Drosophila fasciata Mg. e sulla sua correlazione con il "marciume acido" dell'uva nei vigneti lombardi. Tesi di laurea, Università di Milano, a.a.1983-84.
7. HAAS, E. (1985). Beerenverletzungen fördern das Auftreten von Botrytis und Essigfäule. Obstbau Weinbau, 22, (3),72-74.
8. LACCONE,G., TARANTINO, L., MUROLO, O., and CASILLI, O. (1982). Ulteriori risultati di prove di lotta contro Botrytis cinerea Pers. su uva da tavola in Puglia. Atti Giorn. Fitopat., 2, 345-354.
9. MORANDO, A., NEBIOLO, P., BOSTICARDO, V. and GRASSO, C. (1983). Prove di lotta contro il "Marciume acido" del grappolo. Vignevini,10, 51-55.
10. MORANDO, A., BOSTICARDO, V. and NEBIOLO, P. (1984). Ulteriori prove di lotta contro il "Marciume acido del grappolo" sulle cultivar "Moscato bianco" e "Barbera". Atti Giorn. Fitopat., 2, 127-134.
11. ZIRONI, R., RIPONI, C., FERRARINI, R. and AMATI, A. (1982). Effetti del "Marciume acido" sui costituenti delle uve e sulle caratteristiche dei mosti e dei vini. Vignevini, 9, 39-46.

The phytotoxicity of *Botrytis cinerea*

J.B.Heale

Biology Department, King's College, University of London, Kensington Campus, Campden Hill, UK

Summary

Botrytis cinerea Pers ex Pers is a ubiquitous,necrotrophic, pathogen which invades a wide range of plants through senescent, moribund or wounded tissue.Floral infection and rot of soft fruits such as raspberries and strawberries is also widespread and damaging. In the case of viticulture, the infection of the grape berries by Botrytis has important consequences .If the rate of tissue damage is slow (noble rot), the berries may be used to produce a valuable dessert wine; if it occurs at a high rate, then damage and soft rotting are considerable,and the crop is worthless.Previous studies using a variety of hosts,have pointed to the important role of pectic enzyme secretion by the pathogen in causing maceration and cell death of host tissues.There is also evidence that low mol wt polysaccharides produced by B.cinerea are phytotoxic,but evidence for the role of organic acids is still not convincing. In this review,evidence is presented from a detailed investigation of the phytotoxic fractions secreted in the germination fluid and early mycelial growth phase (up to 48h) by the pathogen, using carrot root tissue as the host system.As well as confirming the presence of a phytotoxic pectic enzyme(characterized as an endopolygalacturonase with a mol. wt. of 49,000 d), and 2 low mol. wt polysaccharides (5000, 10,000d), a phytotoxic, high mol. wt. (63,000d), glycoprotein was identified.The implications of this study are discussed in relation to the mechanisms involved in cell death, host resistance and latency .

1. Introduction

Botrytis cinerea Pers ex Pers is a ubiquitous pathogen which invades a wide range of living,cultivated and wild,host plants through senescent, moribund or wounded tissue; it also survives on dead organic matter and in the soil(Jarvis 1977). This fungus is a major cause of post-harvest and storage losses in fruits and vegetables, including carrots and cabbage(Goodliffe and Heale 1975, 1977, Harding and Heale 1981a, Davies & Heale 1985).

B.cinerea is also particularly important in causing a rot of soft fruits, including raspberries and strawberries (Jarvis 1977). In viticulture, through its diverse effects on the grape berry and its enological potential, the pathogen is of considerable economic significance(Ribéreau-Gayon et al 1980).

As a necrotrophic parasite, B.cinerea has a

requirement for dead plant tissue or exogenously supplied nutrients, which it utilises whilst increasing its inoculum potential. This process is accompanied by the secretion of phytotoxins and macerating enzymes which diffuse ahead of the growing hyphae and result in the death of the tissues , thus producing the characteristic soft rot syndrome and external sporulation associated with "grey mould"" disease.The living host cells situated in advance of the zone of cell death may respond by the accumulation of stress metabolites(phytoalexins)(Goodliffe and Heale 1978,Harding and Heale 1980,1981b, Heale et al 1982 a,b, Hoffman and Heale 1985, Mansfield 1980).Probably of more significance in the primary events determining resistance to infection, is the laying down of wall appositions in areas adjacent to the sites of penetration where cell death occurs. Lignification, callose deposition and suberization have been implicated in such active resistance mechanisms in various host tissues (Friend 1976,Heale and Sharman 1977, Ride 1980,Mansfield 1980, O'Neill and Mansfield 1982). The phytoalexin complex in <u>Vitis</u> <u>vinifera</u>, consisting of the stilbene resveratrol, trans-pterostilbene and a range of viniferins, was described by Langcake et al(1979).

The balance between a susceptible and a resistant response is a complex one and varies with the particular host, cultivar and tissue or organ involved. Other factors interacting to determine the outcome are: (a) age of spore inoculum, and that of the host tissues,(b) inoculum potential,(c) availability of exogenous nutrients eg host wounded tissue, frost damaged tissue etc (d) host water content and relative humidity at the infection court,(e)temperature, (See Jarvis 1977,Goodliffe and Heale 1977,Blakeman 1980, Verhoeff 1980).

Ultimately, the relative rates of host cell death caused by the pathogen on the one hand, and the rate of active responses by the underlying host tissues as described above, will control the outcome of the interaction, as has also been suggested by Mansfield (1980). In some cases, infection of the host remains very limited and is not associated with significant or increasing amounts of host cell death.This frequently occurs in soft fruits as a result of floral infection eg in strawberries (Jarvis 1962) and raspberries (Jarvis 1977), and is associated with frost damage of floral parts. Latent infection in grape flowers has been described by Natal'ina and Svetov (1972) and McClellan and Hewitt(1973); in California the latent presence of the fungus in the necrotic stigma and style tissue can lead to a mid-season rot, even in the absence of the rains typically associated with late-season losses due to this pathogen(Jarvis 1977).Verhoeff (1974) reported that he was able to re-isolate the pathogen from necrotic cells of " ghost spots" in tomato fruits; these small circular haloes on the fruit represent a quiescent state of the pathogen which does not lead to a subsequent rot of the fruit even at maturity, and which may be of considerable importance in understanding the control of expression of phytotoxicity in the host tissues.

Ribéreau-Gayon et al (1980) have summarised the different courses of infection of grape berries by B.cinerea and their significance in enology. The production of sweet dessert wines such as Monbazillac(in the Dordogne region of France) and Sauterne (Gironde) from particular vine cultivars depends upon a limited rot ("Pourriture noble" or noble rot) of a healthy berry with an intact skin,under specific climatic and local edaphic factors peculiar to these wine districts. The much more damaging rot("Pourriture grise", grey rot)is frequently associated with more susceptible cultivars, wounded or burst berries, late-harvesting, heavy rainfall and local conditions.

In all these situations, the expression of phytotoxicity will be governed by mechanisms which control the levels and activity of enzymes and toxins secreted by B.cinerea, thus determining the nature of the rot which develops.The present review is concerned primarily with a study of the nature of the phytotoxic secretions of this fungus and their particular effects on plant cells and tissues.

2.Germination, establishment of infection and the importance of nutrients.

Despite early reports that germinating conidia of B.cinerea penetrate intact host cuticles mechanically (Blackman and Welsford 1916, Brown and Harvey ,1927),it is now established that penetration is achieved with the aid of enzymatic action,including cutin-esterase (McKeen 1974, Shishayama et al 1970). The process of infection from conidia requires the permanent presence of a water film between the spore and host surface (Brown 1915,1916, Jarvis 1962),undoubtedly facilitating the diffusion of phytotoxic secretions from the germ-tubes, as well as the uptake of available nutrients.Death of host cells leads to release of further nutrients at the infection site.

Germination of B.cinerea conidia in pure water is often poor, although it can reach 85% in certain isolates (Blakeman 1975, Yoder and Whalen,1975, Clark and Lorbeer 1977, Sharman and Heale 1979). Blakeman(1980) has suggested that the relatively small spores of B.cinerea (compared with those of the host-specialised forms such as B.fabae etc) contain fewer endogenous reserves, and that this could explain the frequent occurrence of B.cinerea in flowers and fruits where significant amounts of nutrients occur in leachates.Stimulation of both germination and infection by pollen is an additional factor here(Chou and Preece 1968, Blakeman 1980).

Exogenous nutrients are known to enhance germination, appressorial development, and the growth of hyphae over the host surface (Sharman and Heale 1977,1979,Blakeman 1980,Davies and Heale 1985). Several conidia were observed to contribute to the development of dome-shaped infection cushions in 48 h at 12°C on the surface of carrot roots in the presence of a nutrient solution. (Sharman and Heale 1977). Penetration of the suberized periderm layer was then effected by infection pegs arising from swollen hyphal tips on the undersides of infection cushions and enzymatic degradation of suberin lamellae was observed. It may be relevant here to note that Fernando et al (1984) have recently described the production of a suberin-degrading enzyme by _Fusarium solani f. sp. pisi_ grown on suberin, with similar properties to the cutin esterase previously reported by these authors. Infection cushions are structures which increase the inoculum potential of the pathogen , facilitating penetration of resistance barriers and,as discussed below,raising the level of phytotoxic secretions by the pathogen.

Varieties of grapes are more susceptible to infection if they are characterized by high sugar content (Nelson 1951), and resistance falls as sugar content of the berries increases during maturation (Kosuge and Hewitt 1964). Under conditions leading to "Pourriture noble", germ-tubes produce infection hyphae which enter microscopic cracks that develop during growth of the berries (Bulit and Lafon 1970,Ribéreau-Gayon et al 1980). "Noble rot" is typified by the slow development of an intercellular mycelium under the skin,which grows out and conidiates, infecting adjacent healthy berries in the presence of exuded juices. "Pourriture grise" develops in severely wounded grapes and is characterized by a necrotic, brown zone, indicating the rapid diffusion of phytotoxic secretions. The same phytotoxins undoubtedly facilitate the very rapid spread of the "grise" rot to neighbouring berries, which often deteriorate to the "Pourriture vulgaire" (common rot) stage, involving several other soft rot agents and resulting in complete loss (Ribéreau-Gayon et al 1980).

3.Phytotoxic secretions (Table 1).

3.1 Thiourea.
Ovcarov's report (1937), that thiourea produced in culture filtrates caused wilting and chlorosis in tomato plants, was not confirmed by Gentile(1951),who recorded induction of wilting in the absence of thiourea, using similar procedures.

3.2 Organic acids,including Citric acid.
A number of reports have implicated citric acid secretion by _B.cinerea_, both in culture and _in vivo_, as a toxin active in Begonia leaves (Kamoen 1972, Jamart and Kamoen 1972, Kamoen 1976).Oxalic acid, which is normally present in high concentrations in Begonia,was reduced in amount in the chlorotic (yellow) zone surrounding the central necrotic (brown)area after experimental inoculation,indicating that this acid was not important here.Citric acid levels on the other hand, increased significantly over those of the surrounding healthy green tissue, but were low in necrotic tissue. Citric acid,vacuum infiltrated into leaves, resulted in their death within a few days, with symptoms parallel to those occurring in natural infection,and it was considered to be a

vivotoxin. It was further suggested that the fungus in Begonia produced citric acid rather than oxalic acid because of the low pH of the intercelllar fluid of the chlorotic zone.The death of the tissues resulting from citric acid infiltration allowed spontaneous colonization to occur. Kamoen(1976) further suggested that the more rapid spread of lesions under high humidity was due to facilitated diffusion of toxins and enzymes in an increased water film within the intercellular spaces, and that in the chlorotic zone, there was a further increase due to the higher permeability of affected cell membranes. At low humidity, lesion spread was inhibited and the translucent or water-soaked appearance of the tissue, especially the chlorotic zone, diminished.It is necessary in the light of the foregoing to explain the fact that fully turgid carrot slice tissue is resistant to inoculation with 1 x 10^5 spores/ml, whereas at approximately 10% water loss, the tissue gives a susceptible reaction(Goodliffe and Heale 1977, Heale and Sharman 1977). However,the infection test was carried out under high humidity conditions and the limiting plasmolysis induced by severe water loss probably interferes with active resistance mechanisms in these tissues (Hoffman, Roebroeck and Heale 1985).

3.3. Polysaccharides.

Kamoen et al 1978 obtained two major types of polysaccharide from cultures of B.cinerea grown on high glucose media (10%)for 10-30 days ie long term cultures.A high molecular wt. glucan (c. 10^6d) was shown to possess low toxicity only.This (1→ 3:1→ 6) -β-D-glucan is formed by B.cinerea in the grape berry from simple hexoses amd is located between the epidermal cells and the pulp, contributing to the difficulty of clearing wines made from botrytised grapes (Dubourdieu et al 1978, Ribereau-Gayon et al 1980).It appears to function as an extracellular reserve when the fungus is grown in a hexose-depleted medium under which conditions glucanase is induced. The second type was found to be a lower molecular wt fraction (<10^5d), composed mainly of mannose (60-70%), with galactose (20%), and traces of glucose and rhamnose. More recent work shows this fraction to be a variable series of heteropolysaccharides of less than 25,000d, depending upon extraction procedures (Kamoen, personal communication).Kamoen et al (ibid) found the low molecular wt. polysaccharide fraction to be toxic to Begonia and lettuce leaf cells after vacuum infiltration for 24 hrs. Failure to plasmolyse was the criterion adopted and in addition,chloroplasts showed a significant reduction in size. Toxicity occurred both when cells at an incipient state of plasmolysis, and normal turgid cells, were used, unlike the observations when pectic enzymes were employed (see below). Inability to plasmolyse occurred after chloroplast shrinkage, suggesting that the protoplast membrane effect was secondary.The chloroplasts in treated cells showed abnormal contents and exhibited vague boundaries. When the relative diffusion of a pectic enzyme component and the polysaccharide in Begonia parenchyma was compared, they were found to be similar, although the penetration rate of the polysaccharide was significantly higher near the central vein (P=<0.01, t), which agreed with the observation that Botrytis lesions often spread ahead along the leaf veins.No details of the isolation of the polysaccharide from infected Begonia leaves was given, but its secretion in infected plants was reported.

More recently, Bowen and Heale(1985) have reported the presence of two polysaccharide fractions (c5000 and 10,000d) in the cell-free germination fluid of B.cinerea prepared from a short term culture (48 hrs) of a dense spore suspension in low strength medium (1/10th GPSA , Heale et al 1982a), which is probabably chemically related to those investigated by Kamoen and colleagues. They were shown to be individually capable of causing cell death in carrot root secondary phloem parenchyma by the use of Evans blue (membrane dye exclusion test) and by fluorescein diacetate.Time course studies for these individual phytotoxic components are not yet available, but when unfractionated germination fluid was similarly employed, signs of loss of fluorescence using fluorescein diacetate occurred within 2 hrs, whereas first indications of plasma membrane loss of integrity (using Evans blue), did not occur until 4 hrs later. Whilst not being conclusive evidence in itself, this indicates that phytotoxicity primarily results from disturbance of some metabolic target initially, and that loss of protoplast integrity is a later phenomenon - we will return to this when discussing the mechanism of cell death induced by pectic enzymes, also present in germination fluid.

Both polysaccharide fractions , as well as endopolygalacturonase and glycoprotein components were finally characterized after a series of preliminary steps including : (1)ultrafiltration, (2) separation using Sephacryl S-200 (Table 2,3, Fig 1,2).Individual components were tested for their ability to cause cell death and to induce resistance in carrot root slices inoculated 6 hrs later.Previous work had revealed the accumulation of the carrot phytoalexin 6-methoxymellein and other stress metabolites in response to pretreatment of the root slices with heat-inactivated conidia, germination fluid or low levels of live inoculum (Harding and Heale 1978, 1980,1981b). Increased incorporation of tritiated uracil reflected enhanced RNA synthesis and nucleolar volumes in living cells just below the surface layers damaged by the phytotoxic treatments employed, indicative of a requirement for additional ribosomal subunits associated with the expression of specific genes involved in active defence (Heale et al 1982 a,b).They reported further that inducing activity was unaffected during freezing to $-25^{\circ}C$ and rapid thawing, heating to $50^{\circ}C$ for 10 min, but was lost by boiling for 10 mins or by leaving at room temps for 24 hrs.
    All 4 components separately caused cell death and induced resistance as measured by the length of germ-tubes of conidia inoculated on to the surface of pretreated carrot root slices and examined 16 hrs later, as well as by rotting indices

4-5 days after inoculation (Bowen and Heale 1985). No component from germination fluid consistently caused cell death without leading also to a resistance response. Similarly, no fraction induced resistance which was not also phytotoxic. Clearly, we have here cause and effect, strongly indicative of the action of an endogenous elicitor released from cells damaged by the phytotoxins(See 4, this review).

3.4. Pectic enzymes.

Brown(1915) described the phytotoxic nature of extracts of Botrytis cinerea germ-tubes and the relationship between maceration and cell death.He suggested two possibilities: (1)the cells were macerated and killed by the same agent or agents or (2) the cell walls were rendered permeable by the effects of the macerating agent to a separate toxic factor.The macerating factor was attributed to an enzyme,originally termed 'cytase',but subsequently as 'pectinase', in view of its effects in degrading the pectin-rich components of the middle lamella,causing cell separation.
Tribe (1955) investigated the effects of pectin degrading enzymes produced by this pathogen and recorded maximum decrease in viscosity of pectin and pectate solutions between pH 3.5 to 6.0; activity was significantly reduced upon dialysis against distilled water. Plasmolysing concentrations of salts or non-electrolytes greatly retarded the killing action of the enzyme preparations, whilst the effects upon maceration or viscosity reduction were minimal.Isolated protoplasts obtained from plasmolysed tissues were resistant to toxicity.He also recognized the contaminating presence of other, thermostable, toxic factors in crude culture filtrates. In discussing the possibility that cell death was caused by bursting of the protoplasts through the pectinase- weakened walls (apparently the first report of this suggestion), microscopic examination of Sedum spectabilis mesophyll cells treated with a bacterial (Erwinia aroideae) pectinase,failed to show bursting after 1 hr,even though the cells were dead by the neutral red criterion at this time.He concluded that the death of plant cells accompanying the maceration caused by B.cinerea and Erwinia aroideae was explicable by direct hydrolysis of pectic material within them, which was only accessible in unplasmolysed cells.
There have been many detailed studies of the complex of pectic enzymes produced by B.cinerea , including that of Gäumann and Böhni (1947), who described a constitutive endopolygalacturonase (endo PG) and an inducible ("adaptive") pectin methyl esterase (PME).Jarvis (1953) drew attention to the differences in pectate depolymerizing activity on the one hand and maceration on the other,the latter being maximal at pH 2.6 and 6.2 (potato tuber discs),agreeing with the optimum for pectin degradation. Verhoeff and Warren(1972) reported the presence of PME, endo PG, exo PG and a transeliminase (lyase) in petiole stumps of inoculated tomato plants.More recent investigations include those of Urbanek and Zalewska-Sobczak 1975, Di Lenna and Fielding 1983, Marcus and Schejter 1983).
Basham and Bateman (1975) and Bateman (1976) put

forward a generally accepted hypothesis of cell death caused by pectic enzymes produced by soft rot pathogens, following upon their own studies and the earlier work of Tribe(1955) and Hall and Wood (1973).They could find no evidence of any direct injurious interaction between pectic enzymes or their soluble reaction products and plant cells, and concluded that membrane damage in cells treated with endopectic enzymes results from a loss in the ability of the degraded plant cell walls to support the plasma membrane. They had again found the protective effect of plamolysis as described by Tribe (ibid) and that isolated protoplasts in a stable osmoticum were apparently unharmed by pectic enzymes.However, by working only with purified pectic enzyme components, they were not able to investigate other possible phytototoxic agents that might be equally significant in causing cell death in the soft rot syndrome.

Kamoen et al (1978) used an ammonium sulphate precipitated pectin-degrading enzyme from B.cinerea cultures (See 3.3,this review) in phytotoxicity tests on Begonia and lettuce leaf tissue.Treated tissues showed a loss of ability to plasmolyse after 24 hrs, although stomata of lettuce leaves were resistant, as they were to some extent to the polysaccharide toxin.They confirmed Basham and Bateman's (ibid)observations regarding the protective effect of incipient plasmolysis, thus supporting the "bursting hypothesis". More recent studies by Bowen and Heale (1985) involved the characterization of an endo PG (49,000d) in the germination fluid which was shown to be phytotoxic to carrot secondary phloem tissue in roots, and to induce a resistance response(Table 2, 3, Figs 1,2). As mentioned earlier (See 3.3), when unseparated germination fluid (containing endo-PG) was tested over a time course, loss of an enzyme marker (FDA esterase) was observed at 2 hrs, ie 4 hours before evidence of plasma membrane occurred (Evan's Blue dye exclusion test). This suggests that any bursting of cells probably takes place after the cells are dead, thus agreeing with the observations of Tribe (1955) already discussed.This could indicate that some other toxin is involved in the comparatively rapid metabolic disturbance noted here or that pectic enzyme -mediated cell death involves a more rapid effect than could be explained by the bursting of protoplasts through enzyme-degraded and weakened walls. A more satisfactory explanation still has to take account of the well established "protective plasmolysis" effect, and the fact that isolated protoplasts in a stable osmoticum are resistant to attack.

As Byrde (1982) has noted, there is evidence for almost immediate electrolyte leakage from cultured apple cells when treated with an endo-PG from Monilinia fructigena(Hislop et al 1979).Ultrastructural studies indicated a temporary loss of double staining in the plama membrane 5 mins after treatment with a standard 'pectinase' preparation,when no obvious effect could be seen on the walls (Keon 1982). After 25 minutes, paramural vesicles developed, suggesting that the interface between wall and membrane could be a primary site.Cooper (1984) has argued that the plasma membrane may only be damaged when adjacent to the wall, supported by Keon's (ibid) studies indicating the presence of polygalacturonide susbstrate for endo-pectic enzymes in large amounts adjacent to the plasma membrane.There is still no convincing evidence for the direct role of toxic factors solubilized from pectic enzyme-degraded walls (Bateman and Basham 1976),but implications of the release of pectic fragments for elicitation of active resistance responses is discussed in Section 4, this review.Hislop and Pitt(1976) noted an increased acid phosphatase activity in carrot root tissue treated with a PG fraction from S.fructigena , and a similar, but less marked, effect of a transeliminase fraction, but as discussed below, these lysosomal changes are probably the result of cell death, rather than the cause.

The activity of pectic enzymes produced by B.cinerea is known to be regulated by the presence of inhibitors eg Cole (1956) reported that infected apple tissue lacked detectable pectolytic activity, although amounts of pectin did slowly decrease with infection.

Deverall and Wood (1961) suggested an active mechanism of resistance involving a pectic enzyme from B.cinerea and Vicia faba (broad bean). Hydrolysis of host cell walls led to the release of pectic substances activating a latent phenolase.This oxidised phenols which in turn inactivated the pectolytic enzymes of the pathogen, thus limiting ingress.Verhoeff (1970) concluded that the typical symptoms of "ghost spotting" he was able to achieve in tomato fruits by inoculations using a small number of dry conidia , were due to the meristematic activity of the rapidly expanding fruit limiting the growth and enzymatic activity of the pathogen, but failing to do so when conditions were very humid, or when many conidia were employed, when large blisters resulted.Clearly more research is required on the underlying causes of latency or quiescent infections, and in particular to investigate mechanisms determining the limitation of toxin-induced cell death by B.cinerea, including both enzymic and non-enzymic components.Such studies are likely to lead to a better understanding of the marked differences in host and tissue susceptibility to this pathogen.

Whatever the mechanism of pectic enyme-mediated cell death , there is little doubt that the damage caused is an early event in pathogenesis of B.cinerea, as indicated by the report of Verhoeff and Warren (1972), that endo-PG was detected even in ungerminated conidia.

3.5. Phospholipase.

A phospholipase (lacking acyl hydrolase activity, but degrading phosphatidyl choline=lecithin) from culture fluids of B.cinerea, purified 1000 fold, disrupted lysosomal membranes according to Shepard and Pitt (1976a), leading to release of acid phosphatase from lysosome-enriched fractions of potato sprout tissue. There was also evidence for tonoplast membrane damage (betacyanin leakage from treated beetroot discs) but interestingly, only internal cell membranes were affected, and the plasma membrane was apparently undamaged.

Since no maceration of the tissue resulted, and isolated protoplasts survived treatment, the precise role of such fungal phospholipases in subcellular disorganization remains enigmatic.Shepard and Pitt (1976b) also reported two lipolytic enzymes from homogenised potato tuber tissue, one of which showed phospholipase activity and again induced acid phosphatase leakage. The hypothesis that release of lysosomal hydrolases is a primary event determining cell death in host parasite relationships , still lacks convincing evidence (Pitt and Coombes 1968, Wilson 1973, Hislop and Pitt 1976), and it is more likely that decompartmentalisation of the lysosomal complex in host/parasite interactions involving cell death (including the hypersensitive reaction to biotrophic fungi), is a secondary or tertiary event.

3.6. Glycoprotein.
Recent work in this laboratory described earlier (See 3.3 and 3.4 this review) by Bowen and Heale (1985), has involved the separation of those components from the early secretions of germinating spores and young hyphae (48 hr densely-inoculated cultures) of B.cinerea, which showed phytotoxic activity to carrot root secondary phloem parenchyma. In addition to the polysaccharide components and the endo-PG, we identified a glycoprotein (63,000) which caused cell death and induced a resistance response in the tissue(.Table 3, Figs 1,2).The mechanism of toxicity is unknown, and the possibility that it is an enzyme requires further investigation.Another possibility is that the glycoprotein may be similar to the non-specific elicitors described from mycelium and culture filtrates of Cladosporium fulvum (and other biotrophic fungi), that are reported to cause necrosis and to trigger active defence mechanisms in both resistant and susceptible cvs (De Wit and Roseboom 1980).

3.7 Other causes of phytotoxicity.
There are numerous reports of unidentified toxic behaviour by B.cinerea which are difficult to assess and a few cases suggesting that specific factors merit further attention.Of the latter examples, the most obvious perhaps is that of protease where there are conflicting reports.Porter (1966) failed to find evidence for protease activity in fruits of tomato and grape infected by B,cinerea. However protease was detected in petiole stumps of tomato exhibiting rotting symptoms by Verhoeff (1978) and in culture filtrates (Lyr and Novack 1962, Tseng and Lee 1969,Astapovich et al 1972, Shepard and Pitt 1976a). The latter authors reported that four peaks of protease activity separated by isoelectric focussing, were closely associated with 3 phospholipase peaks which made firm conclusions regarding their individual effects on conductivity and electrolyte release from treated potato disc tissue impossible. Kuč (1962) suggested that proteases may act synergistically to increase the macerating activity of PG towards cucumber tissue.

Gentile (1951) reported an unidentified thermostable, non-volatile toxin that was soluble in acetone and alcohol and induced wilting in tomato cuttings, but this was not a good

286

model to test for soft rot toxins which are required to diffuse through parenchymatous tissue in vivo, rather than being carried upwards in the transpiration stream via the xylem.The same criticism can be applied to Purkayastha's reports (1969,1970) of a non-dialysable, partially thermolabile toxin, causing wilting and necrosis of cut shoots of Vicia faba, although here, browning of the primary root as well as softening of the root tip and stem segments was observed.
There are numerous reports of the role of toxins in cabbage tissue infected by B.cinerea by the Russian group led by Rubin (Rubin and Artsikhovskaya 1963,Rubin and Ladygina 1964,Rubin et al 1973). These toxins were reported to include polysaccharide and acidic fractions which caused increases in peroxidase, invertase, cytochrome oxidase and oxidative phosphorylation and to induce enzyme and protein synthesis.It is not clear from these studies whether , as seems likely, that these numerous changes are secondary and follow on from as yet undetermined damage of cellular sites discussed in 3.3 this review.

4.Implications of cell death.
The rapid death of host tissues under the influence of the toxins described above (and others as yet unknown), allows further nutrient availability to the pathogen and stimulates rapid growth and higher levels of toxin secretion to occur, leading to the typical spreading lesions characterizing Botrytis infections of parenchymatous tissue under appropriate humidity conditions. Where the balance of the host/parasite interaction involving B.cinerea is more favourable to the host, eg if humidity is limiting, or if inoculum levels are low, there is evidence that the release of elicitors from toxin-damaged cells leads to a resistant response in the healthy underlying tissues not subject to the diffusion of the toxins. Bowen and Heale (unpublished data) have obtained evidence for the release of a small peptide (approximately 5000d) from damaged carrot tissue,which induced a resistant response leading to reduced germ-tube lengths and rotting index after challenge with live spores. This "endogenous elicitor" was extractable from freeze-thawed and germination fluid-treated carrot root tissue and could also be obtained from homogenised fresh tissue.The detection of the elicitor within 2 h of treatment, suggests that it is released or activated within the early stages of cell damage.Hoffman and Heale (1985) report that the carrot phytoalexin 6-methoxymellein accumulated in carrot tissue in response to several phytotoxic agents,including low levels of live inoculum $(1 \times 10^5$ /ml), 2-10 secs of partial freezing, or 1 mM mercuric chloride. Ethylene did not cause cell death or induce resistance, although it did induce 6 MM, with a similar rate of accumulation as partial freezing.It was therefore concluded that 6MM accumulation is a consequence of cell death, but is not itself a primary determinant in the induced resistance response. The release of rhamnogalacturonan fragments from pectic enzyme-degraded host cell walls may lead to the induced synthesis of a proteinase inhibitor (Ryan et al 1981), and act as an endogenous elicitor of stress metabolites or phytoalexins

(Hahn et al 1981, Nothnagel et al 1983, Walker-Simmons et al 1984).Non-enzymic toxins causing cell damage may activate host pectic enzymes which function in a similar manner.This could be of significance if they diffuse ahead of the pathogen-secreted pectic enzymes in host tissue.

The crucial factors determining the fine balance between resistance (restricted lesions) or susceptibility (spreading lesions) are the rate and extent of cell killing, which will tend to suppress active host responses, and the rate and extent of the latter which includes both cell wall modifications and phytoalexin accumulation.Preformed anatomical barriers, and inhibitors of growth,enzyme and toxin activity,will modify this balance and along with environmental factors , mainly humidity, and temperature, will affect the outcome.Latent infections probably involve additional differences in availability of nutrients (eg sugar levels in ripening fruits),in tissue levels of inhibitors(eg phenol levels in grape berries which fall with maturation, Verhoeff 1980) in the relative insolubility , especially of the pectic material of the middle lamellae, of cell walls (solubility increasing with maturation,Verhoeff 1980), all of which tend to limit or prevent any significant cell damage at the outset. This severely limits nutrient availability to the pathogen and will suppress toxin secretion.

Hutson and Mansfield (1980) in a genetical study reported that the virulence of B.fabae and B.cinerea was closely correlated with cell-killing ability up to 8.5 hrs after inoculation of Vicia faba (broad bean), this being much higher for the host-specialised B.fabae in this tissue.Cooper (1984) has partially attributed the differential sensitivity of broad bean to these Botrytis species(which also rests on other factors such as differential tolerance to the broad bean phytoalexin complex - Mansfield 1980) to much lower levels of endo-PG production by B.cinerea (only 5% of that produced by B.fabae), which is confined to inoculation sites where high levels of wyerone derivative phytoalexins accumulate. It was not clear however, if this difference was partly constitutive, or depended upon the active suppression of PG, as reported earlier by Deverall and Wood (1961).

Bowen and Heale (1985) have shown that at low levels of endo-PG treatment of carrot root slices, where superficial damage (1-3 surface cell layers killed) only occurs, then resistance is induced, whereas at higher concentrations of PG, 10-20 cell layers are killed, and induced susceptibility results. Ride (1975) showed that treatment of wheat leaves wound-inoculated with B.cinerea, with a concentrated solution of cell wall-degrading enzymes, induced lignification of the tissues surrounding the inoculation site,and led to resistance to the pathogen. Induced lignification as a resistant response to wound inoculation with B.cinerea,has also been reported by Friend (1976) in potato and tomato leaves ; lignification is thought to be implicated in the restriction of the pathogen in the "ghost spot" interaction of tomato fruits, but the glycoalkaloid tomatine may also be important here (Verhoeff 1980). Heale and Sharman(1977) reported that the resistance response in carrot callus cells induced by treatment with heat killed conidia or germination fluid, which was later shown to cause cell death (Heale et al 1982a,b),was associated with an increase in peroxidase activity and lignin synthesis in the surface cells. In carrot root tissue on the other hand, an acceleration of suberin deposition was observed over that occurring as a response to slicing (wound healing). In both host systems, there was evidence for induction of phytoalexins, which in carrot root tissue,were shown to include 6-methoxymellein, the polyacetylene falcarinol, and of lesser importance,p-hydroxybenzoic acid (Harding and Heale 1980,1981).

5. Conclusions.

A number of enzymic and non-enzymic toxins secreted by B.cinerea are sufficiently well characterized to identify their importance in causing cell death, but precise understanding of the primary and sequential events involved awaits further biochemical investigation with purified polysaccharíde, glycoprotein, endo-PG components and detailed cytochemical and ultrastructural observations.Differential ability to diffuse through parenchymatous tissue demands further study, as does stability and inactivation mechanisms in different tissues. Such investigations should also include studies of the synergistic effects involved , particularly of the effects of pectic enzymes, proteases and phospholipases on membrane integrity.This may help to explain the very rapid indications of membrane damage, the protective effect of plasmolysis in pectic enzyme-treated tissues,and the resistance of isolated protoplasts to such treatments. The further study of latent and quiescent infections may lead to an elucidation of the differential resistance of host tissues to Botrytis cinerea, and how the potentially harmful effects of toxins are regulated by host and environmental factors.Recent research has revealed the release of rhamnogalacturonan fragments from pectic enzyme-degraded walls , which behave as endogenous elicitors of active resistance responses; in carrot tissue,a peptide apparently functions in a similar manner. Further understanding of the mechanisms involved may lead to direct applications for improved disease control.

REFERENCES.
1. ASTAPOVICH,N.I.,BABITSKAYA,V.R., HREL, M.V. and VIDZISCHCHUK, Z.A. (1972). Biosynthesis of peptolytic enzymes by Aspergillus awamori and Botrytis cinerea. Minsk. Belarus. Akad. Nauk. Vestr. Ser. Biyal.Nauk.4,73-75
2. BASHAM, H.G. and BATEMAN, D.F. (1975). Killing of plant cells by pectic enzymes: the lack of direct injurious interaction between pectic enzymes or their soluble reaction products and plant cells. Phytopathology,65, 141-153.
3. BATEMAN,D.F. (1976). Plant cell wall hydrolysis by pathogens. In "Biochemical Aspects Of Plant Parasite Relationships". Eds. J.Friend & D.R. Threlfall, Academic Press. 79-103.
4. BATEMAN,D.F. and BASHAM, H.G.(1976).Degradation of plant cell walls and membranes by microbial enzymes. In "Encyclopaedia Of Plant Physiology. New Series 4.

Physiological Plant Pathology."Eds. R.Heitefuss & P.H. Williams. Springer-Verlag. 316-355.

5. BLACKMAN, V.H. and WELSFORD,E.J. (1916). Studies in the physiology of parasitism.II.Infection by Botrytis cinerea Annals of Botany, Lond. 30, 389-398.

6. BLAKEMAN,J.P. (1975). Transactions of the British mycological Society,65,239-247.

7. BLAKEMAN,J.P. (1980). Behaviour of conidia on aerial plant surfaces. In "The Biology of Botrytis". Eds. J.R. Coley-Smith, K.Verhoeff & W.R.Jarvis. Academic Press. 115-151.

8. BOWEN, R.M. and HEALE, J.B. (1985).Phytotoxic fractions from germination fluid of Botrytis cinerea inducing resistance in carrot root tissues. (In press).

9. BROWN, W. (1915). Studies in the physiology of parasitism. I. The action of Botrytis cinerea. Annals of Botany, Lond. 29, 313-348.

10. BROWN, W. (1916).Studies in the physiology of parasitism. III. On the relation between the 'infection drop' and the underlying host tissue.Annals of Botany,Lond.30, 399-406.

11. BROWN,W. and HARVEY,C.C. (1927). Studies in the physiology of parasitism. X. On the entrance of parasitic fungi into their host plant.Annals of Botany, Lond. 41, 643-662.

12. BULIT, J. and LAFON, R. (1970). Quelques aspects de la biologie du Botrytis cinerea Pers., agent de la pourriture grise des raisins. Connaiss. Vigue. Vin, 4, 159-174.

13. BYRDE, R.J.W. (1982). Fungal 'pectinases' : from ribosome to plant cell wall.Transactions of the British mycological Society, 79, 1-14.

14. CHOU,M.C. and PREECE, T.F. (1968). The effect of pollen grains on infections caused by Botrytis cinerea. Annals of applied Biology, 62, 11-22.

15. CLARK, C.A. and LORBEER, J.W. (1977). Comparative nutrient dependency of Botrytis squamosa and B.cinerea for germination of conidia and pathogenicity on onion leaves. Phytopathology, 67, 212-218.

16. COLE,J.S. (1956). Studies in the physiology of parasitism XX.The pathogenicity of Botrytis cinerea, Sclerotinia fructigena and S. laxa with special reference to the part played by pectic enzymes. Annals of Botany, Lond., 20, 15-38.

17. COOPER,R.M.(1984).The role of cell wall-degrading enzymes in infection and damage.In " Plant Diseases: infection, damage and loss." Eds. R.K.S.Wood & G.J.Jellis Blackwell. 13-27.

18. DAVIES, R.M. and HEALE,J.B. (1985). Botrytis cinerea in stored cabbage: the use of germ-tube growth on leaf discs as an indicator of potential head rot. Plant Pathology 34,(In Press).

19. DEVERALL, B.J. and WOOD, R.K.S. (1961). Chocolate spot of bean (Vicia faba) - interactions between phenolase of the host and pectic enzmes of the pathogen. Annals of applied Biology, 49, 473-477.

20. DE WITT, P.J.G.M. and ROSEBOOM, P.H.M. (1980). Isolation, partial characterization and specificity of glycoprotein elicitors from culture filtrates, mycelium and cell walls

of Cladosporium fulvum (syn Fulvio fulvum).
Physiological Plant Pathology, 16, 391-408.

21. DI LENNA, P. and FIELDING, A.H. (1983). Multiple forms of
polygalacturonase in apple and carrot infected by isolates
of Botrytis cinerea.
Journal of general Microbiology, 129, 3015-3018.

22. DUBOURDIEU, D., PUCHEU-PLANTÉ, B., MERCIER,M. and
RIBÉREAU-GAYON,P.(1978).C.R.Hebd.Seanc. Acad.Sci.Paris.
Sér. D. 287, 571-573.

23. FERNANDO,G., ZIMMERMANN, W. and KOLATTUKUDY,P.E. (1984).
Suberin-grown Fusarium solani f. sp. pisi generates a
cutin-esterase which depolymerizes the aliphatic
components of suberin. Physiological Plant Pathology,
24, 143-155.

24. FRIEND, J. (1976). Lignification in infected tissue. In
"Biochemial Aspects Of Plant Parasite Relationships".
Eds. J.Friend & D.R. Threlfall. Academic Press. 291-303.

25. GÄUMANN, E. and BÖHNI, E. (1947). Uber adaptiv enzyme
bei parasitischen Pilzen. I.Helv.Chim.Acta., 30, 24-38.

26. GENTILE, A.C. (1951). A study of the toxin produced by
Botrytis cinerea from Exochorda.
Physiologia Plantarum, 4, 370-386.

27. GOODLIFFE, J.P. and HEALE,J.B.(1975). Incipient
infections caused by Botrytis cinerea in carrots
entering storage. Annals of applied Biology, 80, 243-246.

28. GOODLIFFE, J.P. and HEALE, J.B.(1977). Factors affecting the
the resistance of cold-stored carrots to Botrytis
cinerea. Annals of applied Biology, 87, 7-28.

29. GOODLIFFE,J.P. and HEALE, J.B. (1978). The role of
6-methoxymellein in thr resistance and susceptibility of
carrot root tissue to the cold-storage pathogen Botrytis
cinerea.Physiological Plant Pathology, 12, 27-43.

30. HAHN, M.G., DARVILL, A.G. and ALBERSHEIM,P. (1981). Host
pathogen interactions. XIX. The endogenous elicitor, a
fragment of a plant cell wall polysaccharide that elicits
phytoalexin accumulation in soybeans.Plant Physiology,68,
1161-1169.

31. HALL,J.A. and WOOD,R.K.S. (1973).The killing of plant
cells by pectolytic enzymes. In "Fungal Pathogenicity
And The Plant's Response". Eds. R.J.W. Byrde & C.V.
Cutting. Academic Press. 19-38.

32. HARDING,V.K. and HEALE,J.B. (1978). Post-formed inhibitors
in carrot root tissue treated with heat-killed and live
conidia of Botrytis cinerea.
Annals of applied Biology, 89, 348-351.

33. HARDING, V.K. and HEALE, J.B. (1980). Isolation and
identification of the antifungal compounds accumulating in
the induced resistance response of carrot root slices to
Botrytis cinerea. Physiological Plant Pathology, 17,
277-289.

34. HARDING,V.K. and HEALE,J.B.(1981a). A rapid method to
estimate the resistance of cold-stored carrot roots
to Botrytis cinerea. Annals of applied Biology, 99,
375-383.

35. HARDING, V.K. and HEALE, J.B.(1981b). The accumulation of
inhibitory compounds in the induced resistance response
of carrot root slices to Botrytis cinerea.

Physiological Plant Pathology, 18,7-15.
36. HEALE,J.B. and SHARMAN, S.(1977).Induced resistance to
    Botrytis cinerea in root slices and tissue cultures of
    carrot (Daucus carota L.). Physiological Plant Pathology,
    10,51-61.
37. HEALE,J.B.,DODD,K.S. and GAHAN,P.B.(1982a). The induced
    resistance response of carrot root slices to heat-killed
    conidia and cell-free germination fluid of Botrytis cinerea
    Pers ex Pers. I. The possible role of cell death.
    Annals of Botany,Lond. 49,847-857.
38. HEALE, J.B., DODD, K.S.and GAHAN, P.B. (1982b). The induced
    resistance response of carrot root slices to heat-killed
    conidia and cell-free germination fluid of Botrytis
    cinerea.2. Nuclear migration, nucleolar volume and uptake
    of tritiated uracil. Annals of Botany, Lond, 49, 859-872.
39. HISLOP,E.C.,KEON,J.P.R. and FIELDING,A.H.(1979).Effects of
    pectin lyase from Monilinia fructigena on viability,
    ultrastructure and localization of acid phosphatase of
    cultured apple cells.Physiological Plant Pathology, 14,
    371-381.
40. HISLOP, E.C. and PITT,D.(1976). Sub-cellular organization
    in host-parasite relationships. In "Encyclopaedia Of
    Plant Physiology. Ne Series.4. Physiological Plant
    Pathology.Eds R.Heitefuss & P.H.Williams. Springer -
    Verlag. 389-412.
41. HOFFMAN,R.M. and HEALE, J.B. (1985). 6-methoxymellein
    accumulation and induced resistance to Botrytis cinerea
    Pers ex Pers in carrot slices treated with phytotoxic
    agents and ethylene. (In Press).
42. HOFFMAN, R.M.,ROEBROECK, E. and HEALE,J.B. (1985). Ethylene
    biosynthesis regulates 6-methoxymellein production in
    carrots and controls resistance to Botrytis cinerea
    (In Press).
43. HUTSON,R.A. and MANSFIELD,J.W. (1980). A genetical approach
    to the analysis of mechanisms of pathogenicity in Botrytis
    /Vicia faba interactions.
    Physiological Plant Pathology, 17, 309-317.
44. JAMART,G. and KAMOEN, O.(1972).Onderzoekingen over het
    parasitisme van Botrytis cinerea op Knolbegonia.
    Meded. Rijkssstn. Sierpl. Teelt.26,1-66.
45. JARVIS,W.R.(1953).Comparative studies of the pectic enzymes
    of Botrytis cinerea Pers. and Erwinia aroideae
    (Townsend) Stapp.Doctoral Thesis, Univ. of London.
46. JARVIS, W.R. (1962). The infection of strawberry and
    raspberry fruits by Botrytis cinerea Pers.
    Annals of applied Biology, 50, 569-575.
47. JARVIS, W.R.(1977). Botryotinia and Botrytis species:
    taxonomy,physiology and pathogenicity.Research Branch
    Canada Dept. Agriculture. Monograph no. 15, 1-195.
48. KAMOEN, O. (1972).Pathogenesis of Botrytis cinerea on
    tuberous begonia. Doctoral Thesis. Univ. of Gent.
49. KAMOEN, O.(1976). Pathogenesis of Botrytis cinerea
    studied on tuberous begonia. Ministrie van Landbouw
    bestuur voor Landbouwkundig onderzoek-Gent.Verhandelingen
    Rijksstation Voor Plantenziekten. Publicatie nr 23,1-136.
50. KAMOEN,O.,JAMART,G.,MOERMANS,R.,VANDEPUTTE,L. and
    DUBOURDIEU, D. (1978). Comparative study of phytotoxic

secretions of <u>Botrytis</u> <u>cinerea</u>. Mededelingen van de
Faculteit Landbouwwetenschappen Rijksuniversiteit Gent.
43,847-857.

51. KEON,J.P.R. (1982). A study of the effects of extracellular
pectin lyase from <u>Monilinia</u> <u>fructigena</u> on cultured apple
cells. M.Sc. Thesis. Univ. of Bristol.

52. KOSUGE, T. and HEWITT,W.B. (1964). Exudate of grape berries
and their effect on germination of conidia of <u>Botrytis</u>
<u>cinerea</u>.Phytopathology, 54, 167-172.

53. KUĆ, J. (1962). Production of extracellular enzymes by
<u>Cladosporium</u> <u>cucumerinum</u>.Phytopathology,52, 961-963.

54. LANGCAKE, P.,CORNFORD, C.A. and PRYCE,R.J. (1979).
Identification of Pterostilbene as a phytoalexin from
<u>Vitis</u> <u>vinifera</u> leaves.Phytochemistry, 18, 1025-1027.

55. LYR, H. and NOVACK, E.(1962). Vergleichende Untersuchungen
über die Bildung von Cellulasen und Hemicellulasen bei
einigen . Pilzen. Z. Allg. Mikrobiol., 2, 86-98.

56. MANSFIELD,J.W.(1980).Mechanisms of resistance to <u>Botrytis</u>
In "The Biology Of <u>Botrytis</u>. Eds.J.R.Coley-Smith,
K.Verhoeff & W.R.Jarvis. Academic Press. 181-218.

57. MARCUS, L. and SCHEJTER, A. (1983). Single step chromato-
graphic purification and characterization of the endo-
polygalacturonases and pectinesterases of the fungus
<u>Botrytis</u> <u>cinerea</u> Pers. Physiological Plant Pathology,
23, 1-13.

58. McCLELLAN,W.D. and HEWITT,W.B. (1973). Early <u>Botrytis</u> rot
of grapes: time of infection and latency of <u>Botrytis</u>
<u>cinerea</u> Pers. in <u>Vitis</u> <u>vinifera</u> L. Phytopathology,
63, 1151-1156.

59. McKEEN, W.E. (1974). Mode of penetration of epidermal cell
walls of <u>Vicia</u> <u>faba</u> by <u>Botrytis</u> <u>cinerea</u>.
Phytopathology, 64, 461-467.

60. NATAL'INA, O.B. and SVETOV,V.G. (1972). Necrotic generative
organs of grapes as a source of rot infection in berries.
Vinodel. Vinograd. SSSR 4: 49-51.

61. NELSON,K.E. (1951).Fcators influencing the infection of
table grapes by <u>Botrytis</u> <u>cinerea</u>.Phytopathology,41,
319-326.

62. NOTHNAGEL,E., McNEIL,M., ALBERSHEIM, P. and DELL,A.(1983).
Host pathogen interactions XXII. A galacturonic acid
oligosaccharide from plant cell walls elicits phytoalexins
Plant Physiology, 71, 916-926.

63. O'NEILL,T.M. and MANSFIELD, J.W. (1982). Mechanisms of
resistance to <u>Botrytis</u> <u>cinerea</u> in narcissus bulbs.
Physiological Plant Pathology, 20, 243-256.

64. OVCAROV, K.E.(1937). The production of thiourea by fungi.
Dokl.Akad. Nauk. SSSR. 16, 461-464.

65. PITT,D and COOMBES,C.(1968).The disruption of lysosome-
like particles of <u>Solanum</u> <u>tuberosum</u> cells during
infection by Phytophthora erythroseptica Pethybr.
Journal of general Microbiology, 53, 197-204.

66. PORTER, F.M. (1966). Protease activity in diseased fruits.
Phytopathology, 56, 1424-1425.

67. PURKAYASTHA,R.P. (1969). Investigations on phyyotoxicity of
metabolic by-products in the culture filtrates of <u>Botrytis</u>
Spp. Proc. natl. Inst. Sci. India Part B. Biol. Sci. 35,
385-398.

68. PURKAYASTHA, R.P. (1970). The detection of phytotoxicity in *Botrytis*-infected leaves of bean (*Vicia faba*) Sci. Cult., 36, 54-55.
69. RIBÉREAU-GAYON, J., RIBÉREAU-GAYON, P. and SEGUIN,G.(1980) *Botrytis cinerea* in Enology. In "The Biology Of *Botrytis* Eds. J.R. Coley-Smith, K.Verhoeff & W.R.Jarvis. Academic Press. 251-274.
70. RIDE, J.P. (1975). Lignification in wounded wheat leaves in response to fungi and its possible role in resistance. Physiological Plant Pathology, 5, 125-134.
71. RIDE, J.P. (1980). The effect of induced lignification on the resistance of wheat cell walls to fungal degradation. Physiological Plant Pathology, 16, 187-196.
72. RUBIN,B.A., AKSENOVA,V.A. and BRYNZA,A.I.(1973). Protein synthesis in mitochondria from healthy and *Botrytis cinerea*-infected cabbage tissues. Biokhimiya,38,63-68.
73. RUBIN,B.A. and ARTSIKHOVSKAYA,Y.V.(1963).Biochemistry And Physiology Of Plant Immunity.Pergamon Press.1-358.
74. RUBIN,B.A. nd LADYGINA,M.E. (1964).The mechanism of action of toxins of *Botrytis cinerea*. Agrobiologiya, 1964, 443-454.
75. RYAN,C.A., BISHOP,P.,PEARCE,G.,DARVILL, A.G.,McNEILL,M. and ALBERSHEIM, P.(1981). A sycamore cell wall polysaccharide and a chemically related tomato leaf oligisaccharide possess similar proteinase inhibitor inducing activities. Plant Physiology, 68,616-618.
76. SHARMAN,S. and HEALE,J.B.(1977). Penetration of carrot roots by the grey mould fungus *Botrytis cinerea*Pers ex Pers. Physiological Plant Pathology, 10, 63-71.
77. SHARMAN, S. and HEALE, J.B. (1979). Germination studies on *Botrytis cinerea* infecting intact carrot (*Daucus carota* L.)roots. Transactions of the British mycological Society, 73, 147-154.
78. SHEPARD, D.V. and PITT, D. (1976a). Purification of a phospholipase from *Botrytis* and its effects on plant tissues. Phytochemistry, 15, 1465-1470.
79. SHEPARD,D.V. and PITT, D. (1976b). Purification and physiological properties of two lipolytic enzymes of *Solanum tuberosum*.Phytochemistry, 15, 1471-1474.
80. SHISHIYAMA,J., ARAKI,F. and AKAI, S.(1970).Studies on cutin-esterase.II.Characteristics of cutin-esterase from *Botrytis cinerea* and its activity on tomato cutin. Plant and Cell Physiology, 11, 937-945.
81. TRIBE,H.T. (1955). Studies on the physiology of parasitism. XIX.On the killing of plant cells by enzymes from *Botrytis cinerea* and *Bacterium aroideae*. Annals of Botany, 19, 351-368.
82. TSENG,T.C. and LEE,S.L. (1969).Proteolytic enzymes produced by phytopathogens *in vitro*. Bot. Bull.Acad. Sin. (Taipei) 10, 125-129.
83. URBANEK,H. and ZALEWSKA-SOBCZAK,J.(1975).Polygalacturonase of *Botrytis cinerea* E.200 Pers. Biochimica et Biophysica Acta, 377, 402-409.
84. VERHOEFF, K. (1970). Spotting of tomato fruits caused by *Botrytis cinerea*.Netherlands Journal of Plant Pathology 76, 219-226.
85. VERHOEFF, K. (1974). Latent infections by Fungi.

Annual Review of Phytopathology, 12, 99-110.
86. VERHOEFF, K. (1978).
Annls. Phytopath. 10, 137-144.
87. VERHOEFF, K. (1980).The infection process and host
pathogen interactions. In "The Biology of Botrytis"
Eds. J.R. Coley-Smith,K.Verhoeff & W.R. Jarvis.Academic
Press. 153-180.
88. VERHOEFF, K. and WARREN, J.M. (1972). In vitro and
in vivo production of cell wall-degrading enzymes by
Botrytis cinerea from tomato.
Netherlands Journal of Plant Pathology, 78,179-185.
89. WALKER-SIMMONS,M., JIN,D., WEST,C.A.,HADWIGER,L.and
RYAN,C.A. (1984).Comparisons of protein inhibitor
inducing activities and phytoalexin elicitor activities
of a pure endopolygalacturonase, pectic fragments and
chitosans.Plant Physiology, 76, 833-836.
90.WILSON,C.L. (1973). A lysosomal concept for Plant Pathology, Annual Review of Phytopathology,11, 247-272.
91. YODER,O.C.and WHALEN, M.L. (1975). Factors affecting
post-harvest infection of stored cabbage tissue by
Botrytis cinerea.Canadian Journal of Botany,53,691-699.

TABLE 1: REPORTS OF TOXINS AND PHYTOTOXIC ENZYMES PRODUCED BY BOTRYTIS CINEREA

TYPE OF PHYTOTOXIN	CULTURE DETAILS	HOST TISSUE	REPORTS
THIOUREA,	CULTURE FILTRATE	CHLOROSIS AND WILTING IN TOMATO.	OVCAROV,1937.
CITRIC ACID.	CULTURE, INFECTED HOST.	ACTS AS A VIVOTOXIN IN BEGONIA,INDUCING CHLOROSIS, NECROSIS.	KAMOEN,1976.
POLY- SACCHARIDES.	CULTURE, INFECTED HOST	CHANGES IN OXIDATIVE ENZYMES,PROTEIN SYNTHESIS(CABBAGE).	RUBIN & ARTSIKHOV- SKAYA,1963.
a)HIGH MOL.WT. (1000,000d) GLUCAN.	10% GLUCOSE 10-30d.	SLIGHT PHYTOTOXICITY.	KAMOEN ET AL., 1978.
b)LOW MOL.WT. (<25,000d) COMPLEX.	" "	SHRINKAGE OF LETTUCE AND BEGONIA LEAF CHLOROPLASTS;KILLS PLASMOLYSED CELLS.	KAMOEN ET AL., 1978.
PECTIC ENZYMES	GERM-TUBE EXTRACTS	MACERATED POTATO DISCS	BROWN,1917.
" "	CULTURE 4-6d.	MACERATED/KILLED POTATO,CARROT, CUCUMBER TISSUE, ISOLATED PROTOPLASTS RESISTANT.	TRIBE,1955.
" "	10% GLUCOSE	MACERATED/KILLED LETTUCE/BEGONIA LEAF TISSUE.	KAMOEN ET AL., 1978.
PHOSPHO- LIPASE	CULTURE 3d.	DISRUPTS POTATO SPROUT CELL LYSOSOMAL MEMBRANES; NO MACERATION;ISOLATED PROTOPLASTS RESISTANT.	SHEPARD & PITT, 1976.

TABLE 2: PRELIMINARY SEPARATION OF GERMINATION FLUID
         PHYTOTOXIC AND INDUCING FRACTIONS.
         (Bowen and Heale 1985)

MOL WT OF FRACTIONS	PHYTOTOXICITY TO CARROT TISSUE	GERM-TUBE LENGTH μm ON TREATED SLICES+/-SD
(A)ULTRAFILTRATION		
1,000-10,000	+	29.4 +/-3.7 S
10,000-30,000	−	88.3 +/-6.0
30,000-50,000	+	38.2 +/-2.7 S
50,000-100,000	+	51.5 +/-2.1 S
(B)COLUMN CHROMATOGRAPHY (SEPHACRYL S-200)		
5,000-12,000	slight	58.0 +/-5.2
12,000-33,000	+	37.9 +/-1.3 S
33,000-64,000	+	24.1 +/-5.3 S
64,000-200,000	−	77.9 +/-4.2

CONTROL MEAN GERM-TUBE LENGTHS ON UNTREATED SLICES
 WERE FROM 80.9 +/- 8.6 to 112.5 +/- 7.6 μm.
 MEASURED 16H AFTER INOCULATION AT 24°C, 50 REPLICATES.
 S:SIGNIFICANT DIFFERENCE FROM CONTROLS AT P=0.001 (t).
NB. Polysaccharide Fractions 1 and 2 (See Table 3) are grouped
in the 1000-10,000 fraction by ultrafiltration, but appear
in the 12,000-33,000 after Sephacryl separation- this is an
artefact of calibration of the column using protein standards,as
later separative work revealed.

TABLE 3:CHARACTERIZATION OF PHYTOTOXIC FRACTIONS FROM
GERMINATION FLUID. (Bowen and Heale 1985)

FRACTION 1.

    CARBOHYDRATE, 5000d, ACTIVITY RESISTANT TO HEATING
    AT 50 C FOR 15 MIN, RESISTANT TO TRYPSIN.
    DENATURED BY SODIUM PERIODATE AND β-GLUCOSIDASE.

FRACTION 2.

    CARBOHYDRATE, 10,000d, RESISTANT TO HEATING AND
    TRYPSIN; DENATURED BY PERIODATE AND β-GLUCOSIDASE.

FRACTION 3.

    PROTEIN, 49,000d, ACTIVITY REMOVED BY HEATING
    AND TRYPSIN, RESISTANT TO PERIODATE AND β-GLUCOSIDASE,
    CORRESPONDED TO A MAJOR ENDOPOLYGALACTURONASE FRACTION.

FRACTION 4.

    GLYCOPROTEIN?,63,000d,SENSITIVE TO HEATING,AND
    TREATMENT WITH TRYPSIN, PERIODATE OR β-GLUCOSIDASE.

ACTIVITY HERE MEANS BOTH PHYTOTOXIC AND INDUCING ACTIVITY;
LOSS OF PHYTOTOXICITY WAS CORRELATED WITH LOSS OF INDUCING
ACTIVITY, SEE TABLE 2.

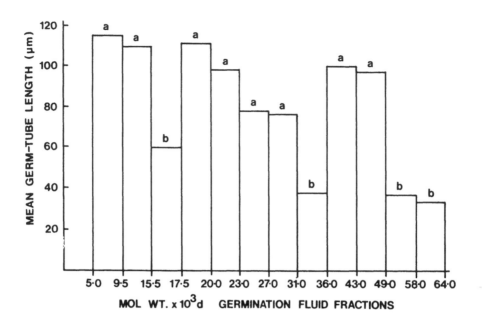

FIGURE 1

Germination fluid fractions from 48 h cultures inoculated at high density with conidia of B. cinerea assayed by their ability to induce resistance in carrot root tissues 6 h after pretreatment.

Germination fluid was separated on Sephacryl S-200 gel; 1 cm^3 fractions were pooled together to give the molecular wt. ranges indicated. Mean germ-tube length is based on 50 observations. Inducing activity was indicated when mean lengths on pretreated slices was significantly reduced (p < 0.001, t test) compared with those on untreated control slices - indicated by (b).

N.B. Molecular wt calibrations are based on protein standards and do not accurately indicate the later-determined values for polysaccharide and glycoprotein fractions.

(Bowen and Heale, 1985)

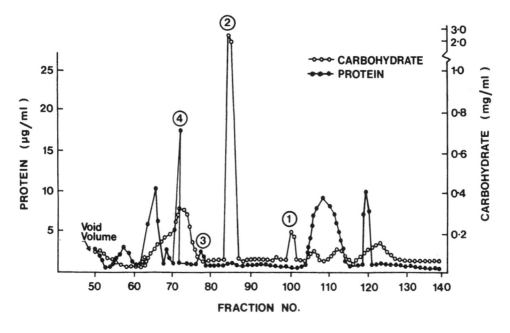

FIGURE 2

Distribution of protein and carbohydrate fractions in germination fluid of B. cinera. Separation as in Fig. 1. Protein assay : Coomassie Brilliant Blue G-250; carbohydrate : Anthrone. Fractions 1-4 correspond to those fractions with both phytotoxic and inducing activities.

(Bowen and Heale, 1985)

Mechanisms of antagonism of *Trichoderma* spp. against *Botrytis cinerea* in grapes

O.Kamoen & G.Jamart
Rijksstation voor Plantenziekten (CLO, Gent), Merelbeke, Belgium
M.Rudawska
Institute of Dendrology, Kórnik, Poland

SUMMARY

Possible mechanisms of the antagonism of Trichoderma harzianum (T.h.) were studied in order to improve its antagonistic activity against Botrytis cinerea (B.c.). A first possible mechanism may be the phytoalexin elicitation by T.h.. Phytoalexins were detected in bean hypocotyls after infiltration with polysaccharides from T.h.. Supply of glucose to T.h. spores will increase the amount of polysaccharide elicitors. This increase may result in an enhanced phytoalexin synthesis. A second possible mechanism may be the mycoparasitism based on enzymatic degradation of the cell wall of the pathogen. The synthesis of the cell wall degrading enzyme 1,3-β-glucanase is inhibited by glucose (catabolic repression) and is induced by a 1,3-β-glucan (substrate induction). Addition of 1,3-β-glucan to T.h. spores may improve the mycoparasitisme. Trying out both mechanisms, elicitation and mycoparasitism, we obtained in experiments with leaves an improved antagonistic activity against B.c. infections.

1. INTRODUCTION

Trichoderma spp. are mentioned in several reports as useful antagonists of Botrytis cinerea (B.c.) (3,8).
Since 1983 we studied aspects of the mechanisms (mode of action) of the antagonism of Trichoderma harzianum. A better understanding of these mechanisms may result in an improvement of the antagonism. At the symposium held at Bordeaux in 1983 concerning "Microbial antagonisms" four hypotheses on this mode of action were proposed by Grosclaude (7). Two of these were put to the test in our research, namely:
a. the antagonism may be based on elicitation of phytoalexins in the host plant tissue.
b. the antagonism may be based on the direct mycoparasitism of Trichoderma to B.c.. These two hypotheses were examined in our experiments.

2. MATERIAL AND METHODS

Fungi and host plants. Trichoderma harzianum Rifai obtained from the Commonwealth Mycological Institute (IMI 206040) was the antagonist used throughout this study. For good sporulation the fungus was grown in petridishes on worth agar half

301

concentration (worth broth 16 g.l^{-1} -Difco 0390-01 - and
agar 15 g.l^{-1}). Conidia obtained from these cultures were
used as sporesuspensions of 10^6 spores.ml^{-1}. An homocaryotic
strain of <u>Botrytis cinerea</u> 69 G 6b obtained from Lauber (13)
was used for infections.Infections on grape leaves were executed with 10 µl drops with spores (10^6.ml^{-1}) in 1/5 diluted
grape juice (12) or in 0,1 M phosphate buffer with pectin.
Infections on beans were carried out with 10 µl drops with
spores (10^6.ml^{-1}) in 0,1 M phosphate buffer with 0,1 M glucose
(16)

Grape plants c.v. Sémillon used for leaf infections were
obtained from Bulit (Gradignan-France). The plants were grown
in pots in the glasshouse and infections were carried out
either on detached leaves in petridishes or on attached leaves
in a moist chamber. Bean plants c.v. Cometa obtained from
Karl Schäffer, Gottingen (Germany) were sown in pots in the
glasshouse. After about a fortnight the plants were removed
to a moist chamber in the lab and infections and treatments
were carried out on the cotyledon leaves.

Growth media for preparation of metabolites. For polysaccharide production the fungus was grown during 10 days as
a mycelium mat in Roux bottles on a liquid medium described
previously (11). The polysaccharides were precipitated by
increased additions of ethanol (0,5-1-2 volumes) to the fil-
trated medium and to the hot water extract of the mycelium
(see also 11). The sugar composition of the polysaccharides
was determined by gaschromatography (17). Column chromatogra-
phy of polysaccharides was described previously (10). The glucanase synthesis by T.h. was studied by growing the fungus on
shake cultures in erlenmeyers with a medium according Del
Rey et al (1) and modifications described by Rudawska and
Kamoen (14)

Chemicals and assays. Laminarin was obtained from Sigma.
An industrial glucanase (Grindazym Fl) a gift from Grindsted-
vaerket (DK-8220 Braband, Denmark) was purified over PD 10
columns (Pharmacia) . Only the high molecular fraction was
used for treatments of B.c. infections. This industrial enzyme
contained several β-glucanases viz: an exo- 1,3-β-glucosidase
(EC 3.2.1.58) an endo-1,3-β- glucanase (EC 3.2.1.6 or
3.2.1.39) a β-glucosidase (EC 3.2.1.21) and an 1,6-β-glucanase
(EC 3.2.1.75) (2).

The assay of 1,3-β-glucanase was based on the release of
reducing sugar groups from laminarin after purification of the
enzymes on PD 10 columns. The reducing sugars were determined
by the method of Nelson - Somogyi.

Elicitation and detection of phaseollin in bean hypocotyls is described previously (9,10).

3. EXPERIMENTAL RESULTS

3.1. Antagonism based on phytoalexin elcitation (hypothesis 1)

We found no information in the literature concerning the
elicitation of phytoalexins by Trichoderma spp.In general,
polysaccharides of fungi are considered as elicitors of
phytoalexins.

3.1.1. Isolation and characterisation of polysaccharides from Trichoderma harzianum

Two polysaccharides were obtained from the medium and from the hot water extract of the mycelium.
1 stly. A glucan, containing mostly glucose, precipitated after addition of half a volume ethanol to the water solution.
2 ndly. A galactomannan, containing mainly mannose, some galactose and glucose, precipitated with equal volumes ethanol added to the water solution. With DEAE analysis of the Trichoderma mannan, we obtained only one fraction retained on the column and eluted with the salt gradient. This was in contrast with the mannan of B.c. which was eluted in two fractions on DEAE (10)

3.1.2 Elicitation experiments on beans

Mannan polysaccharide solutions (2,5 to 10 mg/ml) were succed in by cutted bean hypocotyls (50 µl/hypocotyl). An increase of phaseollin was found after treatment with a mannan from the mycelium (fig. 1). The mannan from the medium was less active. The hypocotyls treated with polysaccharides showed some necroses at the basis, while little necrosis was seen after water treatment.

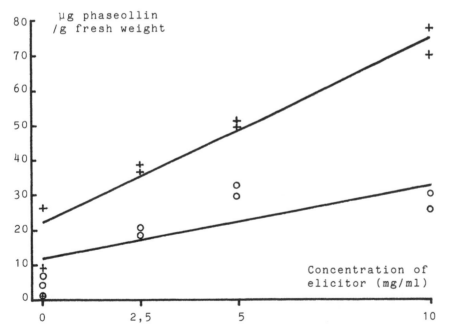

Fig. 1. Increase of phaseollin in bean hypocytyls after treatment with solutions containing increasing concentrations of mannan elicitors (two experiments)

Table 1 Percent infections with and without mannan
pretreatment of bean leaves

Experiment number	With mannan %	Water control %
1	44	85
2	81	100
3	63	88
4	70	95
5	70	90

3.1.3. Pretreatment of bean leaves with mannan polysaccharides of T. harzianum before inoculation with B. cinerea

In a next serie of experiments we examined if a pretreatment of bean leaves with 50 µl drops containg 1 mg/ml mannans
from the T.h. mycelium was able to decrease the Botrytis
infections. Control drops with water were pipetted at the
opposite leaf half. After one or two days, B.c. spores were
added to the partly evaporated drops.
    Very significant less B.c. infections were found in the
drops pretreated with the T.h. mannan (table 1). No attemps
were made to check for the presence of phytoalexins in these
infection experiments.

3.1.4 Pretreatment of grape leaves with mannans from T.h. before inoculation with B.c.

Similar experiments were carried out with grape leaves
pretreated with 50 µl drops containing T.h. mannans. Water
drops were also included as control. After two days B.c. inoculations were carried out either with diluted grape juice or
with pectin-phosphate stimulation. On grape leaves however
there was hardly any decrease of the number of B.c. necroses
after treatment with T.h. mannans (75 % succesfull infections)
in comparision to the water control (85 % drops with necrosis).
In addition no clear resveratrol fluorescence was seen under
the drops with T.h. mannans alone. For phytoalexin precursors
(resveratrol) in grapes may be visulised by fluorescence
under UV at 366 nm. Perhaps these pure polysaccharides did
not penetrate enough in the grape leaf

3.1.5 Antagonism of T.h. with an increased elicitation of phytoalexins on grape leaves

In a next experiment we tried to increase the polysaccharide production in the infection droplets by addition of 1 M
glucose to germinating T.h. spores.Two controls were included:
T.h. spores in water as pretreatment and water drops.Two days
later, 15 µl drops of B.c. (2.10^6) in diluted grape juice
were added to the Trichoderma drops. The pretreatment with
T.h. + glucose resulted in a significant decrease of infection
numbers compared with the T.h.-water pretreatment and in a

very significant decrease compared with the water pretreatments (table II-3). Concerning the size of the succesfull infections there were no significant differences suggesting that T.h. has only an influence on the start of the infection.

3.2 Antagonism based on mycoparasitism (hypothesis 2)

In the literature many authors demonstrated the degradation of cell wall components of parasites by enzymes of Trichoderma. Dubourdieu (4) described the hydrolysis of the B.c. glucan by Trichoderma glucanases. Elad et al (5) showed S.E.M. pictures of cell wall degradation of parasites by Trichoderma and its enzymes. The involvement of 1,3-β-glucanase and chitinase in Trichoderma antagonisme has been suggested by many authors (15,5,6). In our work we examined mainly the regulation of the glucanase in order to favour this aspect of the antagonism.

3.2.1 Effect of purified glucanases on B. cinerea infections in droplets on bean and grape leaves

In a first experiment on mycoparasitism droplets of the industrial glucanase,(Grindazym F4) partly purified by PD 10, were placed on leaves together with B.c. spores in order to obtain an enzymatic degradation of the B.c. germtubes, resulting in a decrease of the infections. However the results were not those that were expected: there was no decrease in infections compared with the control without glucanase. The size of the lesions was even higher with the glucanase treatment than in the control. This may be due to enzymes degrading the plant cells present in the industrial glucanase.
As a consequence of the low succes of the glucanase treatment alone we decided for an other approach viz: to stimulate the glucanase production in germinating spores of Trichoderma in infections droplets.In respect of this problem there are two questions to solve:
1 st question. How can we stimulate the glucanase production in T.h.?
2 nd question. What will be the influence of this stimulated glucanaseproduction of T.h. on the B.c. infections?

Table II Mean lesion surface (in mm^2) and percent succesfull B.c. infections after different pretreatments on grape leaves

Pretreatment	percent succesfull infections	mean lesion surface *
1 Water drops (50 µl)	93	87
2 T.h.spores in water drops	29	12
3 T.h. spores + glucose	0	0
4 T.h. spores + laminarin	0	0

* Mean surface calculated on the succesfull infections and discarding the zero results

305

3.2.2 Stimulation of glucanase in T.h. and effect on infection

In the literature we found conflicting results concerne-
ring the regulation of glucanase (1,6). In our experiments we
found that the 1,3-β-glucanase synthesis is regulated by a
double mechanism viz.: (a) Catabolic repression and (b) substrate induction (14).
-(a)-Catabolic repression means that in the presence of glucose the synthesis of 1,3-β-glucanase is repressed. So if we
want a high glucanase production, glucose and other sugars in
the T.h. droplets should be avoided.
-(b)- Substrate induction means that the enzyme synthesis is
stimulated by the presence of the substrate eg. glucan or
laminarin.
In an infection experiment we treated grape leaves with
T.h. drops supplied with laminarin. Two days later B.c. was
added to the T.h. droplets. As a control T.h. spores without
laminarin and water drops as pretreatment were used.In the T.h.
droplets supplied with laminarin we found a significant decrease of the number of lesions compared with the T.h. + water
and a very significant decrease compared with the water controls (table II-4). The T.h. spores without laminarin were
also effective in controling the B.cinerea infections: a very
significant difference with the water drops pretreatment.

4. DISCUSSION

The conclusions about elicition are als follows.
T.h. produces polysaccharides which are elicitors of phyto-
alexins.
Synthesis of these polysaccharides may be stimulated by providing glucose to the T.h. spores.
The stimulated elicitation may increase the phytoalexin defence
reaction of the plant tissue.
The conclusions about mycoparasitism are.
From our in vitro experiments we found that glucanase synthesis
by T.h. can be increased by avoiding glucose and supplying a
glucan substrate (14).
In our infection experiments partially purified glucanases were
unable to decrease the B.c. infections.
However, addition of laminarin to T.h. spores (substrate induction) resulted in a complete stop of B.c. infection. From
this and other results (4,5,6,15) the involvement of catabolic
enzymes is probable.
At least two mechanisms may be involved in the antagonism
of T.h. Both mechanisms may be stimulated by opposite condi-
tions:phytoalexin elicitation by glucose addition and mycoparasitism by low glucose and substrate addition. Due to this
stimulation by opposite conditions, the question arises what
mechanism should be stimulated by preference. The answer may
be double.
(a) In periods when phytoalexins are less important, enzymatic
mycoparasitism may be stimulated. This may be the case when
flower infections occur or on maturing grapes. In these periods
addition of some glucans and avoiding sugar may be advised.

(b) In periods with more important phytoalexin synthesis, the elicitation may be stimulated. This may be the case on green grapes. Carefull addition of some carbohydrates to the T.h. spores may then favour the elicitation of phytoalexins.

The general conclusion is that according to the vegetation period one of both mechanisms may be activated.

Acknowledgments

The autors are gratefull to ir. J. Van Vaerenbergh for critical revision of the text;for technical assistance to ir. P. Gouwy, ing M. Lambrechts, ing. G. D'Haese, ing Eben and ing. Bos working as students in the lab;to E. Van den Brande for text typing.

REFERENCES

1. DEL REY,F., GARCIA-ACHA,I and NOMBELA, C. (1979)
   The regulation of β-glucanase synthesis in fungi and yeast
   J. General Microbiol. 110:83-89

2. D'HAESE,G. (1984). Polysacchariden en β-glucanasen in verband
   met de bestrijding van Botrytis cinerea .
   Eindstudiewerk, Industriële Hogeschool van het Rijk,C.T.L.-Gent

3. DUBOS, B., ROUDET, J. and BULIT,J. (1983) Influence de la
   température sur les aptitudes antagonistes des Trichoderma
   à l'égard du Botrytis cinerea Pers agent de la pourriture
   grise de la Vigne.
   Les colloques de I.N.R.A. nr 18 (Microbial antagonisms):
   81-87

4. DUBOURDIEU, D. (1982). Recherches sur les polysaccharides
   sécrétes par B. cinerea dans la baie de raisin. Thèse à
   l'Université de Bordeaux II : 251 pp.

5. ELAD,Y., CHET, I., BOYLE, P. and HENIS, Y. (1983). Parasitism of Trichoderma spp. on Rhizoctonia solani and Sclero-
   tium rolfsii-Scanning electron mycroscopy and fluorescence
   mycroscopy.
   Phytopathology. 73 (1) : 85-88

6. ELAD, Y., CHET, I. and HENIS, Y (1982)
   Degradation of plant pathogenic fungi by Trichoderma
   harzianum . Can. J. Microbiol. 28:719-725

7. GROSCLAUDE, C. (1983). Activités du Trichoderma harzianum
   vis a vis du Stereum purpureum.
   Les colloques de l'INRA nr 18 (Microbial antagonisms)
   : 115-118

8. GULLINO, M.L. and GARIBALDI, A.- (1983). Situation actuelle
   et perspectives d'avenir de la lutte biologique et intégrée
   contre la pourriture grise de la vigne en Italie.
   Les colloques de l'INRA nr 18 (Microbial antagonisms)
   :91-97

9.KAMOEN, O. (1984). Secretions from Botrytis cinerea as elicitors of necrosis and defence.
Rev. Cytol. Biol. Végét.-Bot., 7:241-248

10.KAMOEN, O. , JAMART, G., DECLERCQ, H. and DUBOURDIEU, D. (1980). Des éliciteurs de phytoalexines chez le Botrytis cinerea. Ann. Phytopathol., 12:365-376

11.KAMOEN, O., JAMART, G., MOERMANS, R., VANDEPUTTE, L. and DUBOURDIEU, D. (1978). Comparative study of phytotoxic secretions of B. cinerea .
Med. Fac. Landbouww. Rijksuniv. Gent 43 (2):847-857

12.LANGCAKE, P. and PRYCE, R.J. (1976) The production of resveratrol by Vitis vinifera and other members of the Vitaceae as a response to infection or injury.
Physiological Plant Pathology. 9:77-86

13.LAUBER, H.P. (1971). Variabilität und Kernverhältnisse bei Botrytis cinerea.
Scheizerische landwirtschaftliche Forschung. 10(1):1-64

14.RUDAWSKA, M. and KAMOEN, O. (1985). Regulation of β-glucanase production in Trichoderma harzianum. (in preparation)

15.TOKIMOTO, K. (1982). Lysis of the mycelium of Lentinus edodes caused by mycolytic enzymes of Trichoderma harzianum when the two fungi were in an antagonistic state.
Trans. Mycol. Soc. Japan. 23:13-20.

16.VAN DEN HEUVEL, J. (1982). Effect of inoculum composition on infection of French bean leaves by conidia of Botrytis cinerea. Neth. J. Pl. Path. 87:55-64

17.ZANETTA, J.P., BRECKENRIDGE, W.C. and VINCENDON, G.(1972) G.L.C. of Trifluoroacetates of O-methyl glycosides.
J. Chromatogr. 69:291-304

Present situation of grape-vine virus diseases with reference to the problems which they cause in Greek vineyards

A.Avgelis

Plant Protection Institute, Heraklion, Crete, Greece

Summary

The grapevine virus and virus-like diseases occurring in Greece are reported. The spread of the grapevine virus diseases took place mainly after the replacement of the selfrooted virus, which were destroyed by the rapid invasion of Phylloxera vastatrix in 1898, by vines grafted on American rootstooks, both of unknown sanitary conditions. On the basis of the data collected from field surveys carried out for many years, of transmission trials on herbaceous and woody indicators, and of serological tests, the following diseases caused by viruses or by virus-like agents in the areas where grapes are mainly grown in Greece was noticed : Fanleaf, stem pitting, leafroll, enation, "yellows" disease, vein necrosis, corky bark, flat trunk, fleck, tomato black ring virus and Arabis mosaic virus. The indiscriminate marketing of uncertified graftwood and root-stocks has contributed to the dissemination of destructive virus diseases throughout the world. As a result some kind of sanitation procedures for producing virus-free clonal stocks have been established. Heat therapy is a necessary complementation of clonal and sanitary selection. The difficulties to control successfully the nematode Xiphinema index, vector of fanleaf, which is the most important viticultural virus disease in all grape-growing areas, has drawn the attention on research to find tolerance or resistance genes.

1. INTRODUCTION

Vine-growing started in Greece about 3000 a.c. and is one of the main crop of this country as yet. Its development was hardly influenced by the various sociological, economical or political evolutions because it is very well adapted to the local conditions and is closely connected with the traditions of the people.

At the present time vineyards areas are about 175.000 ha. About 54% are wine vineyards, 35% raison vineyards and the remaining 11% table vineyards. Nearly 84% are growing in Southern Greece and the islands, and only 16% in Central and Northern Greece. In the past vine-growers and mainly the raison growers were considered as privileged and their rendition defined the economic vigour of many regions of the country. Condition has gradually changed and nowadays vine-growing is not any more profitable.

Many reasons have contributed to that retrogression. The spreading of Philloxera vastatrix (1898), which in a very short time invaded continental Greece, resulting in a massive replacement of extensive areas. It is of a general acceptance that spreading of viruses and virus-like diseases was immediately followed (7). At that time enormous quantities of vine propagation material of unknown sanitary conditions, was imported, propagated and transported or interchanged in the major viticultural districts of the country. Thus is an environment, where local grapevines probably showed a certain tolerance, at least against some viruses, as a result of natural selection during many centuries, the introduction of new hosts and viruses broke the equilibrium with the result to appear new desease epidemics in the most viticultural districts of the country. The wide use of agricultural machinery during the last decades in the vineyards has contributed to the spread of viruses transmitted by nematodes.

The Greek Department of Agriculture perceived the graveness of the situation to come, but among the various measures which were taken, it did not organize a unit for production of vine propagation material phytosanitary accepted, which could help to prevent up to a certain level the spreading of common grapevine viruses. Virus problems were increasing but only in the early 70s was it taken seriously into account. At the present time the situation is sufficiently better as regards the research on grapevine and in a few years it is expected that "clean" clone stocks will be delivered.

2.1 GRAPEVINE VIRUSES AND VIRUS-LIKE DISEASES IN GREECE

On the basis of the data collected from field surveys, carried out during the last fifteen years, of transmission trials on herbaceous and woody indicators and of serological tests, a good picture of the present situation as regard diseases caused by viruses or by virus-like agents in the areas where grapes are mainly grown in Greece has been obtained.

2.2 SOIL-BORNE VIRUSES TRANSMITTED BY NEMATODES

Grape fanleaf virus : It is the most important virus of grapevine in Greece and is extensively spread (6, 7, 10). It has been noticed either in severe forms with heavy consequences on plant growth and yield or in mild forms with very light symptoms or even symptomless. Deformations of various types on the canes and leaves, dropping off and small berries are frequent. Yellow mosaic is not frequent but causes considerable reduction of the vigour of the infected plants. Vein banding is very limited. Xiphinema index was found in many vineyards affected by fanleaf.

Arabis mosaic virus : It has been found in mixed infections with fanleaf in vineyards of Central Greece.

Tomato black ring virus : It has been isolated from Korinthiaki and Gardinal, mixed with fanleaf in Northern Peloponessus (6, 7)

Enation : It seems to be restricted in Crete and cv Razaki is the most seriously affected (1, 11). The disease was also noticed in the raison grape Sultana and in the wine grape Mandilari. Its symptoms are very similar to those described in other countries except that no stem pitting or fanleaf was found associated with enation-diseased stocks. Attempts to transmit fanleaf virus were unsuccessful. However, several authors consider enation to be a symptom of a particularly virulent strain of fanleaf.

2.3 VIRUSES TRANSMITTED BY SOIL FUNGI OR APHIDS

They were not noticed in Greek vineyards until now.

2.4   VIRUSES WITHOUT KNOWN VECTOR

Leafroll : It occurs wherever grapes are growing with variable incidence (7, 10). Symptoms are most conspicuous on red cultivars. Their leaves become reddish in contrast to the chlorosis which appear on white cultivars. On many rootstocks the desease is latent. The last years a potyvirus and clostervirus-like particles were detected in extracts from grapevines exhibiting leafroll symptoms in a few countries (8).

Stem pitting : It is an important disease widely distributed in Greek vineyards (6, 7, 10, 11). It infects many important cultivars either self-rooted or grafted on American rootstocks. The local cultivars Razaki, Sultana and Korinthiaki are very sensitive and many rootstocks, like R110 which is widely used in Greece, appears severe symptoms of stem pitting. A delayed bud opening and a progressive decline appear on the infected vines. Fanleaf is frequently associated with stem pitting.

Flat trunk and corky bark : They were only occasionally observed in some cultivars of Central and Northern Greece. Corky bark was transmitted on the indicator LN33.

Vein necrosis and fleck : They occur in R110 and St. George rootstocks respectively and in several cultivars of <u>Vitis vinifera</u> as latent (10).

2.5   DISEASES CAUSED BY PROCARYOTES

"Yellows" disease : It exhibits symptoms similar to those attributed to flavescence dorée and black wood. The most susceptible cultivars proved to be Razaki and Roditis, but a limited distribution was observed in some viticultural districts of Central and Northern Greece. Rickettsia-like organisms were isolated from young roots of vines affected by "yellows" (4, 10, 11, 12).

Other viruses reported in several other viticultural countries have not been noticed in Greek vineyards.

3.   PREVENTION AND CONTROL

The indiscriminate marketing of graftwood and rootstocks of unknown sanitary conditions has contributed to the dissemination of grapevine viruses throughout the world. As a consequence a perceptible deterioration of the sanitary conditions of the vineyards have been noticed in many countries and an effective control of these disease became pressing (5). Our knowledge of virus-like diseases of grapevines has markedly advanced in the last years but the nature of the pathogen of many diseases and the control of viruses has not been solved (3). In fact research on virus-grapevine problems was fruitful in recent decades. New data as regard these diseases and their causal agents, epidemiology, vectors and control as well as improved procedures for indexing and diagnosis are frequently presented (8).

In spite of some significant progress the virus control still remains essentially preventive so far as there are not presently available chemicals effective against virus, which are not harmfull to the plants. However, treatments with antibiotics seem to be active against MLO and RLO organisms but its practical significance needs further evaluation.

Production of virus-free plants seems to be at the present time the only available tool for establishing healthy vineyards (5). Some kind of sanitation programmes are used for producing virus-free plants and estimated time for registration of certified virus-free clones is not shorter than 10 years. Procedures generally are based on visual selection, but as it is not trustfull it is supplemented by sanitary selection, heat therapy and technological evaluation. The cooperation of viticulturists, virologists and enologists is indispensably needed. For grapevine viruses which probably have no vector, the above briefly outlined sanitation procedures may give a satisfactory response in the modern viticulture. However, for viruses transmitted by vectors like nematodes, there are serious difficulties. Nematodes are not completely eliminated from grape soils by fumigation and vines are reinfected a few years later (9). For fanleaf, the most important virus disease of grapevine, the inadequate control of their nematode vectors Xiphinema index and X. italiae, have drown the attention on research to find tolerant or resistant genes. The initial results are not disappointing. Very recently Muscadine grapes were found highly resistant to fanleaf transmitted by nematode feeding (2). Thus the development of resistant rootstocks may be an exciting possibility for controlling grapevine viruses in the future (9).

REFERENCES

1. AVGELIS, A. and XAFIS, C. (1978), Presence of enations in Razaki grapevines in Crete (Greece), Phytopath. medit., 17:195
2. BOUQUET, A. (1981), Resistance to grape fanleaf virus in muscadine grapes inoculated with Xiphinema index, Plant Dis., 65:791-793
3. BOVEY, R., GARTEL, W., HEWITT, W.B., MARTELLI G.P. and VUITTENEZ, A. (1980), Maladies à virus et affections similaires de la vigne, Ed. Payot Lausanne, 181 pp.
4. KYRIAKOPOULOU, P.E. and BEM, F.P. (1977). Some virus and virus-like diseases of cultivated plants noticed in Greece during the years 1971 and 1972, Proc. 1st Greek Agric. Res. Symp., B-II: 409-419
5. MARTELLI, G.P. (1979), Identification of virus diseases of grapevine and production of desease-free plants. Vitis, 18:127-136
6. MAVRAGANIS, S.G., THANASOULOPOULOS, C.C. and PANAYOTOY, P.C. (1977), Some observations on virus and virus-like diseases in Greek grapevines. Vitis, 15: 253-257
7. PANAYOTOY, P.C. (1980), Grapevine virus diseases in Greece. Geotecnica, 2:18-24 (in Greek)
8. Proceedings of the 8th conference on viruses and virus diseases of the grapevine (I.C.V.G.), Bari-Sassari, 3-7 September 1984, Abstracts of papers
9. RASKI, D.J., GOHEEN, A.C., LIDER, L.A. and MEREDITH, C.P. (1983), Strategies against grapevine fanleaf virus and its nematode vector, Plant Dis., 67: 335-339
10. RUMBOS, I. (1983), Studies on virus and virus-like diseases of grapevine. Agricultural Research, 7:88-101 (in Greek).
11. RUMBOS, I. and AVGELIS, A. (1984), Natural spread, importance and distribution of yellows, stem pitting and enation disease of grapevine in some viticultural areas of Greece. Proc. 8th Conf. I.C.V.G., Bari-Sassari, Italy
12. RUMBOS, I. and BIRIS, D. (1979), Studies on the etiology of a yellows disease of grapevines in Greece. Z. Pflkrankh. PflSchutz, 86:266-273.

Developments in weed control in viticulture following the Dublin CEC Meeting (June 12-14, 1985) on weed problems

D.W.Robinson
Kinsealy Research Centre, Agricultural Institute, Dublin, Ireland

Summary

The Irish climate is generally unsuitable for grape vines. At Ballygagin, Co. Waterford the 'heat summation' figure (sum of the mean daily temperatures above 10°C for the period April to October inclusive) is usually less than 850. This is well below the desired figure of 1100-2770 degree days. Nevertheless, small vineyards occur in several counties.

The mild, moist climate in Ireland encourages weed growth throughout much of the year. Weed control programmes developed for bush and cane fruits have been adapted for use in vineyards. These are based mainly on the use of overall and spot treatments with herbicides, supplemented occasionally by cultural control measures if appropriate.

Because the weed problem in Ireland is potentially much greater than in southern countries, most growers aim to achieve complete control. This strategy has worked well in fruit crops, an organic mulch being used on gently sloping sites where soil erosion is anticipated. Previously, a strategy based on suppressing weeds until fruit harvest and allowing weeds to develop until the following spring proved unsatisfactory on some occasions. Weed growth during the winter and the development of intractable weeds made control in the spring much more difficult with this system. It is considered that the concept of economic threshold levels is difficult to apply to weeds in fruit plantations because of the long-term nature of these crops and the likely development of a perennial weed flora in such situations.

The Dublin CEC meeting recognised that total weed-free systems carry a risk of undesirable side effects. Research in Ireland using herbicides to achieve weed-freedom has shown no adverse effect on soil structure compared with cultivation. However, more information is required on possible side effects such as the induction of new pest problems through the inhibition of natural mortality factors.

1. Introduction

Ireland lies largely between 52° and 55° N latitute and has a mild, equable, oceanic climate. At the Pomology Research Station, Ballygagin, Co. Waterford (latitude 52°N) mean daily temperature is about 6°0 C in January and 15.5°C in July. Rainfall of about 1000 mm is fairly evenly distributed throughout the year. The 'heat summation' figure (sum of the mean daily temperatures above 10°C for the period April to October

inclusive) only exceeded 850 degree days one year (1976) in the period
1967-78. This is well below the desired figure of 1100-2770 degree days.
Although the Irish climate is generally unsuitable for grape vines, small
vineyards occur in several counties.

2. Research on weed control

The cool, moist climate, so unfavourable for grape vines, enables
weed growth to occur throughout much of the year. Detailed surveys of
weeds in vineyards have not been conducted, but the weed flora is rather
similar to that in blackcurrant, raspberry and gooseberry plantations. A
wide range of species occur in plantations and the most prevalent species
have changed during the last 25 years (Table 1). Weeds in fruit
plantations in Ireland are seldom checked by summer drought. In addition,
many species that were prevalent before herbicides were widely used, such
as Poa annua, Stellaria media, Ranunculus repens and Senecio vulgaris, can
grow slowly during mild spells in the winter.

Table I - Most common weeds in soft fruits in Co. Wexford in 1960 and 1985

1960	1985
Stellaria media	Galium aparine
Chenopodium album	Viola arvensis
Senecio vulgaris	Atriplex patula
Poa annua	Potentilla anserina
Agropyron repens	Epilobium ciliatum
Ranunculus repens	Vicia sativa
Cirsium arvense	Agropyron repens
Rumex spp	Convolvulus arvensis
Sinapis arvensis	Equisetum arvense
Chrysanthemum segetum	Trifolium repens
Capsella bursa-pastoris	Sherardia arvense
Polygonum convolvulus	Hypericum humifusum
Galeopsis tetrahit	Aphanes arvensis
Lamium purpureum	Volunteer crops (cereals,
Polygonum aviculare	potatoes and blackcurrants

Source: N. Rath, Clonroche

Year-round weed growth in fruit plantations in Ireland contrasts with the
normal situation in many south European countries where weeds die back
naturally during the summer.

Because the potential weed problem is much greater in Ireland than in
drier, warmer countries most growers of bush and cane fruits aim to
achieve complete control. Vine growers have adopted the same objective.
Since the early 1960s weed control has been achieved mainly with
herbicides but these are supplemented by cultural treatments (hand forking
or hand pulling of individual weeds) as appropriate.

Simazine is the standard soil-acting herbicide used in bush and cane
fruit plantations and in vineyards. This is applied as an overall
application in February and March. Surviving weeds are controled normally

by directed sprays of paraquat during the growing season. Paraquat is also often used as an overall clean-up spray against winter weeds between November and February. The soil-acting herbicides, diuron and napropamide, are used from time to time as a substitute for simazine, especially where there is a risk of weeds resistant to that herbicide becoming established. Glyphosate and dichlobenil are used as spot-treatments against weeds that are poorly controlled by the other herbicides.

The results of experiments with these herbicides are presented in the annual reports of the Horticultural Centre, Loughgall, Co. Armagh, Northern Ireland and of the Agricultural Institute's Kinsealy Research Centre, Dublin, Republic of Ireland. Research at Loughgall since 1955 and subsequently at Kinsealy, Ballygagin and Clonroche has shown that where weeds can be controlled successfully with herbicides, the optimum method of fruit plantation maintenance, from a yield and management viewpoint, is to eliminate soil tillage completely. Compared with traditional cultivation the use of herbicides without soil-tillage has given increased yields with blackcurrants, gooseberries and apples and no yield difference with raspberries or strawberries. Where herbicides are used and soil tillage eliminated the soil surface becomes compact and smooth and more susceptible to water run-off and erosion on sloping sites. In Ireland, fruit crops are not normally grown on steeply sloping land. On gentle slopes (ca 1 in 10) where erosion can also occur if herbicides only are used for weed control, run-off can be retarded and erosion prevented by the use of a light mulch of organic material such as straw. Where simazine and paraquat are used and cultivation is eliminated under high rainfall conditions, a thick covering of moss usually develops at the soil surface. This moss covering also helps to stabilise the soil surface and reduce erosion.

3.Effect of herbicide use on soil structure

In all fruit plantations treated with herbicides and not cultivated, significant changes were recorded in many soil properties including pore volume, numbers and orientation of pores (Bulfin 1967), bulk density, infiltration capacity, soil strength (Bulfin and Gleeson 1967) and organic matter (Robinson 1975).

In a number of long-term experiments no serious adverse effect on the soil was observed in the absence of cultivation. As a result of this work, soil tillage is normally no longer used in fruit plantations and most growers adopt a non-tillage system of management (Robinson 1964).

The possibility of managing weed populations in perennial soft fruit crops in Ireland has been examined along similar lines to those advocated by Scienza and Mirvalle (1985). Weeds were allowed to develop in the autumn and winter when they are unlikely to cause any damage. They were then controlled in the spring before they could compete with the crop plant. This strategy was satisfactory in situations where perennial weeds did not develop (Allott et al 1971). Usually, however, weeds that were difficult to control such as Agropyron repens and Potentilla anserina became established (Anon 1964). Moreover, periods of prolonged wet and windy weather can delay spraying operations. Timing of herbicide application is sometimes critically important e.g. paraquat shortly before bud-burst. From an economic and management viewpont results in bush and

cane fruit plantations and apple orchards suggested that the most
satisfactory method of soil management is a system in which weeds are
completely controlled by herbicides and the soil is maintained in a clean
condition throughout the year. Although this goal cannot be fully
achieved, weed populations can be kept at a low level and it is possible
to prevent weeds from seeding.

Because of the speed with which the weed flora can change and
resistant species develop, a "weed-free" condition can only be maintained
over a period of years with constant vigilance and continuing effort. The
size of the weed seed population and longevity of weed seeds in the soil,
the ability of the few surviving weeds to replenish that population and
the speed with which "opportunistic" species can exploit rapidly any open
space indicate that weeds will never be completely suppressed.

Where a virtual "weed-free" environment can be achieved, a number of
significant economic benefits can accrue. All weed competition with the
crop is eliminated and crop vigour and yield increase. The spread of weed
biotypes with acquired resistance to herbicides can not occur where weed
seeding can be prevented. In addition a small number of weeds in a fruit
plantation may act as alternate hosts for nematodes and other crop pests.
It seems likely that the aim of complete suppression of weeds, basically
with herbicides (both overall and spot applications) but supplemented with
some hand cultivation, will result in the use of a lower total amount of
herbicides over a 10-year period than a system aimed at achieving
incomplete weed control only. Weed control costs and the amount of
chemical used have been reduced following several years of successful
herbicide treatments in raspberry plantations in Scotland (Lawson 1985).

4. Discussion

While a system of complete weed control has given satisfactory
results in many parts of Ireland, England and Scotland, it is recognised
thta there are situations where this strategy will be impracticable or
undesirable. For example in many situations a light covering of weeds
during the winter would be advantageous provided these weeds could be
controlled easily in the spring. This technique has not always been
successful in Ireland and more detailed studies are required on the
management of weed populations. Where weeds are being managed rather than
controlled, information is required on the feasibility of maintaining
populations of relatively non-competitive annual weeds while suppressing
perennial species. Work is required to determine if it is possible to
prevent the development of difficult weeds including the role of living
and dead mulches.

The Dublin CEC meeting recognised that information is incomplete on
the risk of undesirable side effects associated with weed-free systems.
There is evidence that, in some annual crops, increasing reliance has been
placed on herbicides without sufficient knowledge of the ecological
effects concerned (El Titi 1985). Populations of some weed species can
provide shelter and food for some polyphagous predators or parasites.
There is evidence to indicate that by allowing sugarbeet seedlings to
complete their susceptible establishment phase before removing weeds, the
pressure of pests such as wireworms (Elateridae), Aphis fabae and
Onychiurus fimatus on the crop is reduced (El Titi 1985). However, fruit
crops are established by planting cuttings, runners, vines,bushes or trees

which are likely to be less susceptible to attack by insect pests than
seeds or seedlings. Nevertheless the increasing incidence of vine weevil
(Otiorhynchus sulcatus) in blackcurrant and strawberry plantations in
Ireland in recent years may have been accentuated by a weed-free
environment. While some information is available on the entomophagous
fungi, moulds and insects affecting vine weevil populations (Evanhuis
1978), little is known about the quantitative effect of these natural
control measures on the pest.

The instability of some modern ecosystems has been attributed to
vegetational simplification resulting from the adoption of weed-free
systems without hedges. In Ireland soft fruit plantations and vineyards
tend to be small, units larger than 2 ha being rare. Plantations are
usually surrounded by hedgerows and provide good coverage for certain
insects.

Whether a weed-free fruit plantation is more (or less) at risk from
attack by nematode and insect pests than a weedy one is unknown. The
extent to which an unsprayed hedgerow can substitute as an insect habitat
for weeds growing in the crop is also unknown. Much more information in
these areas is required.

At the Dublin CEC meeting El Titi (1985) rightly indicated that weeds
are only harmful in fruit plantations if they reduce yields or cause
technical difficulties during or after harvest. The weed density at which
level such losses occur is called the economic threshold. Although data
about the economic thresholds for different weed species in certain annual
crops is available, little is known about the practical use of economic
threshold levels for weeds in the long term, particularly in perennial
crops. The concept of economic threshold levels may not be valid in a
perennial crop such as vines or soft fruits. Losses caused by weeds are
influenced by a large number of factors including the time of removal, the
plants or crop infested, weed density, soil moisture content and soil
fertility. These variables make it difficult to predict accurately the
loss caused by weeds in a particular situation. Moreover, weeds do not
usually occur as a single species but in mixtures with probably up to 20
or 30 other species. The behaviour of any one species in a mixture is
often usually different from when that species is grown separately. Hence
the determination of economic threshold levels would need to take into
account the other weed species that are present. Moreover, in perennial
fruit crops the most severe weeds are perennial. These present a special
problem since many individual ramets may be produced from a single
individual weed plant. Further a few plants of a "new" weed species will
not cause any financial loss that year but if allowed to seed will cause
considerable economic loss in subsequent years. The possibility of using
the concept of economic weed threshold levels in fruit crops needs further
examination.

5. Conclusions

The use of herbicides aimed at achieving total control of weeds has
given satisfactory results in fruit plantations in Britain and Ireland
during the last 25 years. It is recognised that total weed control will
reduce the incidence of some pests but the lack of diversity is likely to
accentuate others. Reducing the amount of herbicides used is a desirable
objective but total weed control does not necessarily require the use of

large quantities of herbicides. More detailed studies on the economic and ecological consequences of a 'weed-free' environment are required.

REFERENCES

1. ALLOTT, D.J., ROBINSON, D.W. and UPRICHARD, S.D. (1971). The response of gooseberries to non-tillage systems of management. Hort. Res. 1971, 11: 166-176.

2. Anonymous, (1964). Chemical weed control - gooseberries. Annual Report 1963. Horticultural Centre, Loughgall: 33-36.

3. BULFIN, M. (1967). A study of surface soil conditions under a non-cultivation management system. 11: Micromorphology and micro-morphometrical analysis. Ir. J. agric. Res. 6, 189-201.

4. BULFIN, M. and GLEESON, T. (1967). A study of surface soil conditions under a non-cultivation management system. 1: Physical and chemical properties. Ir. J. agric. Res. 6, 177-188.

5. EVENHUIS, H.H. (1978). Bionomics and control of the black vine weevil Otiorhynchus sulcatus. Med. Fac. Landbouww. Rijksuniv. Gent, 43/2, 607-611.

6. EL TITI, A. (1985). Pest management as a basic foundation for integrated farming systems. Proc. EEC Experts' Meeting 'Weed control in vine and soft fruits', Dublin, Ireland, June 12-14, 1985. (in press).

7. Lawson (1985). Personal communication.

8. ROBINSON, D.W. (1964). Investigations on the use of herbicides for eliminating cultivation in soft fruits. Scientific Horticulture XVI, 1962-1963, 52-62.

9. ROBINSON, D.W. (1975). Some long-term effects of non-cultivation methods of soil management on temperate fruit crops. Proc. XIXth International Horticultural Congress, Warsaw (1975), III: 79-91.

10. SCIENZA, A. and MIRAVALLE, R. New aspects of integrated methods of weed management in Italian vineyards. Proc. EEC Experts' Meeting 'Weed control in vine and soft fruits', Dublin, Ireland, June 12-14, 1985. (in press).

New opportunities for the weeds control in vine

G.Marocchi

Regione Emilia-Romagna, Osservatorio per le Malattie delle Piante, Bologna, Italy

Summary

We relate the reasons and presuppositions for a new and more reasoned weeds control in vine in order to reach economic and agronomic advantages.

Eight years of trials show how the "Soil conservation technique" known as "No Tillage" allows to avoid the weeds' competition and, at the same time, brings remarkable advantages such as : savings, in terms of production costs and working hours, and drought/erosion resistance plus a better soil-bearing.

Vine-growers are afraid of fungus diseases but, on an average they know how to cope with them.

Their knowledge concerning insects, acari and nematodes is not so good and consequently also the more suitable and rational way to control them.

But, perhaps, their knowledge gets even lower when weeds, one of the worst competitor of vine, are concerned.

The growers keep their vineyards free of weeds using several different systems, often without making a balance of the input-output and without any evaluation of the benefits or disadvantages of the system they use.

On a gross average, vine-growers spend 20/30 h/ha/year to till the soil with the main purpose to control the weeds. This means a great amount of manpower, cost, fuel (120-180 l/ha).

Meanwhile, in the most developed countries, "Conservation Tillage" is becoming more and more actual as the answer to the specific needs about conservation of the soil, fuel and so forth.

This technique has been developed in some countries, mainly in France, where "Reduction Tillage" or "Zero-Tillage" are usually practiced in a large part of the vineyards.

To check the opportunities of weed control in vineyards using the "Conservation Tillage" approach, a trial of "No Tillage" has been carried out eight years ago, using several herbicides.

"No Tillage" is an agronomic practice to produce food (grape in this case) without moving the soil by mechanical interventions. Weeds are controlled by 1-2 herbicide applications : one in Winter/Spring against annual weeds and, often one in Summer against perennial ones.

In Italy, despite the benefits showed in France and in other countries, the "No Tillage" is still rather unknown due to the following reasons :

1. Lack of knowledge of the technique
   It is evident that one can put a resistance against such a radical
   change, however, the obstacle has already been overcome by many
   trials, some already started other under procedure. Among these
   there is one trial that was started eight years ago and it is still
   followed up by collecting data, remarks, observations and so forth.

2. Fear to cause damages
   The trials have demonstrated that there is no danger, and more than
   that, there are benefits to the mechanical treatments.

3. Attachment to old traditions
   Farmers have a strong feeling versus giving up the working of the
   soil, but the foreign experience and the results obtained from many
   trials show that it is possible and easy to change the system.

4. Scarce evaluation of the damage caused by weeds
   A difficult evaluation to make, but many figures show that the weeds
   can damage the vineyards' yield at 20-30-50% and even more.

5. Cost of herbicides
   This is an unjustifiable pretext because the main reason which
   induces the grower to use chemicals is exactly that of saving money
   and, a rational usage of herbicides, makes the vineyards' management
   less expensive. It has to be underlined that this technique was
   started in leader farms, the biggest and better managed ones.

   To describe this new and very interesting technique there are
already many components and results obtained by our Italian experiments
and part of them refer to the above-mentioned trial. But there are also
figures, remarks and observations drawn from farms which already use
this technique. All the positive results, whose aspects are useful to be
discussed and investigated, are making a real comparison between our
situation and the French one. The French in fact started this technique
almost twenty years ago and, in many areas, 70-80% or more of their
vineyards are so managed.
   The trial, whose results we are now examining, started in 1978 and
from that time on the weed control was made only by chemicals
alternating, occasionally, Winter applications with Summer operations.
   The remarks and observations made in these eight years can be
summarized as follows :

a. Appearance of the vineyards and grapes quantity-quality production
   The vineyard is now in optimal conditions and with a vegetative
   vigor equal to the ones tilled with traditional means.
   From a visual assessment no signs of suffering can be noted and the
   production/quality of grapes is equal in both sections.
   It is to be added that, without the expensive "tillage" of the soil,
   we also avoided damages to branches and roots.

b. Condition of the soil
   Stratigraphic examinations in various depths of the soil were made
   and the new technique showed a remarkable improvement in the
   physical condition of the soil, such as : stonature, permeability
   and life of microorganisms.

The "Tillage" technique destroys the root-system of vine from the surface down to 20-30 cm, while in the "No Tillage" one, the roots are evenly distributed up to the surface.

The latter is a remarkable advantage both for the mineral nutrition and the drought resistance. In fact, results obtained from the current trials, from observations made in other applications in Italy and, more important, from figures confirmed by the French experiences, the soil of the "No Tillage" technique shows a better endurance to lack of water.

Moreover, it is obvious that, with a root system on the surface of the soil, the vineyard benefits immediately also from a very tiny rainfall.

c. Soil-bearing and erosion resistance

The trial is sited in the lowland, therefore there is not the problem of erosion. An inconvenience, on the contrary, which is well known in the Italian hilly vineyards.

Also in this case we must acknowledge the undeniable benefit which the "No Tillage" technique brings, because it is just in the "Tilled" soil that water makes the biggest erosion and moves the soil downwards.

Another benefit, clearly noted, is the better soil-bearing of the "No Tilled" soil in comparison with the "Tilled" one.

d. Fertilization

A matter always to discuss due to different opinions and various methods used. In the running trial we solved this problem distributing organic substance on the surface and without covering it with earth. In a very short time there was the decomposition with a big benefit both for the earth appearance and for the vine nutrition.

e. The economic aspect

This is the most important point to consider before adopting this new technique. The French figures leave no doubt about it, this technique reduces the cost down to 50%.

Obviously, the comparison between the two techniques needs quite a long time due to the fact that the rate of the herbicide will be reduced in the years that follow the first applications while, on the contrary, the traditional manual or mechanical tillage will have to be done every year and a reduction is unthinkable.

Moreover, the fuel cost is always going up while the cost of herbicides tends to go down. Therefore, on the long run, this new technique will be even more money-saving.

For the running trial we have calculated, on an eight year base, a medium cost of Lit. 280.000 per year for the "No Tillage" plot versus an expenditure of Lit. 450.000 for the "Tillage" one.

FURTHER INVESTIGATION : (in progress)

INTEGRATION BETWEEN ZERO-TILLAGE AND OTHER SOIL/WEEDS TECHNIQUES

The "No-Tillage" system may be married with other techniques. Several trials are placed in Emilia-Romagna to evaluate the most efficacious way.

In the Northern Regions of Italy an integration between spontaneous weeds in Autumn-Spring and a weed control in Spring-Summer using mixture of pre an dpost emergence herbicide, allows a reduction of input (chemical and mechanical) without adverse effect on the crop and more than that, allowing several additional benefits.

In the South, many tillages may be saved using low rate of pre-emergence herbicide + low rate of Glyphosphate after the fall of the leaves.

In flat area, where there is availibility of ground-water close to the surface or in irrigated vineyards, the grassed-down system with chemical weed control under row seems to be one of the best solutions.

Weed management may also be obtained using a weed growth regulator or very low rate of Glyphosphate (0,1 - 0,2 l/ha).

Session 3
Strategies of integrated pest management

Chairman: G.Domenichini

Strategy problems for integrated pest control in viticulture

G.Domenichini

Istituto di Entomologia Agraria dell' Università Cattolica del Sacro Cuore, Piacenza, Italy

Summary

European experience in viticulture over the last few decades indicates the advantage of a strategy in integrated pest control based on biological means understood in the broad sense. Among the factors in favour of such an orientation are: 1) the good power of the plant to bear the infesting agents' attacks, thanks to the abundance of vegetative organs; in addition to that the industrial transformation of the product makes it possible to accept within certain limits those aesthetic faults that in other crops increase the need for treatments; 2) a high biological potential of the vine ecosystem which enables it to hold back pests partially or, sometimes, even totally; 3) the possibility of influencing the habitat of pests through routine operations.
Deeper knowledge of the relations between the micro-climate and the components of the agro-ecosystem is becoming indispensable, and is important for adequate use of cultivation practices,too.
The need to develop an agro-meteorological network in the wine-growing areas is discussed. In particular, the usefulness of extending research into the influence of the wind on the spread of arthropods, on the efficiency of pheromone traps and the epidemiology of plant diseases is pointed out.
The analyses and the use of data in such a complex system of approaches and models demand use of computer technology and steady abandonment of intuitive solutions.

Many experiences in viticulture have matured in Europe since American vines were introduced and deep changes were made in the planting techniques. In particular post-war experiences on the use of pesticides and fungicides have given lots of data on the effects achieved and on the undesired consequences of their use.

In general pesticides are not so frequently used in vines as in other cultivations, but whenever their applications have exceeded certain limits, secondary phytophagous arthropods have appeared very soon. Moreover, some key-species such as grape moths have become even more harmful.

Some fungicides have also shown such side effects on mite pupulations as to upset the balance to the advantage of harmful species.

Although the vine agroecosystem is rather complex as in all perennial cultivations, thus permitting profitable production rates even without too frequent treatments, it undergoes negative changes quite easily when it is exposed to disruptive means, such as chemicals.

For this reason it is advisable in integrated pest control to adopt a strategy based on the use of bioecological means and on the accurate selection of those pesticides which do not interfere with their action.

This is made possible by a number of elements: 1) the good power of
the plant to bear the infesting agents' attacks, thanks to the abundance
of vegetative organs; in addition to that the industrial transformation of
the product makes it possible to accept within certain limits those aesthetic faults that in other crops increase the need for treatments; 2) a
high biological potential of the vine ecosystem which enables it to hold
back pests partially or, sometimes, even totally; 3) the possibility of
influencing the habitat of pests through routine operations.

The plant's prerogatives (see point 1) have already been exploited for
decades in the successful struggle against phylloxera, grafting European
varieties on American hybrids resistant to the root-inhabiting forms of the
aphids. This procedure has been achieved after a wide selection carried out
on American and Euroamerican hybrids. The genetic improvement is going on
in Vitis vinifera L., aiming at finding out single and complex hybrids and
clones, suitable for different pedological and environmental conditions as
well as resistant to diseases, such as downy mildew, oidium, Botrytis, and
so on.

Another factor favourable to a bioecological approach in the integrated pest control in viticulture emerges when we consider the abundance
of vegetative organs produced by vines. The power of the aerial apparatus
to bear unfavourable factors without damages to the production makes it
possible to use slow action means rather than classic pesticides for inter
ventions. Shoots, leaves and the great abundance of inflorescences allow
to fix economic thresholds on much higher infestation levels than for other
crops.

The grape moth attack to inflorescences can be very strong without
causing appreciable damage to the production: the infestation percentages
on the bunches of grapes can be above 5% threshold without heavy consequences provided that the epidemiology of grey mould (Botrytis cinerea
Pers.) in the vine is irrelevant.

Hence comes the possibility of using Bacillus thuringiensis against
these Tortricids taking care that the right moment to apply it is chosen.

It is however necessary to increase researches into biological control
factors of grape moths with a view to finding out suitable means for the
different environments.

As far as point 2) is concerned, we can dispose of a remarkable amount
of papers on the presence and the relevance of the mortality produced by
biological factors of phytophagous arthropods; on the contrary we find that
phytophagous arthropods pullulate when the biological control factors are
modified by classic pesticides. Researches on moulds antagonists to vine
infesting cryptogams are increasing, even if at present they are more inter
esting from the theoretical point of view.

Lobesia causes light damages in many areas of Northern Italy where
more than one insecticide treatment are carried out only on table grapes;
rather than by parasites,larvae are killed by different diseases, such as
e.g. a polyhedric virus.

The problem arises in those areas where some phytophagous species,
favoured by suitable microclimate conditions, become chronically infesting
and cause unacceptable damages. An example for this is given by Planococcus
citri (Risso), an endemic scale insect under muggy conditions which attacks
bunches after multiplying very rapidly on shoots and leaves. The activity
of some specialized parasites,such as Anagyrus pseudococci (Gir.), Leptoma-
stidea abnormis (Gir.), Leptomastix dactylopii How.,and of some predators
like Cryptolaemus montrouzeri (Muls.),whose effectiveness has been experi-
mented in the Mediterranean basin, can be integrated by some farming practices such as summer pruning aiming at obtaining a better and more uniform

exposure to the air and the sun. The evaluation and exploitation of entomophagous insects must be carried out for other infesting arthropods.

As far as point 3) is concerned, modern specialization has brought in
some new techniques as to farming practices such as the use of machines for
working soil, the use of long term manures that can be applied in different
ways; through suitable practices it is possible to modify the ecological
conditions. This problem requires a multidisciplinary approach in order to
better know the effects of different operations on the microenvironment, on
the plant and on the organisms thriving in the cultivation.

In this strategy pesticides find a limited and precise use: they are
tactical instruments for urgent interventions, can be applied in the suitable doses for the target they are addressed to, can be distributed only
in the plots or to the plants where the intervention is really necessary
and applied with all the precautions suitable for limiting the negative
side-effects. Since the agroecosystem can be negatively influenced by biocid
treatments, their use must be sporadic, so that the readjustment of biocenosis in favour of the production is possible.

Hence comes the problem of the availability of the microclimatic data
relevant to the agroecosystem and after all necessary to forecast damages.
This problem is particularly important for viticulture in the Mediterranean
basin. The variety of features, expecially in hilly areas, makes up a
kaleidoscopic picture of ecologic conditions. The vineyard exposure, the
soil nature, windiness, altitude are some of the variables that must be
taken into consideration together with the different cultivating forms, the
characteristics of grafts and rootstocks, the different working techniques.

To formulate simulation models of an ecosystem it is necessary to have
data strictly related to the vineyards under observation having those particular characteristics. For this reason it is extremely important to dispose of an agrometeorologic net able to give data concerning the microclimatic conditions of the different areas of a territory.

The accuracy of information is going to be of great use firstly to
researches and secondly to the method applications. Experimental studies
of Eupoecilia ambiguella Hb. indicate that it is possible to forecast the
duration of each stage and of the whole cycle very precisely referring to
heat unit accumulations. The above is true for Lobesia botrana Schiff. too.
The same accuracy is required for data concerning thermohygrometric and
pluviometric courses in order to define interventions against vine-crypto
gams.

Moreover, the agrometereological net must give indications on the
direction and speed of the wind responsible for the transport of infesting
agents. This subject, often forgotten, must not make us neglect the researches stating not only the diffusion of Phylloxera, of many scale insects
and of mites, but also the responsibility of the wind activity for the
epidemiology of Cryptogams. The catch of few Lobesia by means of pheromone
traps followed by strong infestations in some places raises the question of
possible immigrations from the surroundings favoured by air streams.

Wind direction and speed involve the use of pheromones to monitor grape
moths: as to these elements the location of traps and their effectiveness
ought to be investigated carefully. In my opinion it is useful here to
mention the results of some unpublished experiments that Cravedi and I have
carried out on the wind effects on the attractiveness of pheromone of Cydia
pomonella (L.) in the field.

The catch rate of males released diminishes in proportion to the wind
speed up to 8 km per hour, then it stops beyond this value even if the air
stream moves towards the insects present just some metres off.

The wind action, especially in proximity to coasts and windy areas, cannot be neglected when we consider the activity of Phytoseidae and the transport of pollen, which they eat for lack of victims or as an alternative to them.

The research will investigate the wind activity concerning the transport of adult grape moths and of their parasitoids living on spontaneous plants or present in uncultivated areas around vineyards.

In Northern Italy, Encarsia perniciosi (Tower) has spread in areas which are hundreds of kilometres far from the territories where it had been released, and the same happened for Encarsia berlesei (How.) and for Polynema striaticorne boreum Gir., the parasite of Stictocephala bisonia's Kopp eggs.

The influence of factors related to winds on the infestation of leaf hoppers, whose tendency to migrations and shifts is well-known, is another subject deserving our attention.

We are not far from the concrete implementation of the agrometeorological net, at least in some Italian regions such as Emilia-Romagna, Lombardia and Lazio, where observation posts for downy mildew are already active.

Plans have been made to connect the agrometeorological net with computer centres for damage forecast, which technicians assisting farmers in the integrated pest control can refer to.

There is much work for researchers, but the results will reward them for their efforts provided that specialists of the different disciplines cooperate. To guarantee the successful realization of this plan social bodies and organizations will have to take the financial charge upon themselves so as to give farmers a help destined to yield more and more advantages.

Principles of integrated protection in vine growing

J.P.Bassino

Association de Coordination Technique Agricole (ACTA), Paris, France

Summary

Biological crop-growing cannot respond to the technical and economic needs which guide the producers of table or wine grapes. The only method of progress resides in the intelligent employment of integrated control principles.

Researchers have been perfecting simple and reliable methods for about ten years, in a specialized IOBC working group. The measure of the real risk of damage, on the scale of the small plot or the crop-growing unit, limits the number of treatments.

A rational guide for health protection is proposed and the notion of threshold is defined precisely. Indications are provided for the different means of intervention.

The progressive change of attitude of producers may be seen by the demonstrations which have been carried out on the real scale in the research centres throughout Europe. These Centres, which are united in a network, must be able to act as support to training activities, which are essential.

The material and moral support of national and international institutions is considered to be indispensable.

1. GENERAL SITUATION

In the sector of vine-growing, as in other fields, the producers try to obtain quality products from their soil, in a quantity which is large enough for them to live comfortably. It should be added that the reduction of laborious work is a constant preoccupation in the peasant world.

This very legitimate attitude is not incompatible with research into a maximum limitation of the damaging effects of interventions on the crop-growing habitat and its environment. One possible way is supervised control, which has as immediate consequence a reduction of expenditure by the treasury, which is the easiest to measure.

Biological crop-growing, which has a very fashionable romantic image in Europe, is incapable - in the present state of knowledge - of responding the imperatives which govern the action of producers.

In a more rational, scientific and realistic manner, the best way of reconciling the needs of agricultural production with the preservation of the quality of the environment is to progressively put into practice the principles of integrated control. The most usual definition of this, established by the IOBC in 1973, is the following: "procedure of control against harmful organisms which uses an ensemble of methods which satisfy economic, ecological and toxicological needs, with priority for the thought-out implementation of natural limitation elements and respecting the tolerance thresholds".

Put more simply and in a more operational way, one may say that two essential ideas must guide the activities of researchers (for about ten years numerous scientists of various European countries have been working in the specialized OIBC-WPRS working group), workers for the agricultural Council and also, naturally, farmers :

- a crop is a living environment and natural limitation factors play a considerable role : in other terms and as a result, the richer the environment is in species, the more stable it is;

- everything with an overall effect and, in particular, insecticides which are by nature very efficient, are dangerous for this living environment; obviously they must be used in conditions where the repercussions or the "back-lashes" are as weak as possible.

2.   THE BASES OF INTEGRATED PROTECTION

There are only two possible systems upon which the application of these modes of intervention rests :

a) a programme of treatments established in advance, for a year or more, according to an idea of more or less managed calendar type.

b) an estimation of danger, according to the state of development of the crop and the critical period; in this system one may distinguish two stages in the development towards greater precision and above all towards finer measurement;
- the evaluation of a potential risk, on the scale of a small region
- the evaluation of real risk, on the scale of each plot or each crop-growing unit of the whole which constitutes the agricultural farm.

It is obviously the second way of operating which corresponds best to the spirit and the letter of integrated crop protection.

In practice, the decision to treat or, which is more difficult, not to treat, is taken :

- after having received notification from the agricultural warning station which provides, for a small region or an agriculturl sector, an estimate of potential risk. In fact, in this case the farmer treats the plant and not the enemy of his crop directly.

- after an estimation of real risk in each plot or group of plots (cultural unit); one must see and measure the activity of the pest to take a suitable decision.

This mode of operation - in the spirit of integrated protection - presents advantages and disadvantages.

Advantages	– one knows why one is acting – the number of treatments is reduced to a minimum – one preserves, as far as possible, the useful flora and fauna
Disadvantages	– a considerable length of time is necessary for the observations and the various measurements – the decision is taken, in the best conditions, only on the scale of each crop-growing unit.

To facilitate the decision making, we have drawn up the following guide to reasoning.

QUESTIONS	ELEMENTS OF THE REPLY
1. On the crop considered, is there anything which is dangerous at the economic level?	Main pests and diseases of the crop
2. When? (i.e. at which phenological stages of the plant or at what times?)	Critical periods (potential risk)
3. How is one to know if a risk really exists in the orchard?	Observation – control method (measurement of real risk)
4. What is the importance of the pests or of the diseases?	Tolerance thresholds
DECISION (intervention or no intervention)	
5. If there is a risk, should one intervene?	Choice of means and in particular choice of phytosanitary product

Points 1 and 2 are part of the domain of the general knowledge of the phytosanitary situation and of the environment.

Points 3 and 4 allow preparation of the decision, by reference to the thresholds proposed by the scientists and the agents of agricultural development.

As part of integrated protection, which is the only base acceptable nowadays, the threshold is the basis of all action. It is, with the sampling (or other) method to which it is linked, the most characteristic element of a change in the way of conceiving rational protection taking account of natural limitation factors of the main pests of the crop.

3.   THE GENERAL NOTION OF RISK

This is a complex idea which includes a scale of importance of the decision to be taken (plot, the farm overall, terrain, region..), an economic dimension and also an essential psychological aspect. Is the farmer apt or not apt to act according to an idea other that that of comprehensive insurance, which is obviously very expensive?

An aid to this decision, which would be provided by information from an officiel technical service (e.g. the plant protection service in France) is of considerable help in the intellectual and moral assurance of the farmer.

It cannot go beyond the general environment relative to risk in the region, ... but this is essential!

If the farmes passes to the stage of the measurement of real risk he reasons either on the scale of each plot, or on the scale of each crop-growing unit (group of plots).

The latter approach is particularly interesting because only true knowledge of the general phytosanitary situation allows us to form homogeneous groups which avoid systematic monitoring in each plot.

It is difficult to take account of the economic factor; in fact the basis is fluctuating and uncertain in the crop-growing phase, at the time when the decision must be taken. In reality, this factor is so important that it can only be situated in the overall knowledge of the farmer who decides (or not) to put into practice the principles of integrated control and thus adheres to the philosophy which governs over this type of action.

In fact the operation consists of <u>estimating a technical risk which is immediate or provisional in character</u> :
            - by codified control methods
            - by proceeding through trials.
      Two cases may occur :
- the direct estimation of the risk, for a determined critical period; the control must be simple and rapid for it to be usable,

- the establishment of a provisional risk, based on the biology and the dynamics of development of the pest populations or of the diseases. This makes it possible to manage the control and to intervene, if necessary, at the most favourable moment and, in particular, means the least damage for the group of auxiliary fauna.

4.   THE THRESHOLDS

The aim to be attained is that of obtaining <u>indicative intervention thresholds</u>. The notion of threshold covers different realities. It signifies defining carefully the limits which will be assigned to it, knowing that the main limiting factor is often the lack of a method for estimating a threat.

The classical case in integrated control, illustrated by the following scheme, only takes account of one type of case where the entire problem resides in the best possible evaluation of a density (pest or pathogenic germ populations, etc.) and in the fixing of a benchmark in the "uncertainty zone".

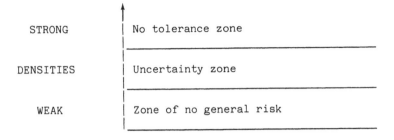

STRONG | No tolerance zone

DENSITIES | Uncertainty zone

WEAK | Zone of no general risk

This threshold, with a benchmark point which is often established empirically, is only a relative value which is closely linked to the sampling method, which itself depends on the means of instruments used : visual inspection, traps, etc. It is also linked to the efficiency, or the power, of the means of intervention available : chemical control, biological control, crop-growing techniques.

The indicative intervention thresholds which one could classify at the "unacceptable risk level" implicitly include :
- an index of harmfulness established by the experimenters,
- the critical period, in relation to the state of greater sensitivity of the plant cultivated,
- the most favourable moment for the control (generally chemical)
- the relative importance of limiting factors which are sometimes crop-growing techniques, but most often the natural predators and parasites.

In many situations the arthropods, which are the farmers' natural friend, seem to have a regulating effect on the pest populations.

There are not yet available, however, simple and rapid methods to quantify the most active predators and parasites (i.e. the auxiliaries). Neither do we know the "efficient ratio" between the pest population and the density of auxiliaries which, at a crucial moment, means the regression of the crop's enemy and consequently, the reduction of the risk of damage.

In summary, the intervention threshold may be written with the following formula :

```
critical   activity      state      activity                  to treat
 period  +  of the   +   of the  +   of the   =   decision
            pest          plant     auxiliaries              not to treat

   ↑          ↑            ↑_____↑
 known    measurable    cannot be measured
                          at present
```

5.   GENERAL METHODS AND MEANS OF INTERVENTION

If one simplifies to the extreme the elements presented in the

preceding sections, one may say that the farmer must, depending on the case : - estimate an immediate risk or a provisional risk,
   - use direct or indirect measurement techniques.

The research work is centred on indirect measurement techniques (such as modelling) to simplify the task of the producer.

If one admits that the so-called "secondary" effects of pesticides are "primordial" when dealing with the preservation of the fauna and the flora of the crop and of its immediate environment (according to the principle that the richer an environment is in species the more stable it is) it must be recognised that the rational choice of intervention techniques is not simple.

As an exemple, we give below the interactions of the treatments against the main pests of the vine.

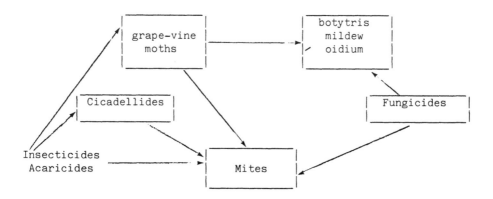

The range of means which can control fungi and parasitic bacteria is not very large and the biological control procedures are continuously being studied in laboratories and experimental fields.

The panoply of chemical and other weapons against the arthropods which are damaging to vines is still very incomplete, as can be seen from the following diagram (after H.G. Milaire).

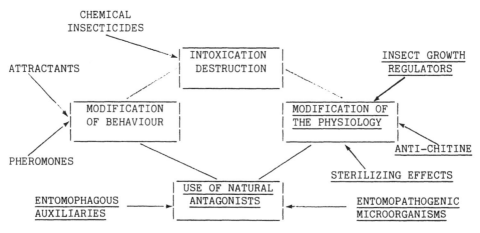

Knowledge still very insufficient

334

6.   THE ASPECTS OF DEVELOPMENT

A change of attitude and of a method of acting is, in general, a generator of stress. The technical basis must, a fortiori, be very sure, and not lend itself to any basic questioning.

Any development can only be progressive and be based on demonstration, i.e. realizations on full scale, in the usual conditions of agricultural practice.

The main difficulties seen by counsellors and farmers appeared clearly when we carried out an enquiry in 1983 in the countries of the European Community. They are summarized below :

Development of integrated protection : THE DIFFICULTIES

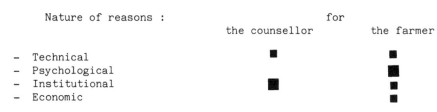

Nature of reasons :	for	
	the counsellor	the farmer
- Technical	■	■
- Psychological		■
- Institutional	■	■
- Economic		■

For the counsellor with a good knowledge of the situation and the farmers of his sector, the difficulties encountered are not psychological and economic but technical (and progress remains to be made) and above all institutional. More precisely, he considers that there are too few counsellors, that assistance and technical training are clearly insufficient and that the encouragement by the State (with public funds) is also insufficient.

For the farmer, the range of difficulties evoked is larger and the zones of uncertainty more diffuse with, however, the psychological factor dominant. This would merit a more precise and in particular a more complete study which could clarify the paths to follow in such an action of vulgarization.

Very briefly, the development of integrated control in Europe requires :

- the demonstration of the reality of this concept of health protection in agricultural centres situated in various European countries;

- the permanent and practical training of counsellors and of farmers, with the support of the technical and scientific personnel of these centres.

To obtain this result it is necessary to form a network to :

- exchange experiments and knowledge and encourage research (IOBC action),

- organize the training activities (progressive training with introduction, practical initiation and perfecting, or an intensive course intended to produce a qualitative jump for those interested).

IN CONCLUSION we can affirm today that integrated protection of the vine is a general source of progress and helps the farmers in dealing with an attack of harmful arthropods or of fungi.

Training must be generalized but progressive.

The "reliability" of technical techniques to be put into practice must be beyond doubt. The task is difficult because one must combine simplicity of methods with certainty of results.

Confidence between counsellor-trainer and farmer is indispensable.

The demonstration of the possibilities of application of the concept of integrated protection, on the real scale, in the technical centres linked together in a network and providing support for training, is an essential activity.

A positive attitude and a desire for encouragement on the part of national institutions could greatly help the work of researchers and technicians; only they can give the necessary impulse and at the same time the material and moral support.

Prediction of risk, control methods and application of integrated pest control

R.Agulhon

Institut Technique de la Vigne et du Vin, Nîmes, France

Summary

 The Working Party on Integrated Vine Pest Control of the
International Organization for Biological Control was set up about
12 years ago and is concerned with the national protection of
vines against pests, diseases and physiological disorders.
 A meeting in Bernkastel (FRG) in June 1985 assessed the
progress made in various projects carried out during recent years
on predicting the riks of major vine diseases and pests,
estimating economic risk, organizing control methods and model
building.
 New protection methods which could replace chemical controls
were discussed by a number of speakers.
 The meeting also examined the work involved in the
application of integrated pest control in various countries, both
in individual vineyards and in areas of various sizes.

1. INTRODUCTION

 The prediction of risk, control methods and the application of
integrated pest control were the subjects of the projects reported on at
the meeting in Bernkastel (FRG) in June 1985 by the Working Party on
Integrated Vine Pest Control of the IOBC (International Organization for
Biological Control), which is run jointly by Mr Augustin Schmid and the
author of this document.
 The Working Party has been in operation for about 10 years and its
task is to study and promote the rational protection of vines against
diseases, pests and physiological disorders.
 At present 80 researchers and technical assistants from a number
of European countries are involved in the work and meet every two years
to discuss the results of their research.
 Six sub-groups have been set up, led by representatives from
various wine-producing countries :
- grape moths and biting insects, R. ROERICH (F) and Laura DELLA MONTA
  (I)
- mites and stinging insects, M. BAILLOD (CH) and W.D. ENGLERT (D)
- fungal and bacterial diseases, W. GARTEL (D) and J. BULIT (F)
- physiological disorders, R. THEILER (CH)
- secondary activities, J. TOUZEAU (F)
- application of integrated control, P. CABEZUELO-PEREZ (E).
 A seventh sub-group on weeds and vegetal encroachment is to be set
up in the near future.

2.   PREDICTION AND ESTIMATION OF RISK

Not having any accurate indication of the risk they face, winegrowers systematically apply a number of treatments, usually chemical. They are also all too often based on an empirical programme which usually involves considerably more treatments than necessary, not always at the right time. This results in pointless expense for both the vineyard and the community.

Only by predicting and estimating the risk can rational protection be given, using not only pesticides but also growing methods which restrict the development of pests. Tolerance levels and control methods, particularly for animal pests, have already been established during earlier projects, but if there is to be a reliable basis for decisions to take action, existing methods must be improved and techniques must be devised to predict the development of the diseases.

2.1   Grape moths

The work on predicting the risk from grape moths is divided into three sections : model building, the use of pheromones and the study of regulating factors.

Two empirical models were presented at the meeting in Bernkastel (Baroni Vita and Caffarelli, Barlattani (I)) which used calculations of temperatures and other meteorological data to predict the main periods of the cycle. They cannot evaluate the risk involved, but they may improve the accuracy of the warnings given to farmers.

Pheromones are a very useful way of assessing variations in population. With Lobesia botrana, although it was shown that differences in stereochemical purity had little effect on the numbers caught, it was also demonstrated that very low doses of pheromones are useful for providing more reliable indications of risk (Roerich).

Future prospects in this fiels include new pheromones for boarmia rhomboidaria and for the web worm Sparganothis pelleriana, which are to be widely tested (Arn-Schmid CH)).

Of the factors which contribute to the reduction of grape moth populations, Coscolla (E) reported that dryness seems to be the most important factor in killing Lobesia botrana eggs.

In conclusion, however, although it was agreed that grape moth larvae cause great damage in summer, particularly as promoter of grey mould, it has not been possible to establish a reliable way of predicting whether or not the damage threshold will be reached. Only negative prediction can be regarded as viable.

2.2   Mites

At present the risk presented by Panonychus ulmi can be successfully estimated by calculating the percentage of leaves occupied. Schruft (D) and Baillod (CH), however, reported that two new methods are ready for application : monitoring eggs in winter, a method developed in Switzerland, and observing symptoms on the leaves during spring, a method developed in Germany.

These two methods replace or simplify the method of calculating the percentage of leaves occupied in the spring for the species Tetranychus urticae and Eotetranychus carpini, but there are still

338

problems in estimating risk as defined in France and Spain. A method must be devised which can be used in practical conditions such as that which Laurent (F) investigated in the South of France, based on exteriorizing the damage.

2.3 Fungal diseases

Strizyk (F) provided further details on the use of model building to predict and estimate the risk of development of certain diseases.

State of Potential Infection (SPI) models were drawn up and used in practical conditions. Around 40 tests on grey mould carried out in Southern France on vines producing tightly clustered grapes showed that the model gave an accurate indication of the degree of risk.

Molot (F) showed that the application of the model both ensured adequate protection and resulted in major economies (two treatments instead of four with the standard method). Experiments using these techniques in various wine growing regions of Europe have produced the same results everywhere. However, in certain situations, the discrepancies between the SPI variations and the mould rate observed during the grape harvest leave room for improvement (Perez-Marin (E)).

As far as mildew (Plasmopora viticola) was concerned, it was agreed that the model currently under examination gave a correct indication of the extent of the risk in early spring, variations in the SPI proving to be in line with the development of the disease on the vine.

Brechbuhler (F) noted a considerable reduction in the resistance to cyclical imides when the number of applications was reduced.

2.4 Physiological disorders

Theiler (CH) appears to have established how to estimate the risk of shanking and the tolerance threshold. He has developed a model which can identify attacks immediately after flowering for certain vines in given locations, having established that there is a close connection between the temperature during flowering and subsequent shanking.

In Alsace, on the other hand, it was shown that the rate of disorders for the Gewürztraminer vine correlated with the total precipitation during the same period. Brechbuhler reported that the rate of disorders for this vine correlated with the precipitation during the 45 days before mid-ripening.

With regard to the tolerance threshold, various reports indicated that vines which were slightly affected sometimes produced higher yields than healthy wines. On the other hand both the quantity and quality of the yield fell with vines which were badly affected.

If a prediction model is to be used a tolerance threshold will have to be fixed, but in the meantime provisional figures have been drawn up for certain types of vine : Gewürztraminer 30%, Chasselas 40%, Müller-Thurgau 40-50%.

3. CONTROL METHOD

Discovering a control method which can be substituted for traditional chemical methods is probably one of the main concerns of all researchers and technical experts working on integrated control. However, the success rate of the methods tested varies greatly for the different pests and diseases and is also dependent on the development of new technologies. A knowledge of the secondary effects of pesticides is also important for traditional treatment programmes.

3.1 Grape moths

Recent projects have concentrated on Trichograms. Remund carried out experiments in Switzerland at a small vineyard, where the insect proved extremely sensitive to climatic conditions despite its resistance to fungicides.

Of the three growth regulators tested in Germany by Schruft on the eggs and larvae of various instars of Eupoecilia ambiguella only one proved to be adequately effective but at three times the recommended dose.

The autocidal method tested in a Swiss vineyard proved to be successful in reducing infestation, but would be economically viable only if no further action were necessary for 12 years, according to the conclusions reached by Boller and Remund.

Sexual confusion was the subject of a number of tests over wide areas in Germany and Switzerland, where considerable quantities of Eupoecilia ambiguella pheromones were spread either manually or by helicopter. Although the results were encouraging not all of the parameters are yet known. Where breeding was heavy the reduction in infestation was not sufficient to provide adequate protection for the crop given that large numbers of fertilized females had been observed. It would appear, however, that increased doses might improve results.

Roerich reported that the effects of sexual confusion on Lobesia botrana were practically nil. Nevertheless the results show that a cheaper way of synthesising the pheromone needs to be developed before the experiments can be taken any further.

Unfortunately, chemical control is still all to often essential and is carried out by applying the insecticides at the most suitable moment. How the products are applied obviously depends on their properties. Marcelin (F) showed that these could easily be broken down into the amount of time the products were active agains newly-hatched caterpillars, their shock effect on already active caterpillars and their ovicidal effect. It is essential to have this information in order to know when to apply the treatment in terms of when the eggs hatch, how the hatching is staggered and the size of the population.

In conclusion, the only control which could replace chemical insecticides is Bacillus thuringiensis, which is difficult to apply and does not always provide adequate protection.

3.2 Mites

The biological control of three species of phytophagous mites, Panonychus ulmi, Tetranychus urticae and Eotetranychus carpini, has made a certain amount of progress with the use of phytoseides. Studies of the populations and distributions of typhlodromes carried out in Italy by Lozzia, Corino and Duverney, and underway in Spain (Garcia) and France (Kreiter) have been accompanied by regular surveys of the population densities as a function of vine treatments.

Typhlodromus pyri is widespread in Germany and Valais and as its population increases it is gradually eliminating red mite even without acaricide treatment. It appears that these predators can be taken to the vineyards and left released there; various aspects of this method have been tested.

A study of their reactions to insecticides in Germany and Zwitzerland indicated that certain strains of Typhlodromus pyri were resistant to a number of phosphoric esters, including parathion in particular.

The use of these predators is thus a new departure in the organisation of the biological control of mite populations and marks new progress in integrated pest control.

3.3 Fungal diseases

Unfortunately no completely successful biological control methods have yet been applied against vine fungi. Only with the fungus Trichoderma viride has there been any measure of success. In Italy Borgo showed that satisfactory results could be obtained with Botrytis cinerea with repeated applications.

The enemy is a ground fungus which is very sensitive to climatic conditions and has great difficulty in establishing itself on hosts in the air.

The control problems most often encountered with the fungal diseases Botrytis cinerea and Plasmopora viticola relate to their resistance. Botrytis has widely developed a resistance to dicarboximides in vineyards in the north, so that it has been necessary to advise producers either to stop applying the product completely or to reduce the number of applications to one per season, which seems to cause regression in the degree of resistance. Unfortunately there are few fungicides currently available which can be substituted for dicarboximides.

The resistance of mildew to anilides, which had been fairly widely observed in certain Franch winegrowing areas, seems to be regressing as a result of reducing the number of treatments with these products.

3.4 Physiological disorders

The methods used to control shanking still involve magnesium-based products. The treatments are 50-80% effective depending on when and how often they are applied, but reports from Austria, Switzerland, Germany and France indicate that they rarely improve yield. However, despite the uncertain results for shanking some of the datapoint in favour of continuing with the treatments since the disorder can cause considerable losses of sugar and acidity.

4.   APPLICATION OF INTEGRATED PEST CONTROL

None of the experiments carried out on the prediction of risk and new control methods could be better applied than in the winegrowing industry where all of the knowledge currently available can be integrated in practice.

For some years now a number of countries have been taking measures which, although different in form, all stem from the same desire to achieve proper integrated pest control in areas both large and small. The Viticulture Group of the IOBC has therefore set up a sub-group responsible for studying the problems involved. A number of the winegrowing countries of Europe are developing various programmes organized in line with the different socio-economic situations of the countries and regions involved. Schmid reported that winegrowers in Valais have set up "integrated protection groups" which are being encouraged to take an active part in monitoring vineyards and decision-making. The groups consist of 5 to 15 people who, once trained, are assisted by viticulture centres which provide information on butterfly catches, interpret meteorological data and recommend which products to use, while leaving the actual decisions to the individuals concerned.

Boller reported that in German-speaking Switzerland there was also an integrated protection plan under which the winegrowers concerned received technical assistance from an Agronomic Research Centre.

In the Federal Republic of Germany there are also centres in direct contact with producers who receive training ans assistance from technical and scientific experts (Schruft).

But it is probably in Spain that the State is most closely involved in the organization of integrated pest control, as Cabezuelo-Perez reported in detail. A state-financed national programme has been set put under which the winegrowers involved receive financial aid to take on technical assistants, who monitor, carry out checks and take decisions in conjunction with the Plant Protection Service. There are currently 28 programmes of this type throughout the country.

In Italy a special organization which has been set up with economic and technical aid from both public and private bodies appoints technical assistants to organize the work of groups of winegrowers (Grande).

In France warning centres and technical institutes such as the ITV* and the ACTA** are setting up programmes to demonstrate the viability of integrated control in the vineyards themselves. Programmes applied in both regions and smaller areas have already produced excellent results.

In conclusion, it would now seem possible to combine all of the knowledge currently available on integrated pest control and to apply it in practice provided that public, professional and private bodies in each of the winegrowing countries of Europe are prepared to give more active, specific and enthusiastic commitment to that end.

---

* Institut technique de la vigne et du vin
** Association de coordination technique agricole

Implementation of integrated pest control over a period of six years in a vineyard in the South-East of France

M.Blanc

Association de Coordination Technique Agricole (ACTA), Manosque, France

Summary

Experiments conducted over six years in a vineyard in the South-East of France show that it is perfectly feasible to implement integrated pest control. This technique makes it possible to reduce the consumption of agro-pharmaceutical products considerably, while remaining within the bounds of cultivation practices. However, vine-growers must acquire more technical skills.

Forecasting of potential damage by certain pests, in particular powdery mildew - Uncinula necator (Burr.) - and the grape moth - Lobesia botrana (Schiff.) - must be made more precise.

1.   INTRODUCTION

The aim of supervised control is to use agro-pharmaceutical products as rationally as possible. This implies a good knowledge, on the one hand, of diseases and pests (economic thresholds, symptoms, location etc.) and, on the other hand, of the direct effects of these products on the pests which attack the crops concerned, as well as any possible indirect (secondary) effects. This concept of pest control, which has been studied in depth by ACTA, the "Association de Coordination Technique Agricole", on fruit-trees, applies to all crops. While it means that the growers must acquire more technical skills, in return it produces economies in plant protection.

With regard to vine growing, the experience with fruit-trees enabled ACTA, together with the ITV (Institut Technique de la Vigne et du Vin), INRA (Institut National de la Recherche Agronomique), SPV (Service de la Protection des Végétaux) and OILB (International Organization for Biological and integrated Control of harmful animals and plants), to define and adapt a rational methodology of plant protection. The task now is to implement these ideas and to test them in practice in a vineyard. This was the aim of the six-year period of experiments carried out by the ACTA Regional Delegation in Manosque, in South-East France.

2.   THE SITE

Since 1979, at the request of the bodies concerned with development (*), this type of rational pest control has been on test in a vine-

---

*   In particular : the GDA Viticulture du Vaucluse, the Union des Vignerons des Côtes du Lubéron (UVCL) and the Association Climatologique Régionale de Carpentras (ACR).

TABLE I
DISTRIBUTION OF THE TREATMENTS CARRIED OUT AGAINST VARIOUS PESTS
IN THE EXPERIMENTAL HOLDING OVER THE SIX-YEAR EXPERIMENTAL PERIOD

A : total number of plots
B : number of plots treated
C : number of treatments per plot

	1979	1980	1981	1982	1983	1984

Mites - Panonychus ulmi Koch
and Tetranychus urticae Koch

	1979	1980	1981	1982	1983	1984
A	28	23	22	24	24	24
B	5	1	0	0	0	0
C	1	1	0	0	0	0

Leaf hoppers - Empoasca vitis Göthe

	1979	1980	1981	1982	1983	1984
A	28	23	22	24	24	24
B	0	3	6	5	0	0
C	0	1	1	1	0	0

Grape moth - L. botrana
and C. ambiguella

	1979	1980	1981	1982	1983	1984
A	28	23	22	24*	24*	24*
B	27	22	22	4	19	19
C	1	1	1	1	1	1

Mildew - P. viticola

	1979	1980	1981	1982	1983	1984
A	28	23	22	24	24	24
B	28	23	22	23	24	24
C	1	5	4	2	4	4

Powdery mildew - U. necator

	1979	1980	1981	1982	1983	1984
A	28	23	22	24	24	24
B	28	23	22	23	24	24
C	3	3	3	4	4 - 5	6

Grey rot - B. cinera

	1979	1980	1981	1982	1983	1984
A	28	23	22	24*	24*	24**
B	17	13	21	4	4	5
C	2	1	1	1	1	2

* 20 in production   ** 21 in production

yard. The vineyard was chosen by the technical experts of the UVCL and the GDA Viticulture of Vaucluse.

The principal criteria governing this choice were :
- the representativity of the vineyard in the production zone under consideration;
- high degree of motivation on the part of the vine-grower.

This second point is very important. The success of these experiments and the chances of repeating them are closely linked to it. Accordingly Mr J.L. Pascal's vineyard was chosen. It is situated in Provence, in South Lubéron, in the municipality of La Tour d'Aygues, in an area which produces quality wines. The total area under vines is 15 ha - 10 for wine grapes and 5 for table grapes, and in 1984 it comprised 24 plots.

3.   BASIC PRINCIPLES

The treatment followed the principles laid down in the ACTA-ITV booklet, i.e. general protection, and periodic checking of the vines. All the information given in this publication is summarized in the check-list of the actions to be carried out (Annex A) and the summary of the thresholds (Annex B).

The information obtained by the sexual trapping of the grape moth (in this case Lobesia botrana Schiff. and the grape berry moth Clysia ambiguella Hb., since the two species cohabit) is only qualitative. It indicates only the beginning and the end of the flights. For the present it is still necessary to check the deposits of eggs on the clusters of grapes, in order to indicate to the vine-grower the action to be taken. The grey rot model, Botrytis cinerea Pers., after it had been adapted to the region, was not used as a hazard indicator until 1982. The mildew model - Plasmopora viticola B and C - is still experimental. The complete chemical suppression of weeds, which was already practised before 1979, was continued, mainly with the aim of preserving maximum load-bearing capacity in the soil, which helps the sprayer to keep moving in unfavourable weather conditions.

4.   AIMS

This study had three main aims, viz.:
- to determine possible changes in the pest control method, in the light of actual practice in the vineyard concerned;
- to determine the nature of the constraints connected with the division and the dispersion of the plots, the vines, the equipment used and the availability of the vine-grower;
- to try to simplify monitoring methods while maintaining the reliability of the measures which must be taken before a decision can be made.

5.   ASSESSMENT OF SIX YEARS OF EXPERIMENTS

Table I gives the number of treatments (*) carried out each year

---

*   This refers to the number of times that commercial products were used and not the number of treatments carried out.

TABLE II
NUMBER OF TREATMENTS CARRIED OUT ON THE EXPERIMENTAL HOLDING
AND THE CONVENTIONAL HOLDINGS

Year	Holdings	
	Experimental	Conventional
1979	7	13
1980	11	18
1981	9	15
1982	7	13
1983	10	15
1984	11	16

The figures show the ratio between the number of plots
treated and the total number of plots forming the holding.

TABLE III
CHEMICAL USED IN THE EXPERIMENTAL HOLDING

Acaricide	Insecticides	Fungicides		
		Powdery mildew	Mildew	Grey rot
Dicofol	Methidathion Deltamethrin B. thuringiensis	Fenarimol Wettable sulphur Sulphur powder	Copper Mancozeb Metalaxyl Phosethyl AL	Procymidone Iprodione

for the main pests listed, and **Table II** a comparison between the
experimental and conventional pest control methods. It should be noted
that the average number of conventional pest control treatments was
obtained by the technical department of the UVCL from a survey on some
25-30 vineyards. The main chemicals used to control the various pests in
the test vineyard are given in **TABLE III**. The very multi-purpose
insecticides such as methidathion or deltamethrin do not seem to affect

346

the environment unduly, since they are applied only once in the season. Bacillus thuringiensis seems to be a very technical product and the question of its use on the grape moth is a delicate one, given the current state of knowledge. Synthetic anti-powdery mildew fungicides such as fenarimol are of interest in cold spring weather, from the D - E stage until the appearance of the grape clusters. The new anti-mildew products, particularly metalaxyl, are really best applied just before the vines flower; copper treatments must be continued at the end of the season, during the second half of August, at the end of vegetative growth. Treatment for grey rot is effective, with one, or even two applications, when specific products are used. This treatment is particularly suitable for table grapes.

Usefulness of the models

- The greey rot model, the oldest and the best known, has been used since 1982 to assess the risk more accurately, and thus to permit fine adjustments to the treatment, particularly in the case of table grape vines, since this disease is very harmful (cutting out clusters is very expensive in terms of manpower).
- The mildew model, which is still in the experimental stage, seems promising.
   The simulations of the potential hazard during three very different years from the epidemiological point of view are as follows :
- 1980 : presence of the disease on very sensitive vines such as Alphonse Lavallée, as from 10-15 June;
- 1981 : total absence of mildew until the beginning of September, general attacks on young shoots thereafter;
- 1982 : total absence of mildew throughout vegetative phase.
   Interpretations of the value of the potentially infectious state (PIS) as from May (when the vine is prone to the disease) are as follows :
- if the PIS is between 0 and -5 the hazard is minimal or very local;
- if the PIS is above 0 the disease is general;
- if the PIS is well below -5 there is no disease. As from 1985 it should be possible to use this model, which has been improved still further during the past two years, in the area under survey for indicating risk.

6.   REMARKS

   Aims

   It appears, in view of the results obtained, that adjustment of the pest control treatment is perfectly possible. The tendency is always towards a reduction, and in fact there is a regular reduction in the use of chemicals each year of 30-40%, which is very encouraging and demonstrates the usefulness of this technique.
   The constraints connected with the structure of the vineyards (the division and dispersion of the plots) are apparently quite easy to overcome, since the vine-grower was able to protect all the plots, under good conditions and at the right time, whatever the vine. The acquisition in 1980 of a towed pneumatic sprayer, equipped with a 300 litre tank, speeded up the process and improved the quality of the treatments.

The six years of observation showed that, in the area under survey, visual checks can be confined to looking for the pests mentioned in **Table IV**, at the stages or times indicated.

Some of the checks in the list of actions to be carried out a certain stages or timers were omitted if :
- the following pests were not found : boarmia - Boarmia gemmaria Brahm. -, cutworms - Scotia sp. -, Sparganothis pilleriana Den et Schiff. -, vine pyralid caterpillars - Arctia caja L -, Black-rot Guignardia bidwellii V and R -, and Roter Brenner - Pseudopeziza tracheiphila Müller-Thurgau;
- at certain periods the pests present were not harmful : e.g. the grape moth in the first generation, leaf hoppers and leaf rollers - Bytiscus betulae L - at the F-G stage;
- the check was very difficult to carry out (rust mites - Phyllocoptes vitis Nal. - at stage A) and accordingly was abandoned as too random.

Differences observed between rational pest control and conventional pest control.

The greater consumption of plant protection products in conventional pest control is mainly due to the fact that there are more treatments for mildew and the grape moth. Early spraying using an anti-mildew product, to treat excoriose - Phomopsis viticola Sacc., is a frequent practice. In addition the treatments are almost always applied to all the plots which make up the vineyard instead of being concentrated on those where they were necessary. This is particularly true of pest control methods against mites and leaf hoppers.

Role of the vine-grower

The role of the vine-grower is not confined to that of a labourer, since he is coming to play an increasingly effective part in the monitoring process. He is also a source of useful information during the various growing stages. He samples suspect wood when it is cut for checking for the presence of Excoriose (*), and he searches for primary spots of mildew or powedery mildew during the work of removing side-shoots and disbudding.

Prospects

These experiments enabled us to achieve the aim initially laid down. Using visual monitoring, which is a simple method, we can indicate the action to be taken against the main pests. Once the vine-growers have been given training on recognition and sites for treatment they should be able to work out the use of insecticides and acaricides themselves.

However, one thing remains obscure regarding treatments for the grape moth. Whereas it is only very seldom that the first generation is harmful, this is not true of the second. But observation of the laying of eggs and their development seems to be the only way of deciding whether and when to spray. These difficult and detailed checks are little used by the professionals. Accordingly an alternative technique must be provided.

The model designed by J. TOUZEAU (SPV) should offer a solution. An initial test carried out in 1984 showed that adaptation to the region was necessary. The experiments on the grape moth will be continued in this respect.

---

* Only one plot had to be protected from excoriose in 1984.

TABLE IV
MAIN REALLY HARMFUL PESTS IN THE REGION UNDER TEST

Stage season	Pest	Other action
A	Excoriose, powdery mildew (pruning stage)	
E	Mites (leaves spread)	
G - J	Mites (leaves in the middle of the branch) Mildew (leaves and grapes) Powdery mildew, excoriose (all branches)	Sexual trapping
to picking	Leafhoppers (early August on leaves in vegetative mass) Grape moths (laying of eggs on grapes in July) Grey rot (grapes at picking time)	

Reference to the tables showing the proportional use of the treatments shows clearly that the general protective system relies very heavily on fungicides. For this reason the information provided by the grey rot and mildew models appears to be indispensable for the calculations for these fungicide treatments. This method of forecasting the risk of infection will be in full use as from 1985. Lastly, it should be noted that the treatment for powdery mildew, although very effective, is somewhat disappointing, since it depends on a high number of regular applications. It is unfortunate that we cannot have a pest control strategy which shows the usefulness of the new synthetic products which have recently appeared on the market.

7. CONCLUSION

These six years of experiments show what can be done in practice in the way of plant protection in the vineyard. The results, which have been achieved in spite of the constraints on the vine-grower, are reassuring. They will certainly be improved still further once the new methods of forecasting the risk of infection, and also, in particular, the models, have been adapted accurately and are in use.

With regard to treatments for grape moth and powdery mildew, these will require more rigorous studies if progress is to be made.

This study must be continued in several vineyards belonging to an association or in areas with the same agricultural and climatic characteristics. It is still essnetial that vine-growers should acquire a certain degree of technical expertise in order to apply treatments rationally.

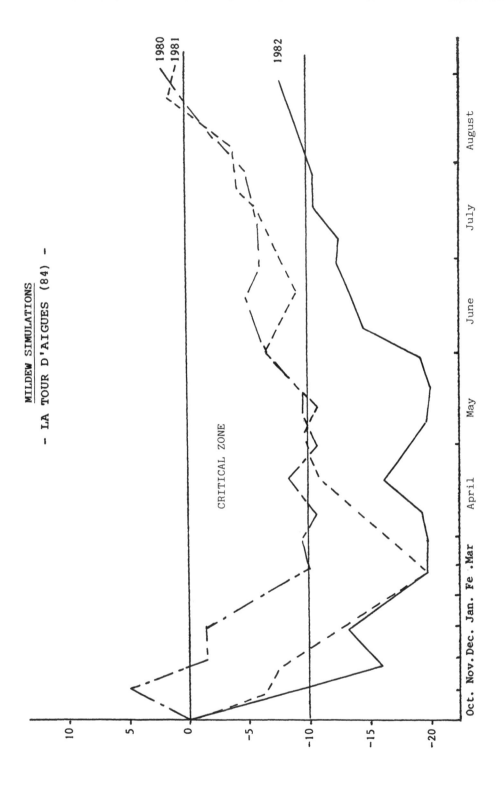

MILDEW SIMULATIONS
- LA TOUR D'AIGUES (84) -

350

CHECK-LIST OF TREATMENT FOR THE MAIN PESTS
(from the 1980 ACTA-ITV handbook)

Stage	Main pests sought for	Organs to be examined (visual monitoring) or other action
A (pruning)	Excoriose (necrosis), powdery mildew (speckled patches) Red spider mite (eggs)	whole of the vine shoots
	Yellow spider mite (adults) Acariasis (adults)	examination of bark under binocular magnifying glass
C and D	Boamie (damage) Noctuidae (damage	developed buds
	Grape moth (damage)	sexual traps
E and F	Red and yellow spider mites (mobile forms) Acariasis, erinose	a fully developed leaf on a young twig *
	Web worm (caterpillars), tiger moth, vine leaf roller (damage)	young branch
	Mildew, grey rot, powdery mildew, excoriose, black-rot, Roter brenner	whole of the stock
	Golden flavescence leaf hopper (larvae)	most advanced leaves on the water shoots
G to J	Red and yellow mites Leaf hoppers (larvae)	a leaf in the middle of a twig in the middle part of the stock
	Grape moth (clusters)	a bunch from among the most advanced on the stock
	Diseases of the foliage and the grapes	whole of the stock
Settina to picking	Red and yellow mites Leaf hoppers (larvae)	a leaf in the middle of a branch in the middle part of the stock
	Grape moth (holes) mildew, powdery mildew, grey rot	a cluster on the stock (inside the vegetable mass for goblet-pruned vines)
	Eutypa disease, esca	the whole of the stock
	Deficiencies	leaf sampling for diagnosis

* Inserted close to the old wood (short and mixed lengths) or located approximately midway along the wood of the previous year (long length)

SUMMARY OF THE TRESHOLDS

(from the 1980 ACTA-ITV handbook)

Red and yellow mites.
° In winter :
  - Red mite : 7 to 20 eggs per bud or 80% of the buds covered by one
    laying or more;
  - Yellow mite : 50% of the stocks covered by 1 female or more.

° In vegetation :
  - Red and/or yellow mite : south-east and south-west of France, 70%
    of the leaves covered from stage E onwards and 30% from mid-August
    onwards; Switzerland, 60% of the leaves covered from stage E on-
    wards and 30 to 45% from mid-August onwards;
  - Red spider mite : 50% of the leaves covered from stage E onwards
    and 30 to 45% from July onwards.

Acariasis
In winter : presence noted at the time of visual monitoring or 1 or 2
            mites detected by soaking.

Grape moth
° First generation : northern areas, 10 to 25 clusters attacked out
  of 100 observed: southern areas : 100 to 200 clusters out of 100
  clusters observed;
° Second and subsequent generation : 1 to 10 clusters with eggs or in
  early stages of infection out of 100 observed, depending on the
  hazard of grey rot.

Web worm
30 to 40 caterpillars on 10 stocks observed.

Green leaf hopper
° Around flowering period : 100 larvae on 100 leaves observed;
° During August : 50 larvae on leaves observed.

Golden flavescence leaf hopper
In the areas affected by the disease - presence of 1 larva.

Problems and prospectives for integrated control in grape cultivation in Piedmont

F.Gremo, M.Pinna & A.Ugolini

Regione Piemonte, Servizio Sperimentazione e Lotta Fitosanitaria, Torino, Italy

Summary

Grape cultivation in Piedmont (N.W.Italy) represents a fundamental sector of the agricultural economy due to the area cultivated (ha 71.000), the size of the crop (t 494.870) and to the quality of the wines produced (40 D.O.C. "Controlled Origin" wines and 2 D.O.C.G."Controlled Origin and Guaranted" wines).
Grapes are beset by many diseases, from the classic fungous disease of plants (common grey mould, grape mildew, downy mildew of grape) to the Mites (Red-spiders, Blister mites), Insects (grape berry moth, vine moth, leaf hoppers, Thrips), virus disease and also minor adversities and physiological disease.
As regards pests, sampling methods for the most harmful phytophagous (red-spiders, grape berry moth, vine moth, leaf hoppers), are being verified, together with "economic thresholds"; useful natural factors (parasites, predators, etc..) are being taken into account.
The some problems have still to be tackeled for others the important phytophagous (Thrips, etc..).
Experimental trials are also being conducted into the efficiency of recently developed chemical and biotechnical active principles.
Research into the operative possibilities of the application of integrated control in grape cultivation is presently being planned; this will of necessity be based on deeper knowledge of agro-ecosystems and of alternative means of control (biological and biotechnical).

1.INTRODUCTION

In 1984 vinyards covered more than 71.000 ha for a total grape crop of 494,870 t and a yield per hectare, where grapes were the main crop, of about 90 q. Although the area cultivated has decreased in recent years, grapes are still therefore one of the fundamental sectors of the agricultural economy of the region, both quantitatively and structurally (area cultivated, size of crop, number of workers) and because the wine produced is of a very good quality, with 40 D.O.C. (controlled origin) and 2 D.O.C.G. (controlled and guaranteed origin) wines, representing between them about 25% of the wine produced.

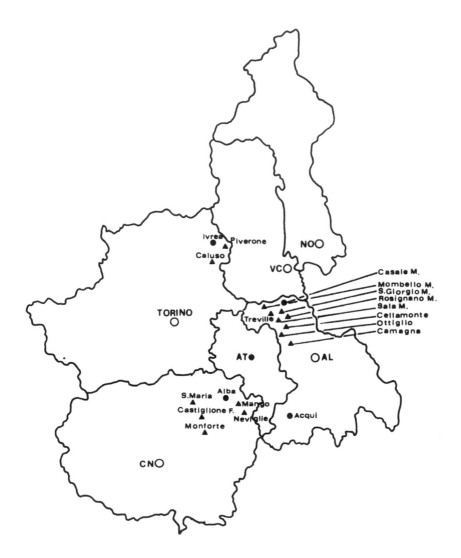

FIG. I

▲ areas controlled
● technical assistance centers

The most widely cultivated vine is still Barbera (about 38.000 ha or 25% vinyards), followed in order of decreasing importance by: Dolcetto, Moscato, Nebbiolo, Freisa, Cortese and Grignolino.

Alongside more general improvements in the cultivation (renewal of structures, development of mechanization, development of more effective agronomic practices etc.), a rational disease control programme is increasingly important; such a programme should guarantee satisfactory yield while fending to avoid unjustified application and waste of

354

pesticides, and favouring the recovery of the preexisting equilibrium between pests and their natural antagonists.

A research group set up some time ago by the Servizio Sperimentazione e Lotta Fitosanitaria della Regione Piemonte and which also offers technical assistance, aims to diffuse integrated pest control methods.

As part of the research programme "Improvement of integrated pest control in vinyards" of the European Community Commission, a survey of the distribution and incidence of various animal parassites attacking the vine was begun, paying particular attention to the study of the biological cycle and the degree of infestation involved. Well known sampling methods and economic thresholds were tested in the field for grape berry moth, grape fruit moth, red spider mites and leafhoppers.

The research was conducted in collaboration with the centres for technical assistance which are already operative in some of the principal grape growing areas of Piedmont; the choice of farms in the various zones of the survey (Fig.I) took the following factors into account:
- the varying intensity of cultivation, accading to geographic location;
- the principal vines present in Piedmont (Barbera, Dolcetto, Moscato, Nebbiolo, Grignolino, Erbaluce) and their differing reactions to the pests studied;
- diversity of approach in pest control paying particular attention to use of fungicides.

From the preliminary survey the following points emerge regarding the principal animal pests:

1.1 Grape fruit moth and grape berry moth

Grape fruit moth (Lobesia botrana Den. et Schiff.) and grape berry moth (Eupoecilia ambiguella Hb.) are the species of Lepidoptera most harmful to grape cultivation. The former is diffused in most of the grape growing area of Piedmont, whereas the latter is confined within areas with more favourable climate, such as the Casale Monferrato (AL) area.

The flight of both these species has been studied using pheromone traps, counting the number of insects captured every two days starting from mid-april. To evaluate the degree of infestation, periodic checks on the bunches were made as follows:
- during the first generation: count of number of bunches occupied by larvae out of a sample of 100;
- during the second generation (starting from the fifth day after the beginning of flight): count of bunches having deposits of eggs, with an intervention threshold of 10%.

First generation infestations were slight in most vinyards; nevertheless on the Erbaluce vine in the Caluso area and on Moscato in the area of Mango d'Alba, more serious attacks occurred (100% of bunches occupied by one or more larvae). In these cases not only damage to the flowers but also perforations of the rachis itself were found.

It is well known that damage to flower bunches from first generation grape berry moth and grape fruit moth is not generally of economic importance; nevertheless, in the case of the Erbaluce vine, which is an early variety, these Lepidoptera attacked prevalently the set berry of grape; in addition the type of training used (pergola) created a microclimate favourable to precocious attacks of common grey mould.

We consequently advised treatment against the first generation in these areas.

Most of the vinyards examined required treatment against the second generation of grape fruit moth; in the Mango d'Alba area and also in the entire Albese area this was carried out on or near July 20th, with about 15% of bunches having eggs and 3% having perforations.

In the Piverone and Caluso areas, on the other hand, insecticide treatments were carried out at the end of July, with about 18% of bunches occupied by larvae and 3% of perforations.

A second treatment was often required, due to the protracted flight and consequent egg laying in August (with a maximum of 30% of bunches occupied), on the expiration of the previous treatment.

In the case of grape berry moth (Eupoecilia ambiguella), the flight was equally intense only in the Casale Monferrato (AL) area, but egg laying was not at a level to justify use of insecticide.

Generally speaking the validity of the pheromone traps is confirmed, both for monitoring the species studied and to determine the most suitable moment to use insecticide; nevertheless this decision cannot depend solely on the data supplied by the traps, which do not give any indication of the ratio between male and female insects of the same species, nor consequently an economic threshold; checks on egg laying are crucial to this end. Further research into these aspects of the problem is clearly within the scope of the research programme; the ratio if any between first and second generation attacks should also be investigated; the arm is to create a model to predict the infestations.

In the years 1983 and 1984 as part of the fight against the grape fruit moth on the grape vine, comparative tests of some biotechnical products were conducted: Bacillus thuringiensis, in various formulations, Beauveria bassiana (an entomopathogenic fungus) and chemical products, including some recently marketed, among which microencapsulated methyl parathion, and also some experimental products. The tests were carried out, in both years, in the same farm situated in Castiglione Falletto (CN).

One single application against the second generation was made for each product being tested, the time of application being established from the number of adult insects caught in the pheromone traps and the course of egg laying.

Data from the successive check of perforations were elaborated by variance analysis and subjected to the Duncan test (tab.I and II).

From the results obtained the following observations may be made: in 1983 the two Bacillus thuringiensis - based formulations (Bacillus CRC and Bactospeine 16) showed a fairly good efficacy, similar to that of the experimental piretrinoid (at a lower dose) and to Bactospeine 16 with pinolene added; it can therefore be said that this additive, used to increase the persistence of the mixture, did not have the expected effect. The addition of sugar to Bactospeine 16 on the other hand did improve the persistence. This showed an efficacy against grape fruit moth similar to that of the best chemical products such as Thiodicarb (Larvin), experimental piretrinoid (at medium and high doses), microencapsuled Methyl parathion (Penncap) and Chlorpyrifos methyl (Tumar).

In 1984, as can be seen from table II, despite a more severe infestation , Bactospeine 16 used as it was confermed a medium activity against grape fruit moth similar to Bactospeine 16 plus an antiperonospore product. The additive had the function of protecting from U.V. and increasing persistance.

Beauveria bassiana (an entomopathogenic fungus) gave satisfactory results, comparable statistically to the Bacillus-based formulations. The microencapsulated Methyl parathion-based chemical products and the experimental piretrinoids confermed their good activity against grape fruit moth, which was better than that of the biotechnical formulations.

In conclusion, the trials showed that the biotechnical products gave satisfactory results, which, in the case of Bacillus, can be improved by adding sugar: it is to be hoped that these products will be improved, especially with regard to their persistence and costance of action in relation to climatic conditions so as to be able to place them alongside the chemical products, and so avoid negative side effects on useful organisms.

1.2 Red spider mites

For over 20 years the phytophagous mites of the grape vine have been of great importance, since they often do considerable damage. Serious infestations began in the late 1950s with the introduction of the new insecticides and copper-free fungicides.

Panonychus ulmi Kock is without doubt the species which causes most problems in grape cultivation; in Piedmont it was found in all the vinyards studied, the population density varying above all in relation to the presence otherwise of Phytoseiidae. Eotetranychus carpini f. vitis Oud. and Tetranychus urticae Koch were observed sporadically, but never in relevant numbers.

Both the phytophages and their predators were sampled every two

Table I – 1983

Pesticides employed	Active substance	Rate g/hl c.p.	Mean perforations/grape	Duncan test
Control	-------	---	1.578	a
Bacillus C.R.C.(Bactucide P)	Bacillus thuringiensis-var.Kurstaki-Serotype 3a-3b	250	0.866	b
Bactospeine 16	Bacillus thuringiensis-var.Kurstaki-serotype 3a-3b	100	0.744	b
Bactospeine 16 + Nu-film 17	Bacillus thuringiensis-var.Kurstaki-serotype 3a-3b + pinolene	100 / 100+	0.733	
Experimental product	Piretrinoid	25	0.577	b c d
Tumar	Chlorpyrifos methyl	150	0.400	c d e
Bactospeine 16 + Nu film 17 + sugar	Bacillus thuringiensis-var.Kurstaki-serotype 3a-3b + pinolene + saccharose	100+ / 100+ / 1000	0.366	d e
Larvin	Thiodicarb (experimental)	150	0.333	d e
Experimental product	Piretrinoid	75	0.177	e
Experimental product	Piretrinoid	50	0.144	e
Penncap	Microencapsuled methylparathion	300	0.044	e

The means having equal letters are not different for p = 0.05

weeks counting the numbers of predator mites and following the evolution of the various generations of Panonychus on samples of 50 leaves per hectare using the frequency of occupation method. This well known method consists of counting the leaves occupied by one or more mobile forms of mites (threshold for treatment was 60-70% of leaves occupied during spring; 30-40% in august-september). Although this method was followed accurately the results cast doubt on its validity (type of sampling and threshold used) when applied to grape cultivation:
- an overestimate of the damage was often found; in fact on reaching the thresholds no bronzing of any type was to be seen on the leaves, not even with 100% occupation;
- the aim of these thresholds is to advise acaricide treatment before the population of phytophagous mites provokes visible damage and becomes no longer controllable, but they do not take into account the ratio between number of mites and leaf surface, nor the potential for infestation, variable in relation to the number of eggs present, nor the influence that populations of predator mites, if present, may have. In fact, from the counts conducted in the various grape growing areas the presence of these predator mites was such that in many cases the infestation of the phytophagous mites could be contained, in particular in those farms where repeated insecticide treatments are not given (Casalese area) or where Dinocap and Ditiocarbammate-based fungicides are not used, preferring to these copper or sulphur-based products.

Populations of Tydeidae were also found in various areas (Piverone, Albese); nevertheless the relationship between the population dynamics of Tydeidae and those of the red spider has not yet been clarified with regard also to the differing localization on the leaf mass of the vine.

In the near future we believe the following themes should be explored:
- since frequency of leaf-occupation does not appear to be a valid sampling method for phytophagous mites on grapes, an economic threshold, based on the number of mobile forms per leaf using the "sequential method", must be established;
- since Phytoseiidae are of such importance in the control of Panonychus ulmi Kock it will be interesting to conduct reintroduction trials in some vinyards, to verify the adaptability of these mites to vinyard conditions in Piedmont.

1.3 Leafhoppers

Damage from the leafhoppers (Typhlocybidae) is becoming more frequent in the vinyard.

The farmers are not always able to recognize the agent of this damage and thus do not act in time to limit the damage.

Many leafhopper species are found on grapes, but the most

Table II - 1984

Pesticides employed	Active substance	Rate g/hl c.p.	Mean of perfora tions/grape	Duncan test
Control	--------	---	2.311	a
Beauveria strain 95	Beauveria bassiana	1000	1.577	b
Bactospeine 16	Bacillus thuringiensis-var.Kurstaki-serotype 3a-3b	100	1.433	b
Bactospeine 16 + antiperonospore product	Bacillus thuringiensis-var.Kurstaki-serotype 3a-3b + antiperonospore product	300	1.211	b
Experimental product	experimental piretrinoid	20	0.300	c
Penncap	Microencapsuled methylparathion	300	0.188	c
Experimental product	Piretrinoid	40	0.066	c
Experimental product	Piretrinoid	30	0.055	c

The means having equal letters are not different for p=0.05

representative, on which we have concentrated our attention, are <u>Empoasca</u> spp. and <u>Zygina rhamni</u> Ferr.

Starting in May, every two weeks the number of nymphs present on 100 leaves was counted, adult flight was observed visually and thus the evolution of the biological cycle was followed. Thresholds for intervention were: 200 nimphs on 100 leaves for the first generation and 50 nimphs for the second.

The most severely infesting species was without doubt <u>Empoasca</u>, whereas <u>Zygina</u> was completely sporadic. In no case was the intervention threshold reached during the first generation, nor did damaging alterations to the colour of the leaves occur. The second generation, on the other hand, caused considerable aesthetic damage in the Monforte d'Alba area (with a maximum of 108 nimphs on 100 leaves) and at San Giorgio Monferrato and Rosignano (85 nimphs on 100 leaves); although the threshold was passed insecticide treatment was not carried out in either zone.

Generally speaking the purpose of the thresholds adopted is avoid the appearance of the first symptoms of attack: leaves turning red or yellow. The possible influence on sugar-content has not yet been quantified however; it is therefore not possible at present to speak of economic thresholds in the real sense, but rather of precautional thresholds. It will therefore be interesting to be able to verify in some sample vinyards the actual effect which various degrees of leafhopper infestation have on the sugar content of the grapes.

1.4 Secondary pests

At present it is difficult to supply a detailed description of the totality of phytophages observed in the grape growing areas of Piedmont. Some of them deserve particular mention, due both to the frequency with which they are found and to the potential future danger.

The Grape Thryps (<u>Drepanathryps reuterii</u> Urel.) was found in almost all the vinyards and in some cases produced deformation and necrosis of terminal buds. Since this phytophagous causes ever more frequent and diffuse damage, a suitable sampling method and intervention threshold must be found.

In spring, at the beginning of the vegetative cycle, infestations of <u>Noctuidae</u> and <u>Geometridae</u> larvae on the primary buds were found in the Casale Monferrato and Mango d'Alba areas, whereas the phylloxera (<u>Viteus vitifolii</u> Fitch) reappeared at Rosignano Monferrato in a vinyard of Malvasia grafted onto Kober 5BB.

Lastly vines affected by rust mite are found with increasing frequency; this condition is caused by <u>Calepitrimerus vitis</u> Nal.; the leaf miner <u>Holocacista rivillei</u> Staint. is completely sporadic.

REFERENCES

1. A.C.T.A.-I.T.V. - 1980 - Protection intégrée, controles périodiques au vignoble.

2. OSSERVATORIO PER LE MALATTIE DELLE PIANTE DI VERONA - 1983 - Parassiti animali e vegetali, difesa e diserbo della vite.

3. A.UGOLINI, F.GREMO, G.MICHELATTI - 1983 - La lotta guidata in frutticoltura e viticoltura in Piemonte: realizzazioni e prospettive. Atti XIII Congresso Nazionale Italiano di Entomologia 1983.

4. C.VIDANO - Alterazioni provocate da Insetti in Vitis osservate, sperimentate e comparate - Annali della Facoltà di Scienze Agrarie dell'Università degli Studi di Torino, Vol.I 1962 - 1966: 513-622.

5. P.GALET - 1982 - Les maladies e le parasites de la vigne. Vol.II.

6. REGIONE LAZIO - ASSESSORATO AGRICOLTURA E FORESTE - 1981 - La difesa integrata della vite.

7. G.CELLI, R.BARBIERI, C.CASARINI - 1975 - Risultati dell'impiego di un preparato a base di Bacillus thuringiensis Berliner contro la Lobesia botrana Schiff. (Lepidoptera Tortricidae). Atti Giornate Fit., 335-339.

8. G.CELLI, R.BARBIERI, C.CASARINI, R.BECCHI - 1980 - Risultati di trattamenti con preparati a base di Bacillus thuringiensis contro la Lobesia botrana Schiff. (Lepidoptera Tortricidae) nel modenese in rapporto al rilievo ferormonico e al danno. Atti Giornate Fit., 431-439.

9. L.CORINO, G. MAGNAGHI - 1982 - Esperienze di controllo delle Tignole dell'uva in Piemonte. Atti Giornate Fit., 197-205.

10. G.C.LOZZIA, M.A. RANCATI - 1984 - La distribuzione delle Tignole della vite in Lombardia. Vignevini, anno XI n° 6-Giugno 1984.

11. A.SCHMID, Ph.ANTONIN - 1977 - Bacillus thuringiensis dans la lutte contre les verse de le grappe, eudemis (Lobesia botrana) e chochylis (Clysia ambiguella) en Suisse romande. Riv.Suis.Vit.Arb.Hort., 9, 119-126.

Integrated control on vine at Bombarral (Lisbon) and Terceira (Azores)

V.Garcia & N.Simões
University of the Azores, Ponta Delgada, Açores, Portugal

Summary

The Applied Ecology Laboratory of the University of the Azores has started
two integrated control programs on vines, one in Continental Portugal (Bom
barral, 50 Kms NE of Lisbon) and another in the island of Terceira (Azores).
The first operation began on 1984,with releases of Trichogramma mass reared in the Azores, and then air mailed to Lisbon. The species used was
Trichogramma embryophagum, against Lobesia botrana and secondarily,against
Eupoecilia ambiguella. In this operation no pesticides were employed. At
Terceira Island, the threat of the japanese beetle, Popillia japonica, a
species introduced in the U.S.Air Force facilities and quite probably with
its origin in the United States is the target of an intensive integrated
control program, to avoid the infestation of neighbouring vineyards.Bacillus
popilliae, Nematodes and entomopathogenic Fungi as well as chemicals, like
Carbaryl, are being used to contain the pest's dispersion.

1.Introduction

Integrated control in vine,with large use of biological control,such as
natural enemies and bacteria,has two examples in the Applied Ecology Laboratory, Division of Biological Control, of the Azores University.

One of these examples is an operation set up on the Portuguese mainland,
at the Bombarral vineyards, 50 Kms North-East of Lisbon. This project uses
mainly entomophagous insects of the oophagous genus Trichogramma,already
employed against the pasture army-worm, Mythimna unipuncta, in the islands
of the Azores. The mass-rearing unit of Trichogramma at work in the University facilities (Department of Biology) rears the natual enemies,which are
then transported to Lisbon by air mail and released at Bombarral.

The other example is consequence of a noxious introduction, about 15
years ago, of a polyphagous pest coming from the Eastern USA, Popillia japonica, the so-called japanese beetle.

Ecological studies started by the end of 1983 and have already shown.the
tendency of the pest to expand, which can become a serious danger to the
vines of Terceira.

This expansion must be stopped and this has to be made, first through
chemical control (spraying of Carbaryl,use of larvicides) and then,in long
term, through biological control.

Biocontrol is to use an entomopathogenic bacteria, already in commerce,
the well-known agent of the "milky disease" Bacillus popilliae. Nematodes
and Fungi are also being tested.

The expert teams of the University and the Regional Directorate for Agri
culture, have commenced their field researches,with emphasis on use of
pheromone trapping (Ellisco Bait Traps, U.S.A.). Larval sampling has alrea
dy given an idea of the strategy and spreading rate of this serious pest.

PLATE I - BIOLOGICAL CONTROL IN GRAPE AT BOMBARRAL

    Treated area .... 1 ha

    Releases per year 5

    Trichogramma (adults per release)............. 100.000

    Treatments design:

        a. every 10 days

        b. 3 treatments in 1st generation

        c. 2 treatments in 2nd generation

PLATE II - POPULATION DYNAMICS OF POPILLIA JAPONICA IN TERCEIRA

    ISLAND

	Infested area (ha)	Mean captures per trap
1974	200	102
1980	1,800	93
1983	6,600	2,493
1984	11,000	3,943

2.Bombarral (vine moths)

In cooperation with DGPPA- General Directorate for Protection of Agricul
tural Production, at Lisbon, the University of the Azores has begun last
year one biological control operation against Lobesia botrana (Eudemis)
and, secondarily, against Eupoecilia ambiguella (Cochylis).
Trichogramma selected was T.embryophagum, and the essays concerned a
trial area of 1 hectare of vineyards.
Releases are realised by packages of 100.000 adults of T.embryophagum.
The releasing design is composed of 5 releases per year, separated by
10 days. Three releases are used against the first moth generation,two
against the second. The scheme of the operation is shown in plate I.

3. Terceira Island (japanese beetle)

In the beginning of the 70's,at the U.S. Air Base of Lajes,the japanese
beetle was first found in some rose gardens. After that and between 1974
and 1984, the area occupied by the beetle has increased from 200 to 11.000
hectares (plate II).

At the moment, the pest does not provoke evident damages, but it repre
sents a serious danger to the other Azorean islands.This has influenced
the Government of the Azores to declare Terceira under partial quarantine
measures. Several control means were settled, since the University took on
the coordination of the programs.

3.1. Chemical control

Some previous sprayings,using Malathion and Carbaryl for several years,
did not show impressive results.

This year (1985) one spraying operation of Carbaryl was carried out
by helicopter, over the zones of higher population densities.This has cau
sed a sharp dropping effect, soon followed by a total recover of the popu
lation density (plate III).

We have treated 405 hectares of pasture and hedges, the equivalent of
2000 ha if only pastures were to be considered.

The drop in adult beetles' population was clear, since the first day
after spraying,falling from 100 to 50 p.100.

By the 4th. day, densities approached 10 p.100,to recover again the
100 p.100 level on the 8th. dau after treatment.

3.2. Biological control

We include here phero-traps,nematodes, fungi,and bacteria.

Concerning the phero-traps, Ellisco Bait Traps were used. This type of
trap is in commerce in the USA, since 1974. It uses a feeding attractant
composed of PEP, Eugenol and Geraniol (1:2:1). As a sexual attractant,the
pheromone (Japonilure).

On 4,2 hectares Bacillus popilliae was introduced, on a basis of 2,5 Ki
los of commercial powder with spores per hectare. These spores are produced in the U.S.A. but its experimental production is to be considered in
the Azores by the end of 1986, if the results of the tests are positive.

Use of Nematodes seems to present good perspectives. Species under study are Neoaplectana carpocapsae and Neoaplectana glaseri. The results of
the essays already made are presented on plate IV. A neat reduction of lar
val numbers down to 40 p.100 on N. carpocapsae and 79 p.100 on N.glaseri
is shown.

4. Conclusions

4.1. Biological control of the vine moths by Trichogramma embryophagum is
still in a preliminary phase. It will be necessary to increase the treated
surfaces and to determine the rate of parasitism to draw out some valid
answers.

4.2. Integrated control of Popillia japonica, in the island of Terceira
(Azores) is to prevent the pest spreading to other islands, with an adapta
tion to vines. This could lead to introductions of already vine-adapted
beetles on the mainland.

4.3. Chemical control only results during a limited time. It must be follow
ed by long- term biological control operations,including microbiological
control. This strategy seems to be the only feasible to stop spreading of
the japanese beetle in the Azores.

PLATE III - TREATMENTS WITH PESTICIDE BY HELICOPTER - 1985

Treated area (indirectly).............. 2000 ha
            (directly) .... 405 ha

Active ingridient
       Carbaryl 50% .... 1.4 Kg/ha

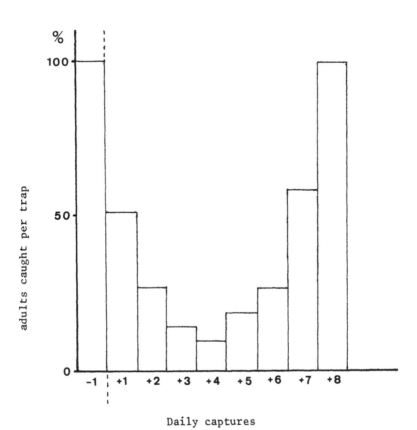

PLATE IV - REDUCTION OF LARVAE SURVEY (%) AFTER TREATMENTS WITH

N. CARPOCAPSE AND N. GLASERI (FIELD ASSAYS)

	MARCH (13.2ºC)	MAY (17.5ºC)
Control	0	0
N.carpocapse	0	40
N.glaseri	16	79

BIBLIOGRAPHY

1- SIMÕES,N.,A.Martins. - Preliminary note on the infestation of Popillia japonica Newman (Coleoptera:Scarabacidae)in Terceira Island - Azores. (in press).

2- MARTINS,A.,N.Simões. - Population Dynamycs of the Japanese beetle (Coleoptera: Scarabaeidae) in Terceira Island - - Azores. (in press).

3- GUENNELON G. & D'ACIER F., 1972 - Piégeage sexuel de l'Endémis de la vigne, Lobesia Gotrana SCHIFF.(Lepidoptera,Tortricidae)dans la région d'Avignon.Rev.Zool.Agric. et Path.Vég., 2- 61-77

4- SOROKINA A.P., 1977 - Trichogramma embryophagum (Hymenoptera,Trichogrammatidae)in the USSR. Zoolog. Jorn.,56 1112-1115.

The contribution of grape-vine breeding to integrated pest control

G. Alleweldt
Federal Research Centre for Grapevine Breeding Geilweilerhof, Siebeldingen, FR Germany

1. Summary

Plasmopara viticola, Oidium tuckeri and Botrytis cinerea are the most important fungus diseases in Central European viticulture. In the feral forms of Vitis native to North America and Eastern Asia resistance genes against these fungus diseases are present which can be used for crossing purposes with the cultivars of Vitis vinifera.

The paper describes the results of a breeding program, initiated by Professor Husfeld, more than 50 years ago. Meanwhile a wide range of fungus resistant strains could be selected which possess the yield and wine quality potential of V. vinifera cultivars and which are at the moment officially tested in order to decide if their viticultural ability satisfies the demand of German viticulturists.

Further breeding success could be achieved by developing phylloxera-resistant varieties with high fungus disease resistance, high yields and high wine quality.

Emphasis is given to the observation that experiments have indicated that the population pressure of spider mites within the vineyards planted with fungus resistant vines is reduced, probably due to the simultaneous increase in the number of predatory mites.

2. Introduction

In the Federal Republic of Germany, the manifold efforts to introduce an ecological or biological viticulture proved to be rather successful as far as mineral nutrition, soil fertility and vinification are concerned, and led to numerous impulses for a prospective viticultural management. In consequence, a reduced application of mineral fertilizers and herbicides could be achieved, accompanied by an increased green manure and cover crops. Within this scope it was early recognized that the limits of this ecological viticulture are given in an effective control of fungus diseases and other pest diseases by means of an integrated pest control, dominated by the use of "biological" pesticides (10). Particularly, the control of fungus diseases, like powdery mildew, downy mildew and grey mould causes per se serious difficulties which are obviously not to be solved by an integrated pest control itself. Furthermore, the introduction of systemic fungicides, especially in controlling Botrytis, resulted in new fungus races, resistant to the chemicals (4, 5, 15). This picture is a circulus vitiosus: New chemicals are developed which create new fungus races without a great chance to untie the Gordon knot by biological means unless breeding efforts create fungus resistant varieties.

As far as insect control is concerned, the use of phylloxera-tolerant rootstocks eliminated damages caused by phylloxera. The initiated tests to monitor grape moth caterpillars by Bacillus thuringiensis, pheromones and other means respectively, are very promising (3, 6, 7, 17). It could also be demonstrated that cover crops reduce the population of spider mites in viticulture by increasing the predators (2, 13).

No doubt, breeding efforts to develop fungus and pest resistant or tolerant cultivars, are time-consuming and of considerable expenditure but the results seem to be most effective. Thus it seems worthwhile to present a report on the status of breeding pest resistant vines and give an outlook on existing and so far not used gene resources.

In the Federal Republic of Germany, breeding for fungus-resistant grapevines was initiated more than 50 years ago. Ever since it has been continued mainly at the Federal Research Centre for Grapevine Breeding Geilweilerhof. In the following I would like to give a survey on the breeding results as an essential part of an integrated pest control.

3. Fungus diseases

3.1 Plasmopara viticola

It is well known that the introduction of the so-called Bordeaux mixture protected the grapevines against the attack of downy mildew for more than 8 to 10 decades very effectfully. However, the frequent application of copper during so many years led to a steady enrichment in copper of the soils, making some of them already unfit for producing other crops than vines. This, at least, was one reason to apply organic fungicides. But meanwhile first reports (4) indicate the existence of fungus races resistant to some fungicides. Under these circumstances one ought to consider a breeding program as an effective tool to solve the problems.

The genetic resources for resistance are mainly found in the American gene centre of the genera Vitis. Numerous native grape species, e.g. V. riparia, V. rupestris, V. aestivalis, V. lincecumii and others, are carriers of complete resistance against P. viticola. Worldwide known hybrid breeders in France and the United States have made use of these resources. Russian geneticists and breeders found also genes for resistance in Asiatic species, especially in V. amurensis (8).

Generally, breeding progress depends on the number of genes determining resistance, the development of suitable test systems and breeding strategy. Since Husfeld developed a method to test the degree of resistance against powdery mildew in the early thirties (11), it was possible to initiate a breeding program which resulted in a number of strains with a high degree of resistance in combination with an acceptable yield and wine quality potential. One example of the latest breeding results is shown in Table 1.

Table 1

Yield and must quality of the new cultivar O R I O N (Ga-58-30) (Figures represent the average data of results obtained from 1977-1984)								
y i e l d			sugar content			a c i d i t y		
Riesling kg/ar	Orion kg/ar	rel.(%)	Riesling °Oe	Orion °Oe	rel.(%)	Riesling g/1	Orion g/1	rel.(%)
114.3	177.7	156	66.0	76.5	116	16.5	9.2	49

This new cultivar "Orion" for which plant variety protection has been applied, is very productive, winter hardy and is highly resistant to Plasmopara viticola. Any chemical protection against powdery mildew is not necessary. The wines are neutral to fruity, mild and pleasant; a foxy flavour cannot be tasted. The wine is undistinguishable from vinifera wines.

Although the genetic and biochemical causes of the resistance of vines to P. viticola are still unknown, infected vines react hypersensitive. The fungus obviously penetrates into the plant cells, but spreading within the tissue is blocked by an active mechanism, probably by phytoalexins. The results are differently sized necrotic spots, their size depends on the degree of resistance. The reaction of leaves is largely identical with the resistance of the berries.

It is remarkable that the genetic resistance to P. viticola has been proved to be stable in all experiments and under detrimental conditions. It can be concluded from this fact that up to now no symptoms for the presence of fungus races are existing which are able to overcome the genetic barriers of resistance of these grapevines. Thus grapevine varieties can be developed, possessing complete resistance to Plasmopara viticola in combination with wine quality and yield.

3.2 Oidium tuckeri (Powdery mildew)

Modern viticultural management, resulting in higher yields by applying organic fungicides, using mineral fertilizers and very vigorous rootstocks, led to a high susceptability of the cultivars to powdery mildew. Although the control by means of sulphur is most effective and within the frame of environmental protection, organic fungicides have been applied. The occurrence of aggressive and fungicid resistant Oidium races (5) call for new methods and strategies of control.

Again breeders are asked to develop Oidium-resistant varieties. As with P. viticola, the genetic resources are present in American native species. The partial resistance of V. amurensis against Oidium tuckeri may not be sufficient to develop fully resistant varieties (8).

Breeding for resistance against Oidium tuckeri is characterized by the lack of a suitable test method for determination of the resistance of seedling populations. Although there exists a high correlation between Plasmopara- and Oidium resistance, particularly in the F_1 of American wild species and V. vinifera, genetically and frankly speaking this correlation is not very close. Nevertheless, some highly resistant cultivars could be developed.

Only recent results with in vitro culture of vines showed new perspectives concerning the breeding of Oidium-resistant grapevines (16). At the moment, the infection of leaf with in vitro cultured fungal spores is tested with promising results. Independent of this screening method, breeding strains with high resistance to powdery mildew were obtained. Table 2 shows an example. The variety is highly resistant to downy mildew and partly resistant to Botrytis. Wines of this cultivar are neutral, somewhat similar to Silvaner, a very important variety grown in Germany and in other vinegrowing countries.

Table 2

Yield and must quality of the new cultivar S I L V A (Ga-54-14)
(Figures represent the average data of results obtained from 1977-1984)

yield			sugar content			acidity		
Riesling kg/ar	Silva kg/ar	rel.(%)	Riesling °Oe	Silva °Oe	rel.(%)	Riesling g/l	Silva g/l	rel.(%)
114.3	162.4	142	66.0	68.4	104	16.5	10.2	62

The biochemical background for resistance to powdery mildew is not
fully elucidated. A hypersensitivity exists - similar to Plasmopara -
possibly due to the presence of phytoalexins. Also the thickness of leaf
and berry cuticula seems to be very important. Perhaps this might be the
reason for the differences in the degree of resistance of leaves and berries. All tests have shown that resistance to powdery mildew is very steady
and not overcome by eventually present, aggressive fungal races.

Thus breeding efforts led to cultivars resistant to powdery mildew.

Most of these new cultivars are also resistant to downy mildew.

3.3 Botrytis cinerea

The control of Botrytis cinerea causes very sincere problems. The
change of copper-containing to organic fungicides, the increase of yield by
fertilizers, the introduction of new cultivars, mostly susceptible to Botrytis and most of all the high mutagenety of the fungi led to considerable
difficulties in fungus control by chemical compounds. This resulted in the
occurrence of fungicide resistant Botrytis races in quick sequences calling
for new and effectful fungicides. Although it is possible to reduce fungus
infestation by cultural methods, for instance by reducing the amount of N
fertilizer, however, these procedures might not be sufficient to produce
mature and healthy grapes for winemaking.

Breeding of vines with a partial or full resistance to Botrytis causes
no problem any more (Table 3). This new cultivar develops a very pronounced
muscat flavored wine which so far has been accepted by the consumers. The
berries are highly resistant to summer rot of Botrytis, i.e. only very mature berries with more than 15 % sugar become susceptible to Botrytis. This
very late attack causes the desired noble rot.

The genetic resources for resistance to Botrytis are partly present
in V. vinifera and sufficiently in American wild species. Also V. amurensis possesses a fairly high resistance to Botrytis. Thus within the breeding
program of mildew resistance one obtains simultaneously seedlings with all
degrees of leaf and berry resistance to Botrytis. Whereas all viticulturists are highly interested in vines with a complete leaf and bud resistance
to Botrytis, winemaking demands a release of berry resistance in order to
produce wines from fully matured noble rot berries. Therefore, breeding

Table 3

Yield and must quality of the new cultivar P H O E N I X (Ga-49-22)
(Figures represent the average data of results obtained from 1977-1984)

yield			sugar content			acidity		
Riesling kg/ar	Phoenix kg/ar	rel.(%)	Riesling °Oe	Phoenix °Oe	rel.(%)	Riesling g/l	Phoenix g/l	rel./%)
114.3	181.7	159	66.0	70.7	107	16.5	10.1	61

efforts are under this premise directed to the development of vines which
possess a pronounced stage resistance to Botrytis. This means that the resistance to Botrytis has to be active as long as the attack of Botrytis can
lead to summer rot. This possibility can be realized by breeding work.

The causes of resistance to Botrytis are nearly clarified. Under the
influence of a toxin, formed by the Botrytis fungus, the plant produces
biochemically phytoalexins, the so-called viniferins. The capability of producing viniferins is positively correlated with the Botrytis resistance.
Besides that, a morphological character for resistance is existing, i.e.
the porosity of cuticula of berries.

It is a known fact that berry resistance to Botrytis reduces with increasing berry maturity. While searching for reasons for this phenomenon,
the microporosity of the cuticula was discovered. Very dainty pores were
stained and proved under UV-radiation (Fig. 1). Resistant varieties are
characterized by only a few and very small pores, susceptible ones, however,
by a lot more and larger pores.

W e   c a n   t h u s   s o l v e   t h e   s e r i o u s   p r o b -
l e m   o f   B o t r y t i s   c o n t r o l   b y   d e v e l o p i n g
g r a p e v i n e   v a r i e t i e s ,   r e s i s t a n t   t o
B o t r y t i s .

3.4 Other fungus diseases

Other occuring fungus diseases of the grapevine do not play an economic role in the Federal Republic of Germany. We can furthermore antici-
pate that corresponding resistance genes are present in the species V. vinifera for endemic fungus as for instance Pseudopeziza tracheifila (9).

In other vinegrowing regions, mainly in subtropical and tropical areas
a series of important diseases occur. Corresponding genes for resistance
are present. In this connection the high resistance of Muscadinia species
has to be mentioned (Table 4). Perhaps future methods of protoplasm fusion
could solve the existing crossing barriers between Muscadinia and Euvitis
species and thus initiate intensive breeding programs.

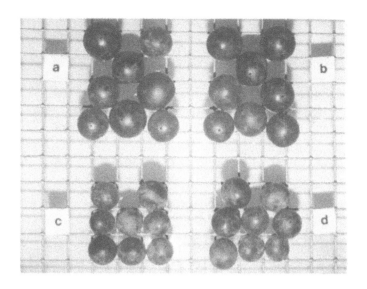

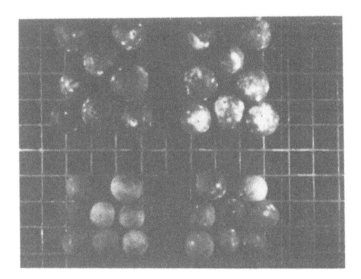

Fig. 1: The perforation of the berry cuticula of 4 grapevine cultivars.
The berries were dipped into aqueous solution of
fluorescine and photographed under UV-light.

a = Müller-Thurgau, b = Foster's White Seedling
c = Pollux,   d = Ga-54-14

Table 4

Vitis rotundifolia
as a gene resource for resistance

Insect/disease	degree of resistance

Phylloxera	resistant
Meloidogyne incognita	tolerant
Meloidogyne javanica	tolerant
Pierce's disease	tolerant
Oidium tuckeri	tolerant
Plasmopara viticola	resistant
Gloeosporium ampelophagum	resistant
Phomopsis viticola	resistant
Botrytis cinerea	tolerant
Guignardia bidwellii	tolerant

OLMO (14)

---

4. Grape pests

4.1 Phylloxera

The destruction of grapes in Europe by Phylloxera are well known. It was caused by the formation of rudimentary root galls on vines of V. vinifera which became necrotic and led to the decline of vines.

By introducing Phylloxera-tolerant grapevines from America and by using these vines as rootstocks, the damage of Phylloxera could be averted.

Numerous breeding efforts were initiated to develop Phylloxera-tolerant wine grapes, worldwide known as "direct producers". Today, breeding for fungus resistant grapevines follows two lanes. The first one leads to fungus resistant vines which behave to Phylloxera like V. vinifera varieties. These varieties have to be grafted on Phylloxera-tolerant rootstocks. The second lane leads to Phylloxera-tolerant direct producers with a high fungus resistance and satisfactory yield and quality performance. For an example, the new breeding strain Pollux of the Federal Research Centre for Grapevine Breeding can be mentioned. Recently, KING and RILLING (12) could prove the

375

existence of different Phylloxera races which are obviously capable to attack even those varieties which had been considered to be Phylloxera-resistant. The future will show the eventual need of new rootstocks, tolerant or resistant to all Phylloxera races.

4.2 Other pests

Genes for resistance to many pests, e.g. nematodes, have been found in several Vitis species. We can suppose that the genetic diversity as far as pest resistance is concerned is not exhaustively evaluated. The breeding profit of this potential is still at the beginning.

We can, however, presume that it is possible to develop grapevines, resistant to very many pests, thus leading to a reduced application of chemical substances.

In this connection experiments with fungus resistant cultivars demonstrated that the lack of fungicide treatments, the lack of herbicides, and the use of cover crops improved the predatory mites's population, thus keeping the damage impact by mites below the economical threshold. This aspect has to be considered as an essential part of an integrated pest control.

5. Final remarks

The aim, to create a viticulture without or with a very reduced use of chemicals, can be realized. Prerequisites, however, are necessary changes in viticultural management and the development of pest resistant varieties which replace very susceptible cultivars. This gradual displacement of traditional varieties, however, is not impossible, theoretically spoken, but knocks against a wall of mistrust, and may induce invincible marketing problems. Nevertheless, viticulturists in Western Germany have accepted new grapevine varieties, raised from V. vinifera, since they have recognized their yield and quality potential. The next step is the introduction of new hybrids with, as far as wine quality is concerned, the benefits of the European grape and the high pest resistance of American Vitis species. These new breeding strains are not easily accepted, independent of their productivity and efficiency and independent of the efforts to maintain a productive viticulture without the necessity of chemicals. But long-sighted, the only way which leads to a biologically well-balanced and healthy viticulture is the growing of cultivars resistant to the main regional diseases. Thus, breeding is an essential part of an effective integrated pest control.

REFERENCES

1. BLAICH, R., STEIN, U. und WIND, R. (1984). Perforationen in der Cuticula von Weinbeeren als morphologischer Faktor der Botrytisresistenz. Vitis 23, 242-256

2. BOLLER, E. (1978). Abschätzung des Spinnmilbenrisikos und Schonung der Raubmilden im ostschweizerischen Weinbau. Schweiz. Z. Obst- und Weinbau 114, 257-264

3. BOLLER, E. (1980). Umweltfreundliche Schädlingsbekämpfung im ostschweizerischen Weinbau. Stand der Arbeiten im Jahre 1980. Schweiz. Z. Obst-Weinbau 116, 723-730

4. BOSSHARD, E. und SCHUEPP, H. (1983). Variabilität ausgewählter Stämme von Plasmopara viticola bezüglich ihrer Sensibilität gegenüber Metaloxyl unter Freilandbedingungen.
Z. f. Pflanzenkrankheiten und Pflanzenschutz 90, 449-459

5. CLERJEAU, M. (1985). Evolution de la lutte contre le Mildiou de la vigne. Consequences de l'utilisation des produits systemiques. Bull. OIV 58, 403-415

6. DANKO, L. and JUPP, G.L. (1983). Field evaluation of pheromone-baited traps for monitoring grape berry moth (Lepidoptera:Olethreutidae). J. Econ. Entomol. 76, 480-483

7. EGGER, E. and BORGO, M. (1983). Monitoring flights of grape moths with specific pheromone traps in the Piave flats and Conegliano hills. Riv. Viticoltura e Enologia, Conegliano 36, 51-70

8. GOLODRIGA, P.J. et SOUYATINOV, I.A.: Vitis amurensis. Habitat, aptitudes et technologiques, variabilité de l'espèce. Bull. OIV 54, 971-982

9. HAHN,H. (1957). Eine Auslesemethode für die Resistenz gegen den Roten Brenner (Pseudopeziza tracheiphila Müller-Thurgau). Vitis 1, 32-33

10. HOLZ, B. (1983). Über die Wirksamkeit der Pflanzenpflegemittel "Bio-S" und "Algifert" gegen Peronospora, Oidium und Botrytis im Weinbau. Wein-Wiss. 38, 126-140

11. HUSFELD, B.(1932). Über die Züchtung plasmoparawiderstandsfähiger Reben. Gartenbauwiss. 7, 15-92

12. KING,P. D. and RILLING, G. (1985). Variations in the galling reaction of grapevines. Evidence of different phylloxera biotypes and clonal reaction to phylloxera. Vitis 24, 32-42

13. OBERHOFER, H. (1983). Spinnmilben - ein Folgeschädling unserer Pflegemaßnahmen. Obstbau-Weinbau, Mitt. Südtiroler Beratungsring Bozen 20, 15-21

14. OLMO, H.P. (1971). Vinifera x rotundifolia hybrids as wine grapes. Amer. J. Enol. Viticult. 22, 87-91

15. PEARSON, R.C. and RIEGEL, D.G. (1983). Control of Botrytis bunch rot of ripening grapes: Timing applications of the dicarboximide fungicides. Amer. J. Enology Viticult. 34, 167-172

16. STEIN, U., HEINTZ, C. und BLAICH, R. (1985). Die in-vitro-Prüfung von Rebsorten auf Oidium- und Plasmopara-Resistenz. Z. Pflanzenkrankheiten u. Pflanzenschutz 92, 355-369

17. STEINER, H. (1983). Biologische Traubenwicklerbekämpfung mit Bacillus thuringiensis. Rebe und Wein 36, 304-305

Closing session

Chairman: R.Cavalloro

Conclusions and recommendations

<u>Insects and mites</u>

The large number of papers presented in this session showed that considerable research has being carried out in European countries on such pests of <u>Vitis vinifera</u>, as the grape berry moths, <u>Auchenorrhyncha</u>, the grape phylloxera, <u>Drosophila</u> spp. associated with sour--rot of grapes, and phytophagous mites and their natural enemies.

1. With respect to grape berry moth a number of papers reported the population fluctuations of male moths, as judged by pheromone trap captures. In some regions, the first generation larvae do not usually cause such damage as would require control measures. Thus, one insecticidal treatment against the second generation larvae is sufficient for most cultivars. In some regions, pheromone trap captures, if low or medium, are not reliable for the estimate of attack and therefore for determining when to treat. Larval hole counts in the berries were then found to be more reliable. The mortality during the first (flower-feeding) generation was suggested as an estimate of population density of the second generation.

    More than one speaker reported that the populations of both species of berry moths fluctuate widely and that the reasons are not known. Egg mortality was noticed under high summer temperatures.

    A faster larval development and lower mortality was noticed when <u>Lobesia botrana</u> (Denis et Schiff.) larvae fed on Botrytis-infected grape berries in the laboratory. The phenomenon awaits verification under field conditions.

    A number of data on the occurrence and abundance of <u>Lobesia botrana</u> and <u>Eupoecilia ambiguella</u> (Hb.) are different from older information. This suggests to need for more basic work to better know the life history of those two species.

2. The <u>Hemiptera Auchenorrhyncha</u> found on grapevines in Italy were recorded and classified according to their feeding habits and relationship to the plant. One of them has recently entered Italy from America and is causing damage in the absence of effective

natural enemies. Those potentially hazardous as disease vectors were suggested. Effective natural enemies of some of those Hemipteran pests of the vines were identified.

3.   A review of the situation concerning the grape phylloxera, Viteus vitifoliae (Fitch), during the past 100 years was presented as well experimental data, showing: a) no correlation between density of foliar infestation of Vitis vinifera (L.) and root population and symptoms density; b) that certain sandy soils were highly unfavorable for the insect; c) that infestation of Vitis vinifera foliage changes with time and the cultivar; d) that the population structure in leaf galls varies with the maturity of the leaves and therefore the type of sampling must depend on the purpose of the work.

4.   Sour-rot of grapes in northern Italy was found to be correlated with the presence of Drosophila spp. adults and larvae in the berries.

5.   As far as phytophagous mites are concerned, it was pointed out that by adopting an intervention threshold higher than before, the number of acaricide applications was considerably reduced.

It was agreed that spider mites become major pests only as a result of the excessive and/or improper use of pesticide, which eliminate useful predatory arthropods such as the Phytoseiidae. Integrated control, which by its very nature aims at protecting useful arthropods should, therefore, be developed and applied wherever possible.

What is even more encouraging is that by the use of selective fungicides and insecticides, in conjunction with high intervention thresholds for spider mites, and the presence of Phytoseiidae, satisfactory natural control of spider mites has been achieved in many Italian vineyards. Certain Phytoseiidae such as Kampimodromus aberrans (Oud.) can keep phytophagous mites in vineyards to satisfactorily low levels.

6.   It is underlined that member countries and the C.E.C. pay more attention to pests newly introduced to Europe, such as Popilia japonica Newm., and support eradication programs.

Diseases and Weeds

The participation of phytopathologists and weed-experts is proving increasingly important in meetings on integrated control, and thus during this meeting interesting papers have been given. It has turned out to be a session of quality thanks to valuable presentations.

If pathologists did not in the past feel motivated by the notion of integrated control it was only because the concept did not appear very clear to them. In fact, integrated and biological control have been assimilated for a long time and many pathologists, particularly those who study the interactions between the various cultural parameters on the development of diseases, are researchers who - without knowing it - are interested in integrated control.

The major conclusions of this session on diseases and weeds underline the following points:

1. Biological control

One has mostly considered the interesting prospect of using Trichoderma (which has been talked of for a long time), but will this happen in practice? The problem rests on the need for complementary studies between laboratories with true collaboration and division of work such as on the selection of strains, and the ecology of fungi. This is a theme of work which the C.E.C. could stimulate, in the same way as vine growing plays the role of model in this field.

Furthermore, the antagonistic possibilities of other agents or the study of hypovirulent strains have not been sufficiently exploited. The time has come to urge on the development of the screening of useful biological agents.

2. Herbicides

Although reports lead to the conclusion that technically there are good reasons to use herbicides one must henceforth gof much further, i.e. to appreciate the various biological influence of the use of these products such as the action on insect or virus-harbouring plants, on the microflora-fauna of the soil, on the vegetal organs which shelter cryptogamic parasites, etc. Thanks to this type of study one could certainly speak of integrated control for herbicides also.

Strategies of I.P.M.

The application of Integrated Pest Control in vineyards in various European countries indicates clearly that in practice very good production results can be obtained on reducing the use of chemical pesticides.

Damage prediction studies require more detailed knowledge of the biotic agents which restrict phytophagous arthropods and pathogenic fungi, of the abiotic factors which influence these and of the organism/environment interrelationships.

It is recognised that at both experimental and the application stages, agrometeorological services should be widely available as should means of computerizing data.

Study on the influence of crop growing action on agro-biocoenosis should be encouraged taking account of the changes which they cause in the microclimate.

The participants aknowledge with satisfaction the interest of the C.E.C. in the phytosanitary aspects of viticulture, which aim at improving the quality of the well-known wines and table grapes of Europe.

It was evident that certain potential or secondary pests of the grape-vine become major pests as a result of the improper use of chemical pesticides, which eliminate useful parasite or predator species.

The grape berry moths continue to be major pests. Important work has been done in some areas, which allows the reduction of the number of insecticide applications. Yet, now data reported at this meeting and elsewhere show that much more basic knowledge on the moths' biology and ecological associations is needed, in order to learn the causes of population fluctuations and improve integrated pest control in vineyards.

More basic knowledge is needed also for a number of other vine major or potential pests. The neglection of such standard practices as using Phylloxera-resistant rootstocks may cause serious problems in some regions.

It is, therefore, recommended that the C.E.C. supports basic research aiming at Integrated Pest Control, and it reducing the adverse effects of current pest control methods. In this respect, natural enemies, including insect-pathogenic microorganisms and selective pesticides should be paid attention to.

Where the development of Integrated Pest Control has progressed to a desired degree, a large-scale application to demonstrate its value to the growers should be encouraged. It is recommended that the C.E.C. supports such an application.

The Commission of the European Communities is asked to encourage research and actions as well as to stimulate multidisciplinary and international collaboration in the sector, with particular reference to the Member countries.

International collaboration is increasingly necessary and can only develop through mutual respect. Exchanges and specific meetings like this one are thus essential: the C.E.C. must be aware of this and is called upon to intensify and develop the possibilities of exchange of scientists within the Member countries.

It is considered useful to have the next large meeting on Integrated Pest Control in viticulture preferably in a Mediterranean country: the C.E.C. is asked to select the country and to realize the meeting.

List of participants

Belgium

    KAMOEN, O.
    Rijksstation voor Plantenziekten
    Burg. van Gansberghelaan, 96
    9220 Merelbeke

F.R. Germany

    ALLEWELDT, G.
    Federal Research Institute for Grape Vine Breeding
    Geilweilerhof
    6741 Siebeldingen

    SCHRUFT, G.
    Staatliches Weinbau Institut
    Merzhauserstrasse, 119
    7800 Freiburg-i-Brisgau

France

    AGULHON, R.
    Institut Technique de la Vigne et du Vin
    Domaine de la Grande Bastide - route du Général
    3000 Nîmes

    BASSINO, J.P.
    Association de Coordination Technique Agricole
    A.C.T.A.
    149, rue de Bercy
    75595 Paris

    BLANC, M.
    Association de Coordination Technique Agricole
    A.C.T.A.
    Manosque

CLERJEAU, M.
Station de Pathologie Végétale
I.N.R.A.
Domaine de la Grande Ferrade
Centre de Recherches de Bordeaux
33140 Pont-de-la-Maye

Great Britain

HEALE, J.B.
Biology Department - King's College - University
Campden Hill Road
London WB 7 AH

Greece

AVGELIS, A.
Plant Protection Institute
71306 Heraklion (Crete)

SAVOPOULOU-SOULTANI, M.
Department of Plant Protection
Laboratory Applied Zoology and Parasitology - University
54006 Thessaloniki

STAVRAKI, H.
"Benaki" Phytopathological Institute
Kiphissia (Athens)

TZANAKAKIS, M.
Department of Plant Protection
Laboratory Applied Zoology and Parasitology - University
54006 Thessaloniki

Italy

ANTONELLI, R.
Istituto di Entomologia Agraria - Università
Via San Michele degli Scalzi, 2
56100 Pisa

BAGNOLI, B.
Istituto Sperimentale per la Zoologia Agraria
Via di Lanciola
50125 Cascine del Riccio (Firenze)

BELCARI, A.
Istituto di Entomologia Agraria - Università
Via San Michele degli Scalzi, 2
56100 Pisa

BERTONA, A.
Sandoz S.p.A.
Via Arconati, 1
20135 Milano

BISIACH, M.
Istituto di Patologia Vegetale - Università
Via Celoria, 2
20133 Milano

BORGO, M.
Istituto Sperimentale per la Viticoltura
Viale XXV Aprile, 1
31015 Conegliano

CASTAGNOLI, M.
Istituto Sperimentale per la Zoologia Agraria
Via di Lanciola
50125 Cascine del Riccio (Firenze)

CHIAPPINI, E.
Istituto di Entomologia Agraria
Università Cattolica del Sacro Cuore
Via Emilia Parmense, 84
29100 Piacenza

CORINO, L.
Istituto Sperimentale per la Viticoltura
Corso Luigi Einaudi, 60
14100 Asti

CRAVEDI, P.
Istituto di Entomologia Agraria
Università Cattolica del Sacro Cuore
Via Emilia Parmense, 84
29100 Piacenza

CROVETTI, A.
Istituto di Entomologia Agraria - Università
Via San Michele degli Scalzi, 2
56100 Pisa

DALLA MONTA', L.
Istituto di Entomologia Agraria - Università
Via Gradenigo, 6
35100 Padova

DEL BENE, G.
Istituto Sperimentale per la Zoologia Agraria
Via di Lanciola
50125 Cascine del Riccio (Firenze)

DELRIO, G.
Istituto di Entomologia Agraria - Università
Via De Nicola
07100 Sassari

DE SILVA, J.
Istituto Sperimentale per la Zoologia Agraria
Via di Lanciola
50125 Cascine del Riccio (Firenze)

DOMENICHINI, G.
Istituto di Entomologia Agraria
Università Cattolica del Sacro Cuore
Via Emilia Parmense, 84
29100 Piacenza

DUSO, C.
Istituto di Entomologia Agraria - Università
Via Gradenigo, 6
35100 Padova

FELLIM, F.
E.S.A.T.
Via Rosmini, 42
38100 Trento

FONTANELLI, G.P.
Sandoz S.p.A.
Via Ugo Foscolo, 19
50124 Firenze

GAMBARO, P.
Villafontana
37132 Verona

GARIBALDI, A.
Istituto di Patologia Vegetale - Università
Via P. Giuria, 15
10126 Torino

GENDUSO, P.
Osservatorio Regionale per le Malattie delle Piante
Via N. Garzilli, 3
90100 Palermo

GENTINI, U.
Azienda Autonoma di Cura, Soggiorno e Turismo
dell'Isola d'Elba
Calata Italia, 26
57031 Portoferraio

GIROLAMI, V.
Istituto di Entomologia Agraria - Università
Via Gradenigo, 6
35100 Padova

GRANDE, C.
Osservatorio Regionale per le Malattie delle Piante
Via Tevere, 5
00198 Roma

GREMO, F.
Osservatorio Regionale per le Malattie delle Piante
Corso Grosseto, 71
10126 Torino

GRIECO, L.
Azienda Autonoma di Cura, Soggiorno e Turismo
dell'Isola d'Elba
Calata Italia, 26
57031 Portoferraio

GULLINO, M.L.
Istituto di Patologia Vegetale - Università
Via P. Giuria, 15
10126 Torino

LIGUORI, M.
Istituto Sperimentale per la Zoologia Agraria
Via di Lanciola
50125 Cascine del Riccio (Firenze)

LORSUZO, G.
Azienda Autonoma di Cura, Soggiorno e Turismo
dell'Isola d'Elba
Calata Italia, 26
57031 Portoferraio

LOZZIA, G.
Istituto di Entomologia Agraria - Università
Via Celoria, 2
20133 Milano

MANCA, M.
Azienda Autonoma di Cura, Soggiorno e Turismo
dell'Isola d'Elba
Calata Italia, 26
57031 Portoferraio

MAROCCHI, G.
Osservatorio Regionale per le Malattie delle Piante
Via di Corticella, 133
40029 Bologna

MAURI, F.
Istituto di Entomologia Agraria - Università
Via Gradenigo, 6
35100 Padova

MESCALCHIN, E.
E.S.A.T.
Via Rosmini, 42
38100 Trento

MINERVINI, G.
Istituto di Patologia Vegetale - Università
Via Celoria, 2
20133 Milano

PALMIERI, M.
Azienda Autonoma di Cura, Soggiorno e Turismo
dell'Isola d'Elba
Calata Italia, 26
57031 Portoferraio

PARDI, G.
Sindaco di
57031 Portoferraio

PAVAN, F.
Istituto di Entomologia Agraria - Università
Via Gradenigo, 6
35100 Padova

PEGAZZANO, F.
Istituto Sperimentale per la Zoologia Agraria
Via di Lanciola
50125 Cascine del Riccio (Firenze)

PICONE, P.
Istituto di Entomologia Agraria - Università
Viale delle Scienze
90100 Palermo

PINNA, M.
Osservatorio per le Malattie delle Piante
Corso Grosseto, 71
10100 Torino

QUAGLIA, F.
Istituto di Entomologia Agraria - Università
Via San Michele degli Scalzi, 2
56100 Pisa

RASPI, A.
Istituto di Entomologia Agraria - Università
Via San Michele degli Scalzi, 2
56100 Pisa

ROSSI, E.
Istituto di Entomologia Agraria - Università
Via San Michele degli Scalzi, 2
56100 Pisa

SAGILOTTO, G.B.
Consorzio Vini
Via Vittorio Veneto, 13
30020 Pramaggiore

SANCASSANI, G.P.
Osservatorio Regionale per le Malattie delle Piante
Lungadige Capuleti, 11
37132 Verona

SANTINI, L.
Istituto di Entomologia Agraria - Università
Via San Michele degli Scalzi, 2
56100 Pisa

STRAPAZZON, A.
Istituto di Entomologia Agraria - Università
Via Gradenigo, 6
35100 Padova

SUSS, L.
Istituto di Entomologia Agraria - Università
Via Celoria, 2
20133 Milano

TRANFAGLIA, A.
Istituto di Entomologia Agraria
Università di Napoli
Via Università, 100
80055 Portici

VACANTE, V.
Istituto di Entomologia Agraria - Università
Via Valdisavoia, 5
95100 Catania

VERCESI, A.
Istituto di Patologia Vegetale - Università
Via Celoria, 2
20133 Milano

VIDANO, C.
Istituto di Entomologia Agraria - Università
Via P. Giuria, 15
10126 Torino

ZANGHERI, S.
Istituto di Entomologia Agraria - Università
Via Gradenigo, 6
35100 Padova

ZIEROCK, R.
Istituto di Coltivazioni Arboree - Università
Via Celoria, 2
20133 Milano

Ireland

ROBINSON, D.W.
An Foras Taluntais - The Agricultural Institute
Kinsealy Research Centre
Malahide Road
Dublin 17

Portugal

SIMOES, N.
Universidade dos Azores
Rua Mae de Deus
9500 Ponta Delgada (Azores)

C.E.C.

CAVALLORO, R.
Commission of the European Communities
"Integrated Plant Protection" Programme
Joint Research Centre
I-21020 Ispra

ROTONDO', P.P.
Commission of the European Communities
Directorate General XIII/A2
Plateau du Kirchberg - B.P. 1907
L-2920 Luxembourg

Index of authors